Nature's Teachings

Human Invention Anticipated by Nature

John George Wood

CAMBRIDGE
UNIVERSITY PRESS

CAMBRIDGE UNIVERSITY PRESS

Cambridge New York Melbourne Madrid Cape Town Singapore São Paolo Delhi

Published in the United States of America by Cambridge University Press, New York

www.cambridge.org
Information on this title: www.cambridge.org/9781108000710

This edition first published 1877
This digitally printed version 2009

ISBN 978-1-108-00071-0

CAMBRIDGE LIBRARY COLLECTION

Books of enduring scholarly value

Religion

For centuries, scripture and theology were the focus of prodigious amounts of scholarship and publishing, dominated in the English-speaking world by the work of Protestant Christians. Enlightenment philosophy and science, anthropology, ethnology and the colonial experience all brought new perspectives, lively debates and heated controversies to the study of religion and its role in the world, many of which continue to this day. This series explores the editing and interpretation of religious texts, the history of religious ideas and institutions, and not least the encounter between religion and science.

Nature's Teachings

Nature's Teachings, first published in 1877, was one of many books on natural history by J.G. Wood, a Victorian clergyman who was hugely influential in popularising the subject, as well as being the editor of 'The Boy's Own Magazine'. Here he examines the close parallels between nature and human inventions in areas including seafaring (the raft, paddle and oar), war and hunting (barbs, poisons and projectiles), architecture, tools, optics and acoustics, as well as 'useful arts' including sewage disposal. His text contains over 750 figures and illustrations, and he argues that future great discoveries could be made as a result of careful observations of nature. Although a contemporary of Darwin, Wood largely ignored the evolution debates and focussed on communicating his enthusiasm for the natural world to a non-scientific audience. His successful publications still make fascinating reading for those interested in Victorian culture and the history of education.

Cambridge University Press has long been a pioneer in the reissuing of out-of-print titles from its own backlist, producing digital reprints of books that are still sought after by scholars and students but could not be reprinted economically using traditional technology. The Cambridge Library Collection extends this activity to a wider range of books which are still of importance to researchers and professionals, either for the source material they contain, or as landmarks in the history of their academic discipline.

Drawing from the world-renowned collections in the Cambridge University Library, and guided by the advice of experts in each subject area, Cambridge University Press is using state-of-the-art scanning machines in its own Printing House to capture the content of each book selected for inclusion. The files are processed to give a consistently clear, crisp image, and the books finished to the high quality standard for which the Press is recognised around the world. The latest print-on-demand technology ensures that the books will remain available indefinitely, and that orders for single or multiple copies can quickly be supplied.

The Cambridge Library Collection will bring back to life books of enduring scholarly value across a wide range of disciplines in the humanities and social sciences and in science and technology.

NATURE'S TEACHINGS

The HOME

NATURE'S TEACHINGS

HUMAN INVENTION
ANTICIPATED BY NATURE

BY THE
REV. J. G. WOOD, M.A., F.L.S., ETC.

AUTHOR OF "HOMES WITHOUT HANDS,"
"MAN AND BEAST, HERE AND HEREAFTER," ETC.

LONDON
DALDY, ISBISTER & CO.
56, LUDGATE HILL
1877

PREFACE.

A GLANCE at almost any page of this work will denote its object. It is to show the close connection between Nature and human inventions, and that there is scarcely an invention of man that has not its prototype in Nature. And it is worthy of notice that the greatest results have been obtained from means apparently the most insignificant.

There are two inventions, for example, which have changed the face of the earth, and which yet sprang from sources that were despised by men, and thought only fit for the passing sport of childhood. I allude, of course, to Steam and Electricity, both of which had been child's toys for centuries before the one gave us the fixed engine, the locomotive, and the steamboat, and the other supplied us with the compass and the electric telegraph.

In the course of this work I have placed side by side a great number of parallels of Nature and Art, making the descriptions as terse and simple as possible, and illustrating them with more than seven hundred and fifty figures. The corollary which I hope will be drawn from the work is evident enough. It is, that as existing human inventions have been anticipated by Nature, so it will surely be found that in Nature lie the prototypes of inventions not yet revealed to man. The great discoverers of the future will, therefore, be those who will look to Nature for Art, Science, or Mechanics, instead of taking pride in some new invention, and then finding that it has existed in Nature for countless centuries.

I ought to mention that the illustrations are not intended to be finished drawings, but merely charts or maps, calling attention to the salient points.

CONTENTS.

NAUTICAL.

CHAPTER I.

THE RAFT.

CHAPTER II.

THE OAR, THE PADDLE, AND THE SCREW.

CHAPTER III.

SUBSIDIARY APPLIANCES.—Part I.

CHAPTER IV.

SUBSIDIARY APPLIANCES.—Part II.

CHAPTER V.

SUBSIDIARY APPLIANCES.

Part III.—The Boat-hook and Punt-pole.—The Life-buoy and Pontoon-raft.

WAR AND HUNTING.

CHAPTER I.

THE PITFALL, THE CLUB, THE SWORD, THE SPEAR AND DAGGER.

CHAPTER II.

POISON, ANIMAL AND VEGETABLE.—PRINCIPLE OF THE BARB.

CHAPTER III.

PROJECTILE WEAPONS AND THE SHEATH.

CHAPTER IV.

CHAPTER V.

CHAPTER VI.

THE HOOK.—DEFENSIVE ARMOUR.—THE FORT.

CHAPTER VII.

SCALING INSTRUMENTS.—DEFENCE OF FORT.—IMITATION. —THE FALL-TRAP.

CHAPTER VIII.

CONCEALMENT.—DISGUISE.—THE TRENCH.—POWER OF GRAVITY.—MISCELLANEA.

ARCHITECTURE.

CHAPTER I.

THE HUT, TROPIC AND POLAR.—PILLARS AND FLOORING.—
TUNNEL ENTRANCE OF THE IGLOO.—DOORS AND HINGES.—
SELF-CLOSING TRAP-DOORS.

CHAPTER II.

WALLS, DOUBLE AND SINGLE.—PORCHES, EAVES, AND
WINDOWS.—THATCH, SLATES, AND TILES.

CHAPTER III.

THE WINDOW.—GIRDERS, TIES, AND BUTTRESSES.—THE
TUNNEL.—THE SUSPENSION-BRIDGE.

CHAPTER IV.

TOOLS.

CHAPTER I.

OPTICS.

CHAPTER I.

THE MISSIONS OF HISTORY.—THE CAMERA OBSCURA.—LONG AND SHORT SIGHT.—STEREOSCOPE AND PSEUDOSCOPE.—MULTIPLYING-GLASSES.

The Camera Obscura.—Telescopes, Microscopes, and Spectroscopes, and their separate Objects.—Structure of the Camera Obscura.—The Double Convex Lens.—Its Use as a Burning-glass.—The Meridian Gun in Paris.—

CHAPTER II.

THE   WATER-TELESCOPE.—IRIS   OF   THE   EYE.—MAGIC LANTERN.—THE SPECTROSCOPE.—THE THAUMATROPE.

USEFUL ARTS.

CHAPTER I.

PRIMITIVE MAN AND HIS NEEDS.—EARTHENWARE.—BALL-AND-SOCKET JOINT.—TOGGLE OR KNEE JOINT.

CHAPTER II.

CRUSHING INSTRUMENTS.—THE NUT-CRACKERS, ROLLING-MILL, AND GRINDSTONE.—PRESSURE OF ATMOSPHERE. —SEED DIBBLES AND DRILLS.

CHAPTER III.

CLOTH-DRESSING. — BRUSHES AND COMBS. — BUTTONS, HOOKS AND EYES, AND CLASP.

CHAPTER IV.

THE STOPPER, OR CORK.—THE FILTER.

CHAPTER V.

THE PRINCIPLE OF THE SPRING.—THE ELASTIC SPRING. —ACCUMULATORS.—THE SPIRAL SPRING.

CHAPTER IX.

ARTIFICIAL WARMTH.—RING AND STAPLE.—THE FAN.

CHAPTER X.

WATER, AND MEANS OF PROCURING IT.

CHAPTER XI.

AËROSTATICS.—WEIGHT OF AIR.—EXPANSION BY HEAT.

ACOUSTICS.

CHAPTER I.

PERCUSSION.—THE STRING AND REED.—THE TRUMPET.
—EAR-TRUMPET.—STETHOSCOPE.

The Science of Sound.—Rhythmical Vibrations.—The Drum.—Primitive
Drums.—The Solid and the Hollow Log.—The Bass Drum and Kettle-
drum.—African Drums.—Gnostic Gems and the Ashanti Drum.—Tym-
panum, or Drum of the Human Ear, and its Mechanism.—An artificial
Tympanum.—The String.—The Bow and the Harp.—The Harpsichord
and the Zither.—The Bow and the Violin.—The Cricket.—The
Vibrator, or Reed.—The Jew's Harp and Harmonium.—The Cicada
and its Song.—Harmonics upon Strings.—The Æolian Harp.—Har-
monics upon the Trumpet.—The Trombone.—Trachea of the Swan.
—The Ear-trumpet.—The Sea-shell.—The Stethoscope.—Savage Food.
—The Aye-aye.—The Siren and its Uses.—Echo and Whispering
Gallery 513

NAUTICAL.

CHAPTER I.

Poetry and Science.—The Paper Nautilus and the Sail.—Montgomery's " Pelican Island."—The Nautilus replaced by the Velella.—The Sailing Raft of Nature and Art.—Description of a Velella Fleet off Tenby.—The Natural Raft and its Sail.—The Boats of Nature and Art.—Man's first Idea of a Boat.—The Kruman's Canoe and the *Great Eastern.*—Gradual Development of the Boat.—The Outrigger Canoe a Mixture of Raft and Boat.—Natural Boats.—The Water-snails. — The Sea-anemones.— The Egg-boat of the Gnat.—The Skin-boat of the same Insect.—Shape and Properties of the Life-boat anticipated in Nature.—Natural Boat of the Stratiomys.

THE RAFT.

IT has been frequently said that the modern developments of science are gradually destroying many of the poetical elements of our daily lives, and in consequence are reducing us to a dead level of prosaic commonplace, in which existence is scarcely worth having. The first part of this rather sweeping assertion is perfectly true, but, as we shall presently see, the second portion is absolutely untrue.

Science has certainly destroyed, and is destroying, many of the poetic fancies which made a part of daily life. It must have been a considerable shock to the mind of an ancient philosopher when he found himself deprived of the semi-spiritual, semi-human beings with which the earth and water were thought to be peopled. And even in our own time and country there is in many places a still lingering belief in the existence of good and bad fairies inhabiting lake, wood, and glen, the successors of the Naiads and Dryads, the Fauns and Satyrs, of the former time. Many persons will doubtless be surprised, even in these days, to hear that the dreaded Maelström is quite as fabulous as the Symplegades or Scylla and

Charybdis, and that the well-known tale of Edgar Poe is absolutely without foundation.

Perhaps one of the prettiest legends in natural history is that of the Paper Nautilus, with which so much poetry is associated. We have all been accustomed from childhood to Pope's well-known lines beginning—

"Learn of the little Nautilus to sail,"

and some of us may be acquainted with those graceful verses of James Montgomery, in his "Pelican Island:"—

"Light as a flake of foam upon the wind,
Keel upward, from the deep emerged a shell,
Shaped like the moon ere half her horn is filled.
Fraught with young life it righted as it rose,
And moved at will along the yielding water.
The native pilot of this little bark
Put out a tier of oars on either side,
Spread to the wafting breeze a two-fold sail,
And mounted up and glided down the billow
In happy freedom, pleased to feel the air,
And wander in the luxury of light.
 * * * * * * *
It closed, sank, dwindled to a point, then nothing,
While the last bubble crowned the dimpling eddy
Through which mine eye still giddily pursued it."

So deeply ingrained is the poetical notion of the sailing powers attributed to the nautilus, that many people are quite incredulous when they are told that there is just as much likelihood of seeing a mermaid curl her hair as of witnessing a nautilus under sail. How the creature in question does propel itself will be described in the course of the present chapter; and the reader will see that although one parallel between Nature and Art in the nautilus does not exist, there are several others which until later days have not even been suspected.

It is, therefore, partially true that science does destroy romance. But, though she destroys, she creates, and she gives infinitely more than she takes away, as is shown in the many late discoveries which have transformed the whole system of civilised life. Sometimes, as in the present instance, she discovers one analogy while destroying another, and though she shatters the legend of the sailing nautilus, she produces a marine animal which really does sail, and does not appear to be able to do anything else. This is the VELELLA, a figure of which, taken from a specimen in my collection, is given in the illustration, and drawn of the natural size.

It is one of that vast army of marine creatures known familiarly by the name of "jelly-fishes," just as lobsters, crabs, shrimps, oysters, whelks, periwinkles, and the like, are lumped together under the title of "shell-fish." As a rule, these creatures are soft, gelatinous, and, in fact, are very little more than sea-water entangled in the finest imaginable mesh-work of animal matter; so fine, indeed, that scarcely any definite organs can be discovered. The Velella, however, is

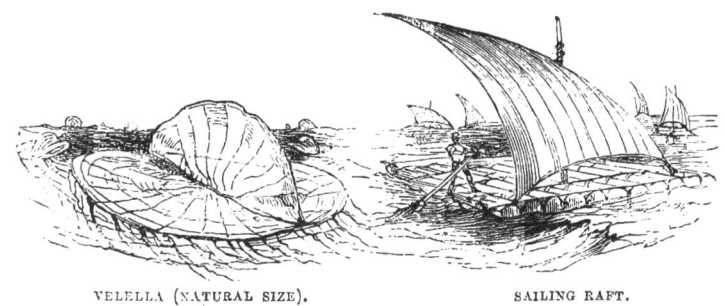

VELELLA (NATURAL SIZE).   SAILING RAFT.

remarkable for having a sort of skeleton, if it may be so called, consisting of two very thin and horny plates, disposed, as shown in the illustration, so as to form an exact imitation (or perhaps I should say a precursor) of a raft propelled by a sail. Indeed, the Latin name Velella signifies a little sail.

How well deserved is the name may be seen by the following graphic account of a Velella fleet sent to me by a lady who takes great interest in practical zoology :—

"The specimens which I send came from Tenby, a very rough sea having driven a large living fleet of them on that coast.

"When in life, they are semi-transparent, and radiant in many rainbow-tinted colours. They came floating towards me in all their fragile beauty on the rough sea waves. I succeeded in capturing some of them, and preserved the only portion available for my collection.

"They are extremely tender, and by no means with which I am acquainted can be preserved more than these skeleton-like cartilaginous plates. They soon dissolve in either spirits of wine or water, and lose every vestige of their shape and substance. The upright, thin, pellucid plate has the appearance of a fairy-like miniature sail, and apparently acted as such when the creature was floating with its long and many-tinted tentacles pendent from its lower surface.

"Although widely distributed, they are seldom seen on our own coast, although sometimes driven there from the warmer regions by stress of wind and waves.

"These little creatures had never before been seen at Tenby, but when I asked a native bathing-woman whether she knew their name, she immediately replied, 'Sea-butterflies.' Although the name was evidently of her own invention, it was most appropriate and poetical. I have always found the Welsh people abound more than any other nation in pretty and characteristic synonyms."*

In answer to a letter in which I asked the writer for some further information concerning the Velella, sending also an outline sketch of the animal, which I asked the writer to fill in with the proper colours, I received the following reply:—

"I will do my best to answer your questions, and to give you what information I can concerning the creatures.

"When seen at Tenby, they were all floating on the surface of the sea, the tentacles only being submerged. My specimens floated for a very short time after capture, death following so quickly that I was obliged to set to work at once with camel's-hair brush and penknife to take away the gelatinous part. Indeed, decomposition took place so rapidly, that Velellas and myself were simultaneously threatened with extermination.

"Both raft and sail were equally enveloped in a soft, gelatinous covering, certainly not more than the sixteenth of an inch in thickness, except under the centre of the raft, where it became slightly thicker. The covering of the sail was exceedingly thin, and like a transparent and almost invisible soft skin. The sail is very firmly attached to the raft, as they did not separate when decomposition began.

"The tentacles were entirely composed of the same soft, jelly-like substance as that of the envelope, and every part was iridescent in a sort of vapoury transparent cloud of many-tinted colours, blue and pale crimson predominating. I have

* By sailors the Velella is popularly known by the name of "Sally-man;" *i.e.* Sallee-man.

filled up to the best of my memory the little sketch, and only
wish you could have seen the Velellas as I did, in their full life
and beauty."

Two of the specimens here mentioned are in my collection,
and beautiful little things they are. The two plates are not
thicker than ordinary silver paper, but are wonderfully strong,
tough, and elastic. The oval horizontal plate, or raft, if it
may be so called, is strengthened by being corrugated in con-
centric lines, and having a multitude of very fine ribs radiating
from the centre to the circumference. It is slightly thickened
on the edges, evidently for the attachment of the tentacles.

The perpendicular plate, or sail, does not occupy the larger
diameter of the raft, but stretches across it diagonally from
edge to edge, rising highest in the centre and diminishing
towards the edges, so that it presents an outline singularly like
that of a lateen sail. It is rather curious that the magnifying
glass gives but little, if any, assistance to the observer, the
naked eye answering every purpose. Even the microscope is
useless, detecting no peculiarity of structure. I tried it with
the polariscope, scarcely expecting, but rather hoping, to find
that it was sensitive to polarised light. But no such result
took place, the Velella being quite unaffected by it.

The corresponding illustration is a sketch of a raft to
which a sail is attached. Such rafts as this are in use in many
parts of the world, the sail saving manual labour, and the
large steering oar answering the double purpose of keel and
rudder. In the Velella, the tentacles, though they may not
act in the latter capacity, certainly do act in that of the former,
and serve to prevent the little creature from being capsized in
a gale of wind.

The Boat.

There is no doubt that the first idea of locomotion in the
water, independently of swimming, was the raft; nor is it
difficult to trace the gradual development of the raft into a
Boat. The development of the Kruman's canoe into the *Great
Eastern,* or a modern ironclad vessel, is simply a matter of
time.

It is tolerably evident that the first raft was nothing more
than a tree-trunk. Finding that the single trunk was apt to

turn over with the weight of the occupant, the next move was evidently to lash two trunks side by side.

Next would come the great advance of putting the trunks at some distance apart, and connecting them with cross-bars. This plan would obviate even the chance of the upsetting of the raft, and it still survives in that curious mixture of the raft and canoe, the outrigger boat of the Polynesians, which no gale of wind can upset. It may be torn to pieces by the storm, but nothing can capsize it as long as it holds together.

Laying a number of smaller logs or branches upon the bars which connect the larger logs is an evident mode of forming a continuous platform, and thus the raft is completed. It would not be long before the superior buoyancy of a hollow over a solid log would be discovered, and so, when the savage could not find a log ready hollowed to his hand, he would hollow one for himself, mostly using fire in lieu of tools. The progress from a hollowed log, or "dug-out," as it is popularly called, to the bark canoe, and then the built boat, naturally followed, the boats increasing in size until they were developed into ships.

Such, then, is a slight sketch of the gradual construction of the Boat, based, though perhaps ignorantly, on the theory of displacement. Now, let us ask ourselves whether, in creation, there are any natural boats which existed before man came upon the earth, and from which he might have taken the idea if he had been able to reason on the subject. The Paper Nautilus is, of course, the first example that comes before the mind; but although, as we have seen, the delicate shell of the nautilus is not used as a boat, and its sailing and rowing powers are alike fabulous, there is, as is the case with most fables, a substratum of truth, and there are aquatic molluscs which form themselves into boats, although they do not propel themselves with sails or oars.

Many species of molluscs possess this art, but we will select one as an example of them all, because it is very plentiful in our own country, and may be found in almost any number. It is the common WATER-SNAIL (*Limnæa stagnalis*), which abounds in our streams where the current is not very strong. Even in tolerably swift streams the Limnæa may be found plentifully in any bay or sudden curve where a reverse current is generated, and therefore the force of the stream is partially neutralised. These molluscs absolutely swarm in the Cherwell, and in the multitudinous ditches which drain the flat country about Oxford into that river as well as the Isis.

Belonging to the Gasteropods, the Water-snail can crawl over the stones or aquatic vegetation, just as the common garden snail or slug does on land. But it has another mode of progression, which it very often employs in warm weather. It ascends to the surface of the water, reverses its position so

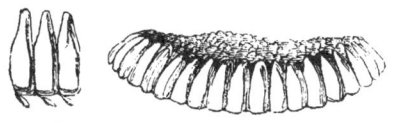

GNAT-EGG BOAT AND THREE EGGS.

"DUG-OUT" BOAT OF VARIOUS PARTS OF THE WORLD.

SEA-ANEMONE ACTING AS BOAT.

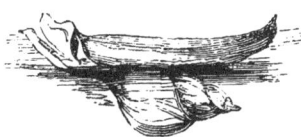

WATER-SNAIL ACTING AS BOAT.

BIRCH-BARK CANOE.

PUPA SKIN OF GNAT ACTING AS BOAT.

that the shell is downward, spreads out the foot as widely as possible, and then contracts it in the centre, so as to form it into a shallow boat.

The carrying capacity of this boat is necessarily small, but as the shell and nearly the whole of the animal are submerged, and therefore mostly sustained by the water, a very small amount of flotative power is sufficient for the purpose. Sometimes, on a fine day, whole fleets of these natural boats may be seen floating down the stream, thus obtaining a change of locality without any personal exertion.

In perfectly still water, where no current can waft the Limnæa on its easy voyage, it still is able to convey itself from one place to another. By means of extending and contracting the foot, it actually contrives to crawl along the surface of the water almost as readily as if it were upon the under side of some solid body, and, although its progress is slow, it is very steady. Another very common British water-snail, the Pouch-shell (*Physa fontinalis*), has almost exactly the same habits. Reference will be made to the Pouch-shell on another page.

The capacity for converting the body into a boat is not confined to the molluscs, but is shared by many other animals. Take, for example, the well-known marine animals, called popularly SEA-ANEMONES. As they appear when planted on the rocks, they look as incapable of motion as the flowers whose names they bear. Yet, by means of the flattened base, which they use just as a snail uses its feet, they can manage to glide along the rocks in any direction, though very slowly.

The base is capable of extension and contraction, and by elongating one side of it, fixing the elongated portion, and then raising the remainder of the base towards it, the animal makes practically a series of very slow steps. This mode of progression may often be seen in operation on the glass front of an aquarium.

The same property of expansion and contraction enables the Sea-anemones to convert their bodies into boats, and float on the surface of the water. When one of these animals wishes to swim, it ascends the object to which it is clinging— say the glass of the aquarium—until it has reached the air. It then very slowly, and bit by bit, detaches the upper part of the base from the glass, allowing itself to hang with its tentacles downward. These, by the way, are almost wholly withdrawn when the animal is engaged in this business. By degrees the whole of the base is detached from the glass except a very tiny portion of the edge. The base is next contracted in the middle into the form of a shallow cup, and, when this is done, the last hold of the glass is released, and the animal floats away, supported by its hollowed base.

Entomologists are familiar with the following facts, and were this work addressed to them alone, a simple mention of the insect would be sufficient. But as this work is intended for the general public, it will be necessary to give a description, though a brief one, of the wonderful manner in which an insect, which we are apt to think is only too common, plays the part of a boat at its entrance to life and just before its departure from this world, not to mention its intermediate state, to which reference will be made under another heading.

The insect in question is the common GNAT (*Culex pipiens*), which makes such ravages upon those who are afflicted, like myself, with delicate skins, and can have a limb rendered useless for days by a single gnat-bite.

In this insect, the beginning and the end of life are so closely interwoven, that it is not easy to determine which has the prior claim to description, but we will begin with the egg.

With very few exceptions, such as the Earwig, which watches over its eggs and young like a hen over her nest and chickens, the insects merely deposit their eggs upon or close to the food of the future young, and leave them to their fate. The eggs of the Gnat, however, require different treatment. The young larvæ, when hatched, immediately pass into the water in which they have to live, and yet the eggs are so constituted that they need the warmth of the sun in order to hatch them. The machinery by which both these objects are attained is singularly beautiful.

The shape of the egg very much resembles that of a common ninepin, and the structure is such that it must be kept upright, so that the top shall be exposed to the air and sun, and the bottom be immersed in the water. It would be almost impossible that these conditions should be attained if the eggs were either dropped separately into the water or fixed to aquatic plants, as is the case with many creatures whose eggs are hatched solely in or on the water.

As is the case with many insects, each egg when laid is enveloped with a slight coating of a glutinous character, so that they adhere together. And, in the case of the Gnat, this material is insoluble in water, and hardens almost immediately after the egg is deposited. Taking advantage of these peculiarities, the female Gnat places herself on the edge of a floating leaf or similar object, so that her long and slender hind-legs rest on the water. In some mysterious way, the eggs, as they are successively produced, are passed along the hind-legs, and are arranged side by side in such a manner that they are formed into the figure of a boat, being fixed to each other by the glutinous substance which has already been mentioned.

It is a very remarkable fact, which assists in strengthening the theory on which this book is written, that the lines of the best modern life-boats are almost identical with those of the Gnat-boat, and that both possess the power of righting themselves if capsized. In all trials of a new life-boat, one of the most important is that which tests her capability of self-righting; and any one who has witnessed such experiments, and has tried to upset a Gnat-boat, cannot but be struck with the singular similitude between the boat made by the hand of man and that constructed by the legs of an insect, without even the aid of eyes.

Push the Gnat-boat under water, and it shoots to the surface like a cork, righting itself as it rises. Pour water on it, and exactly the same result occurs, so that nothing can prevent it from floating. Then, when the warm air has done its work in hatching the enclosed young, a little trap-door opens at the bottom of the egg, lets the young larvæ into the water, and away they swim.

Now we come to another phase of existence in which the Gnat forms a boat. Every one knows the little active Gnat larvæ, with their large heads and slender bodies, much like tadpoles in miniature. When they have reached their full growth, and assume the pupal form, their shape is much changed. The fore part of the body is still more enlarged, as it has to contain the wings and legs, which have so great a proportion to the body of the perfect Gnat. And, instead of floating with its head downwards, and breathing through its tail as it did when a larva, it now floats with the head uppermost, and breathes through two little tubes.

Even in its former state the creature had something almost grotesque in its aspect, the head, when magnified, looking almost as like a human face as does that of a skate. But in its pupal state it looks as if it had put on a large comical mask much too large for it, very much like those paper masks which are enclosed in crackers, and have to be worn by those who draw them.

In process of time the pupa changes to a perfect Gnat within this shelly case, able to move, but unable to eat. The body shrinks in size, and the wings and legs are formed, both being pressed closely to the body. When the Gnat is fully developed, the pupal skin splits along the back, and opens out into a curiously boat-like shape, the front, which contains the heavier part of the insect, being much the largest, and consequently being able to bear the greatest weight.

By degrees, the Gnat draws itself out of the split pupal skin, resting its legs on it as fast as they are released. It then shakes out its wings to dry, and finally takes to the air.

It is a really wonderful fact that the insect which, for three stages in life—namely, an egg, larva, and pupa—lived in the water, should in the fourth not only be incapable of aquatic life, but should employ its old skin to protect it from that very element in which it was living only a minute or two before.

Should the reader wish to examine for himself either the egg or skin boat of the Gnat, he can easily procure them by searching any quiet pond, or even an uncovered water-butt. They are, of course, very small, averaging about the tenth of an inch in length, and are nearly always to be found close to the side either of pond or tub, being drawn there by the power of attraction.

I may here mention that there are other dipterous insects belonging to the genus Stratiomys, which undergo their metamorphosis in a very similar fashion. In these insects, the larva breathes through the tail, and when it attains its pupal condition, the actual insect is very much smaller than the pupal skin, only occupying the anterior and enlarged part. Indeed, the difference of size is so great, that several entomologists believed the future Stratiomys to be but a parasite on the original larva. The beautiful Chameleon-fly (*Stratiomys chamæleon*) is a familiar example of these insects.

NAUTICAL.

CHAPTER II.

THE OAR, THE PADDLE, AND THE SCREW.

Propulsion by the Oar.—Parallels in the Insect World.—The "Water-boatman."
—Its Boat-like Shape.—The Oar-like Legs.—Exact mechanical Analogy
between the Legs of the Insect and the Oars of the human Rower.—
"Feathering" Oars in Nature and Art.—The Water-boatman and the
Water-beetles.—The Feet of the Swan, Goose, and other aquatic Birds.—
The Cydippe, or Beroë.—The Self-feathering Paddle-wheel.—Indirect Force.
—The Wedge, Screw, and Inclined Plane.—"Sculling" a Boat.—The
"Tanka" Girls of China.—Mechanical Principle of the Screw, and its
Adaptation to Vessels.—Gradual Development of the Nautical Screw.—
Mechanical Principle of the Tail of the Fish, the Otter, and the sinuous
Body of the Eel and Lampern.—The Coracle and the Whirlwig-beetle.

THE Boat naturally reminds us of the Boatman. In the two
gnat-boats which have been described there is no propelling power used or needed, the little vessel floating about at
random, and its only object being to keep afloat. But there
are many cases where the propelling power is absolutely
essential, and where its absence would mean death, as much
as it would to a ship which was becalmed in mid ocean without
any means of progress or escape. There are, for example,
hundreds of creatures, belonging to every order of animals,
which are absolutely dependent for their very existence on
their power of propulsion, and I believe that there is not a
single mode of aquatic progression employed by man which has
not been previously carried out in the animal world. There
are so many examples of this fact that I am obliged to select a
very few typical instances in proof of the assertion.

Taking the Oar as the natural type of progression in the
water, we have in the insect world numerous examples of the
very same principle on which our modern boats are propelled.

And it is worthy of notice, that the greater the improvement in rowing, the nearer do we approach the original insect model.

The first which we shall notice is the insect which, from its singular resemblance to a boat propelled by a pair of oars, has received the popular name of WATER-BOATMAN. Its scientific name is *Notonecta glauca*, the meaning of which we shall presently see. It belongs to the order of Heteroptera, and is one of a numerous group, all bearing some resemblance to each other in form, and being almost identical in habits. Though they can fly well, and walk tolerably, they pass the greater part of their existence in the water, in which element they find their food.

Predacious to a high degree, and armed with powerful weapons of offence, it is one of the pirates of the fresh water, and may be found in almost every pond and stream, plying its deadly vocation.

Its large and powerful wings seem only to be employed in carrying it from one piece of water to another, while its first and second pairs of legs are hardly ever used at all for progression. The last pair of legs are of very great length, and furnished at their tips with a curiously constructed fringe of stiff hairs. The body is shaped in a manner that greatly resembles a boat turned upside down, the edge of the elytra forming a sort of ridge very much like the keel of the boat.

When the creature is engaged in swimming, it turns itself on its back, so as to bring the keel downwards, and to be able to cut the water with the sharp edge. From this habit it has derived the name of Notonecta, which signifies an animal which swims on its back. The first and second pairs of legs are clasped to the body, and the last pair are stretched out as shown in the illustration, not only looking like oars, but being actually used as oars.

Now, I wish especially to call the reader's attention to the curiously exact parallel between the water-boatman and the human oarsman. As the reader may probably know, the oar is a lever of the second order, *i.e.* the power comes first, then the weight, and then the fulcrum. The arm of the rower furnishes the power, the boat is the weight to be moved, and the water is the fulcrum against which the lever acts.

I have more than once heard objections to this definition, the objectors saying that the water was a yielding substance,

and therefore could not be the fulcrum.   This objection, however, was easily refuted by taking a boat up a narrow creek,
and rowing with the oar-blades resting on the shore, and not in
the water.

Now, the swimming legs of the water-boatman are exact
analogues of the oars of a human rower.   The internal muscles
at the juncture of the leg with the body supply the place of the
rower's arms, the leg itself takes the office of the oar, and the

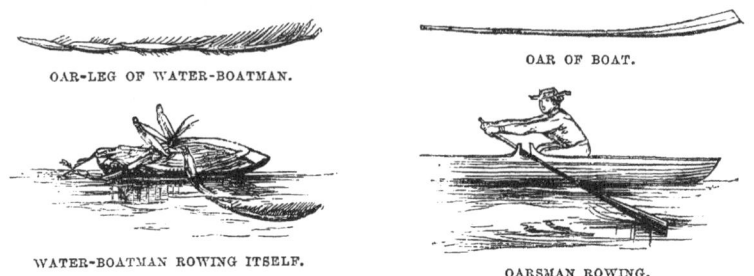

OAR-LEG OF WATER-BOATMAN.

OAR OF BOAT.

WATER-BOATMAN ROWING ITSELF.

OARSMAN ROWING.

body of the insect is the weight to be moved, and the water
supplies the fulcrum.   Even the broad blade at the end of the
oar is anticipated by the fringe of bristles at the end of the leg,
and its sharpened edge by the shape of the insect's limb.

Besides these resemblances, there is another which is worthy
of notice.   All rowers know that one of their first lessons is to
"feather" their oars, i.e. to turn the blade edgewise as soon
as it leaves the water.   Nothing looks more awkward than
for a boatman to row without feathering.   (We all must
remember the eulogy on the "Jolly Young Waterman," who
"feathered his oars with skill and dexterity.")   In the first
place, he must lift his oar very high out of the water, and, in
the second, he will be impeded by any wind that happens to
come against the blades.

The Water-boatman, however, does not lift its legs out of the
water after every stroke, as a human boatman does, and therefore it has no need to feather in the same way.   But there is
even greater need for a feathering of some kind in the insect's
leg, on account of the greater resistance offered by water than
by air, and this feathering is effected by the arrangement of the
blade-bristles, which spread themselves against the water as
the stroke is made, and collapse afterwards, so as to give as
little resistance as possible when the stroke is completed.

In Art we have invented many similar contrivances, but I believe that there is not one in which we have not been anticipated by Nature. Putting aside the insect which has just been described, we have the whole tribe of water-beetles, in which the same principle is carried out in an almost identical manner. In the accompanying illustration, the oar, the rower, and the boat are placed above one another, and next to them are seen one of the oar-legs of the water-boatman and the insect as it appears when swimming on its back.

Then, there is the foot of the duck, goose, swan, and various other aquatic birds, in which the foot presents a broad blade as it strikes against the water, and a narrow edge as it recovers from the stroke. Some years ago, a steam yacht was built and propelled by feet made on the model of those of the swan. She was a very pretty vessel, but art could not equal nature, and at present the swan-foot propeller, however perfect in theory, has not succeeded in action. Perhaps, if some nautical engineer were to take it in hand, he would procure the desired result.

Almost exactly similar is the mode of propulsion employed by the lobster, the prawns and shrimps, their tails expanding widely into a fan-like shape as they strike against the water, and then collapsing when the stroke is withdrawn, so as to allow them to pass through the water with the least possible resistance.

The same principle is to be seen in the lively little Acaleph, for which there is unfortunately no popular name, and which we must therefore call by its scientific title of CYDIPPE, or Beroë, these names being almost indifferently used. When full grown, it is about as large as an acorn, and very much of the same shape. It is as transparent as if made of glass, and, when in the water, is only visible to practised eyes.

En passant, I may remark that the familiar term of "water," when applied to diamonds, is owing to their appearance when placed in distilled water. Those which can be at once seen are called stones of the second water. Those which cannot be seen, because their refractive powers are equal to those of the water, are called " diamonds of the first water," and are very much more valuable than the others.

As the Cydippe is, in fact, little more than sea-water,

entangled in the slightest imaginable and most transparent tissue of animal fibre, it is evident that the water and the Cydippe must be of almost equal refracting power, and that therefore the acaleph must be as invisible as diamonds of the "first water." Indeed, I have often had specimens in a glass jar which were absolutely invisible to persons to whom I wished to show them.

But an experienced eye detects the creature at once. Along its body, at equal distances, are eight narrow bands, over which the colours of the rainbow are, though very faint, perpetually rippling. This appearance is caused by the machinery which impels the body, and which seems never to cease. Each of these bands is composed of a vast number of tiny flaps, which move up and down in regular succession, so as to cause the light to play on their surfaces. And, as they move as if set on hinges, they of course offer no resistance to the water after their stroke is made.

Now let us compare these works of nature with those of art. We have already seen the parallels of the oar, and we now come

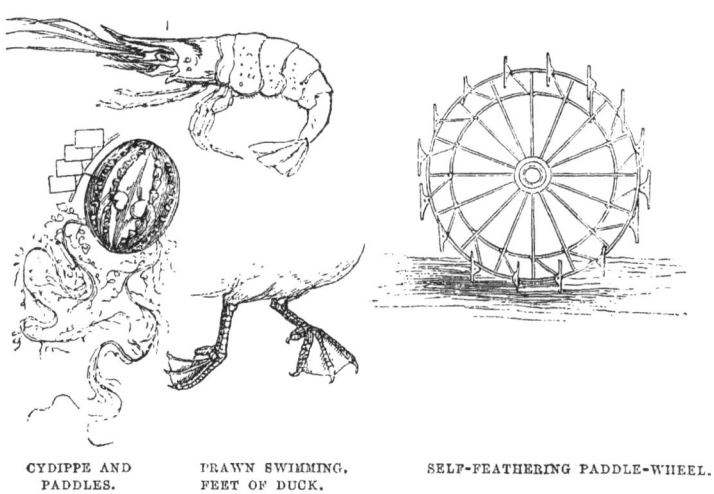

CYDIPPE AND PADDLES.   PRAWN SWIMMING, FEET OF DUCK.   SELF-FEATHERING PADDLE-WHEEL.

to those of the paddle-wheel. When paddle-steamers were first invented, the blades were fixed and projected from the wheel, as if they had been continuations of its spokes. It was found, however, that a great waste of power, together with much inconvenience, was caused by this arrangement. Not only was a considerable weight of water raised by each blade after it passed the middle of its stroke, but the steam power was given nearly as much to lifting and shaking the vessel as to propulsion.

A new kind of paddle-wheel was then invented, in which the blades were ingeniously jointed to the wheel, so that they presented their flat surfaces to the water while propelling, and their edges when the stroke was over. This, which is known by the name of the "Self-feathering Paddle-wheel," was thought to be a very clever invention, and so it was; but not even the inventors were likely to have known that if they had only looked into the book of Nature, they might have found plenty of self-feathering paddle-wheels, beside the few which my limited space enables me to give.

If the reader will look at the illustration, he will see that on one side is represented the self-feathering paddle-wheel of Art, with its ingenious arrangement of rods and hinges. On the other side there comes, first, the common Prawn, shown with its tail expanded in the middle of its stroke.

Just below it is a Cydippe of its ordinary size, showing the paddle-bands, one of which is drawn at the side much magnified, so as to show the arrangement of the little paddles. As to the tentacles which trail from the body, we shall treat of them when we come to our next division of the subject of the work.

Lastly, there is a representation of the self-feathering feet of the Duck, the left foot expanded in striking the water, and the right closed so as to offer no resistance when drawn forward for another stroke. The swan's foot shows this action even more beautifully than does that of the duck.

WE now come to another mode of propulsion, namely, that which is not due to direct pressure of a more or less flat body against the water, but to the indirect principle of the screw, wedge, or inclined plane.

Space being valuable, I will only take two instances, namely, the well-known mode of propelling a boat by a single oar working in a groove or rowlock in the middle of the stern, and the ordinary screw of modern steamers.

Most of my readers must have seen a sailor in the act of

C

" sculling " a boat.   A tolerably deep notch is sunk in the
centre of the stern, and the oar is laid in it, as shown in the
central illustration, on the right-hand side.   The sailor then
takes the handle of the oar, and works it regularly backwards
and forwards, without taking the blade out of the water.   The
boat at once begins to move forward, and, when the oar is

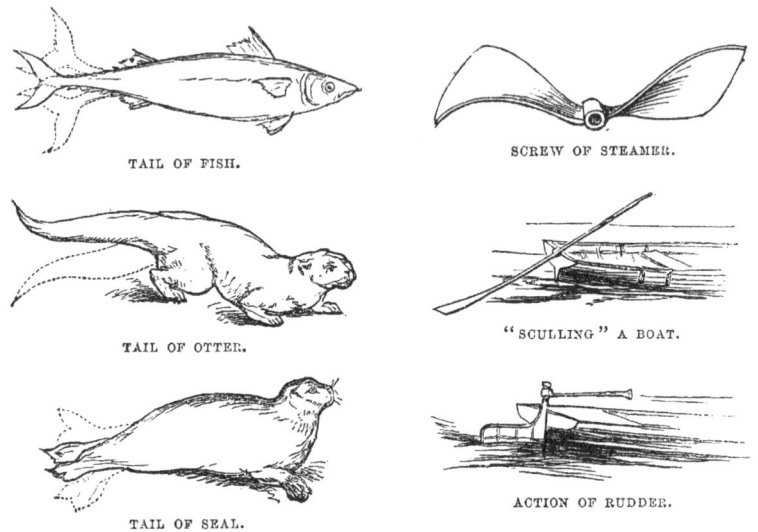

TAIL OF FISH.

SCREW OF STEAMER.

TAIL OF OTTER.

"SCULLING" A BOAT.

TAIL OF SEAL.

ACTION OF RUDDER.

urged by a strong and experienced man, can be propelled with
wonderful speed.   The well-known " Tanka " boat-girls of
China never think of using two oars, a single oar in the stern
being all-sufficient for the rapid and intricate evolutions
required in their business.

The mechanical process which is here employed is nothing
more than that of the inclined plane, or rather, the wedge, the
oar-blade forming the wedge, and the force being directed
against the stern of the boat, and so driving it through the
water.

The Rudder affords another example of a similar force,
although it is used more for directing than propelling a vessel.
Still, just as the scull is used not only for propelling, but for
steering the boat, the rudder, when moved steadily backwards
and forwards, can be used for propulsion as well as steerage.
In the absence of oars, this property is most useful, as I can
practically testify.

So different in appearance are the screw and the inclined plane, that very few people would realise the fact that the screw is nothing but an inclined plane wound round a cylinder, or rather, is a circular inclined plane. The ordinary corkscrew is a good example of this principle, the cylinder being but an imaginary one.

Now, if the screw be turned round, it is evident that force is applied just on the principle of the wedge, and this principle is well shown in the various screw-presses, of which the common linen-press is a familiar example, as was the original printing-press, which still survives as a toy for children.

We all know the enormous force exerted by screws when working in wood, and how, when the screw-driver is turned in the reverse direction, the instrument is forced backwards, though the operator is leaning against it with all his weight. In fact, a comparatively small screw, if working in hard wood or metal, so that the threads could not break, could lift a heavy man.

Substitute water for wood or metal, and the result would be the same in principle, though the resistance would be less. As the loss of power by friction would prevent a large vessel from being propelled by a stern oar moved like a scull, the idea was invented of applying the same kind of power by a large screw, which should project into the water from the stern of the vessel. This modification, moreover, would have the advantage of forcing the vessel forward when the screw was turned from left to right, and drawing it back when turned in the opposite direction, whereas the sculling oar would only drive it forward.

The principle was right enough, but there was at first a great difficulty in carrying it out. Firstly, several turns of a large screw were used, and were found to need power inadequate to the effect. Then the screw was reduced to four separate blades, and now only two are used, as shown in the illustration, these saving friction, being equally powerful for propulsion, and running less risk of fouling by rigging blown overboard or other floating substances.

So much for Art. Now for the same principle as shown in Nature, of which I can take but a very few instances.

The first and most obvious example is that of the Fish-tail, which any one may observe by watching ordinary gold fish in a bowl. Their progression is entirely accomplished by the movement of the tail from side to side, exactly like that of the sculling oar, and moreover, like the oar, the tail acts as rudder as well as propeller.

The force with which this instrument can be used may be estimated by any one who is an angler, and knows the lightning-like rush of a hooked trout, or who has seen the wonderful spring with which a salmon shoots clear out of the water, and leaps up a fall several feet in height. This is not done, as many writers state, by bending the body into a bow-like form, and then suddenly straightening it, but by the projectile force which is gained by moving the tail backwards and forwards as a sculler moves his oar.

Perhaps some of my readers have seen the wonderful speed, ease, and grace with which an Otter propels itself through the water. As the otter feeds on fish, and can capture even the salmon itself, its powers of locomotion must be very great indeed. And these are obtained entirely by means of the tail, which is long, thick, and muscular, and can be swept from side to side with enormous force, considering the size of the animal. The legs have little or nothing to do with the act of swimming. The fore-legs are pressed closely against the body, and the hind-legs against each other. The latter act occasionally as assistants in steering, but that is all.

Then there are the various Seals, whose hind-legs, flattened and pressed together, act exactly like the tail of the fish, that of the otter, the oar of the sculler, or the screw of the steamer. Also, the eel, when swimming, uses exactly the same means, its lithe body forming a succession of inclined planes; so does the snake, and so does the pretty little lampern, which is so common in several of our rivers, and so totally absent from others.

I can only now give a short description of the woodcut which illustrates these points.

On the right hand Art is shown by the screw-blades of the modern steamer. In the middle is the ordinary mode of sculling a boat by an oar in the stern, and below it is the rudder, which, like the sculling oar, may be used either for propulsion or direction.

On the left hand we have three examples of the same

mechanical powers as shown in Nature. The uppermost figure represents a fish as in the act of swimming, the dotted lines showing the movement of its tail, and the principle of the wedge. In the middle is an otter, just preparing to enter the water, and below is a seal, both of them showing the identity of mechanism between themselves and the art of man. I need not say that the mechanism of art is only a feeble copy of that of nature, but nothing more could be expected.

WHILE we are on this subject I may as well mention two more applications of the screw principle. The first is the windmill, the sails of which are constructed on exactly the same principle as the blades of the nautical screw. Only, as they are pressed by the wind, and the mill cannot move, they are forced to revolve by the pressure of the wind, just as the screw of a steamer revolves when the vessel is being towed, and the screw left at liberty.

Moreover, just as the modern screws have only two blades, so, many modern windmills have only two sails, the expense and friction being lessened, and the power not injured.

Again : some years ago there was a very fashionable toy called the aërial top. It was practically nothing but a windmill in miniature, rapidly turned by a string, after the manner of a humming-top. The edges of the sails being turned downwards, the instrument naturally screwed itself into the air to a height equivalent to the velocity of the motion.

A similar idea has been mooted with regard to the guidance of balloons, or even to aërial voyaging without the assistance of gas, but at present the weight of the needful machinery has proved to be in excess of the required lifting power.

In fine, the application of the inclined plane, wedge, or screw as a motive power, is so wide a subject that I must, with much reluctance, close it with these few and obvious examples.

IT is worth while, by the way, to remark how curiously similar are such parallels. I have already mentioned the very evident resemblance between the water-boatman, the water-beetles, and the human rower, the body of the insect being shaped very much like the form of the modern boat. I must now draw the attention of the reader to the similitude between the very primitive boat known by the name of Coracle, and the common Whirlwig-beetle (*Gyrinus natator*), which may be found in nearly every puddle. The shape of the insect is almost identical with that of the boat, and the paddle of the

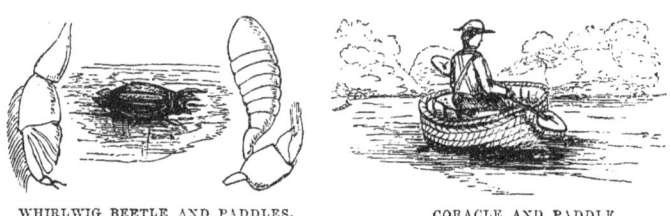

WHIRLWIG BEETLE AND PADDLES.   CORACLE AND PADDLE.

coracle is an almost exact imitation of the swimming legs of the whirlwig. And, as if to make the resemblance closer, many coraclers, instead of using a single paddle with two broad ends, employ two short paddles, shaped very much like battledores.

NAUTICAL.

CHAPTER III.

SUBSIDIARY APPLIANCES.—Part I.

General Sketch of the Subject.—The Mast of Wood and Iron.—Analogy between the Iron Mast and the Porcupine Quill.—The Iron Yard and its Shape prefigured by the same Quill.—Beams of the Steam-engine.—Principle of the Hollow Tube in place of the Solid Bar.—Quills and Bones of Birds.—Wheat Straws and Bamboos.—Structure of the Boat.—The Coracle, the Esquimaux Boat, and the Bark Canoe.—Framework of the Ship and Skeleton of the Fish. —Compartments of Iron Ship and Skull of Elephant.—The Rush, the Cane, and the Sugar-cane.—" Stellate " Tissue and its Varieties.

HAVING now treated of the raft, the boat, the ship, and their various modes of propulsion and guidance, we come to the subsidiary appliances to navigation, if they may be so called in lack of a better name.

First in importance is necessarily the mast; and the yards, which support the sails, are naturally the next in order. Then there come the various improvements in the building of vessels; namely, the substitution of planks fastened on a skeleton of beams for a mere hollowed log, and the subsequent invention of iron vessels with their numerous compartments, giving enormous strength and size, with very great comparative lightness.

Then we come to the various developments of the ropes or cables, by which a vessel is kept in its place when within reach of ground, whether on shore or at the water-bed. Next come the different forms of anchors which fasten a vessel to the bed of the ocean, of grapnels by which she can be made fast to the shore, or of " drags," which at a pinch can perform either office, and can besides be utilised in searching for and hauling up objects that are lying at the bottom of the sea.

Next we come to the boat-hook, which is so useful either as a temporary anchor, or as a pole by which a boat can be propelled by pushing it against the shore or the bed of the water ; and then to the "punt-pole," which is only used for the latter purpose.

Lastly, we come to the life-belt and life-raft, which are now occupying, and rightly, so much of the public attention. These subjects will be treated in their order in the present chapter, and I hope to be able to show the reader that in all these points nature has anticipated art.

I presume that most, if not all, of my readers are aware of the rapidly extending use of iron in ship-building, not only in the standing rigging, but in the material of the vessel. First there came iron "knees," *i.e.* the angular pieces of wood which strengthen the junctions of the timbers. Formerly these were made of oak-branches, and, as it was not easy to find a bough which was naturally bent at such an angle as was required for a "knee," such branches were exceedingly valuable. Iron, however, was then employed, and with the best results. It was lighter than the wooden knee, was stronger, could be bent at any angle, and took up much less space.

By degrees iron was used more and more, until vessels were wholly made of that material. Then the masts, and even the yards, were made of iron, and, strange as it may appear, were found to be lighter as well as stronger than those made of wood. Of course, the masts and yards were hollow, and it was found by the engineers that in order to combine lightness with great strength, the best plan was to run longitudinal ridges along the inside of the tube.

A section of one of these masts is given at Fig. B, and taken from the drawings of one of our largest engineering firms. The reader will see that the mast is composed of rather slight material, and that it is strengthened by four deep though thin ribs, which run throughout its length.

When I first saw this mast I was at once struck with the remarkable resemblance between it and the quill of the Porcupine. These quills, as all anglers know, are very light, and of extraordinary strength when compared with their weight. Indeed, they are so light that they are invaluable as penholders to those who are obliged to make much use of their pen. I have used nothing else for a very long time, and the drawing of the Porcupine quill which is here given at Fig. A was made from a small piece cut from the top of the penholder which I have used for some fifteen years, and with which all my largest and most important works were written, including the large "Natural History," "Homes without Hands," "Man and Beast," &c., &c. A portion of the same quill is also shown of its real size.

If the reader will cut a Porcupine quill at right angles, make a thin section of it, and place it under the microscope, or even under an ordinary pocket lens, he will see that the exterior is composed of a very thin layer of horny matter, and the interior filled with a vast number of tiny cells, which are formed much on the same plan as the pith of elder and other plants. The analogies of the pith will be treated in another page.

But were the quill merely a hollow tube filled with pith, it would be too weak to resist the strain to which it is often

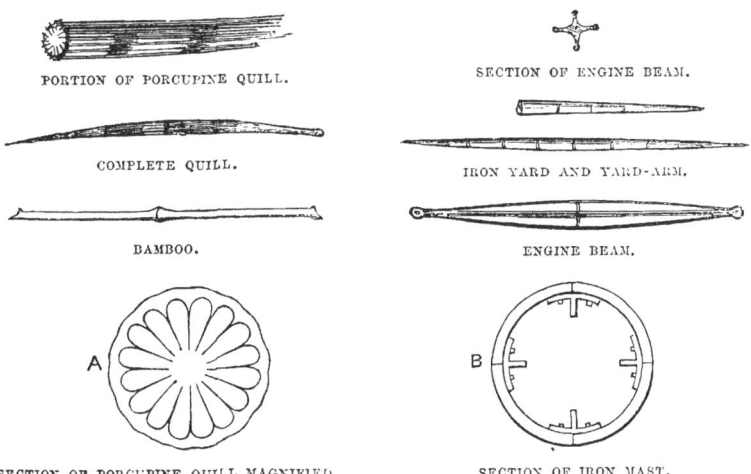

PORTION OF PORCUPINE QUILL.

SECTION OF ENGINE BEAM.

COMPLETE QUILL.

IRON YARD AND YARD-ARM.

BAMBOO.

ENGINE BEAM.

A

B

SECTION OF PORCUPINE QUILL MAGNIFIED.

SECTION OF IRON MAST.

liable. Consequently it is strengthened by a number of internal ribs, composed of the same horny material as the outer coat, and arranged in exactly the same way as those of the mast.

There are yet other points in the structure of the Porcupine quill which might be imitated with advantage in the mast. In the first place, the internal ribs are much more numerous than those of the mast, but they are very much thinner, and taper away from the base, where the greatest strain exists, to the end, where they come to the finest imaginable edge. This modification of structure enables the outer shell of the quill to be exceedingly thin and light, and, moreover, gives to the whole quill an elasticity which is quite wonderful, considering its weight and strength.

Then, in the iron mast the exterior is quite smooth, whereas in the Porcupine quill it is regularly indented, exactly on the principle of the corrugated iron, which combines great strength with great lightness. And I cannot but think that our iron masts might be made both lighter and stronger if the shell were thinner, the internal ribs made like those of the Porcupine quill, and the shell corrugated instead of being quite smooth. The internal cells of the quill are, of course, not needed in the mast, as they are intended for nutrition, and not for strength.

BEING on this subject, we may take the shape of the Porcupine quill, and compare it with that of the ship's yard. It will be seen that the two are so exactly similar in form that the outline of one would answer perfectly well for the other. The only perceptible difference is, that in the ship's yard both ends are alike, whereas in the Porcupine quill the end which is inserted in the skin is rounded and slightly bent, while the other end is sharply pointed.

The principal point to be noticed in the form of both quill and yard is, that they become thicker in the centre, that being the spot on which the greatest strain comes, and which, in consequence, needs to be stronger than any other part. While holding and balancing the pole which Blondin uses to preserve his balance when walking on the high rope, I was struck with the fact that the pole, which is heavily weighted at each end, had to be strengthened in the middle, exactly on the principle of the Porcupine quill and the ship's yard. It could not, of course, be thickened, as the hands could not grasp it, but it had to be furnished with additional strengthening. And the necessity of such strengthening is evident from the fact that on one occasion the pole did break in the middle, so that any one of less nerve and presence of mind must have been killed.

Bearing in mind, then, that in a rod or pole the centre is the part which most requires to be strengthened, we can see, in cases too numerous to mention, how art has followed, though perhaps unconsciously, in the footsteps of nature. Take, for example, the beam of a steam-engine, such as is given in the sketch, and for which the great engine at Chatham acted as model. The reader will observe that in this case the beam is gradually thickened towards the centre, the ends, where the strain is slightest, being comparatively small.

Another point also must be noticed. Equal strength could have been obtained had the beam been solid, but at the expense of weight, and consequent waste of power. Lightness is therefore combined with strength by making the beam consist of a comparatively slight centre, but having four bold ridges, as shown in the section given in the accompanying illustration. This plan, as the reader will see, is exactly the same as that which is adopted in the iron mast and porcupine quill, except that the ridges are external instead of internal. The same mode of construction is employed in ordinary cranes, the principal beam of which is almost identical in form with that of the engine, both being thickest in the centre, and both strengthened with external ridges.

There are also other analogies between the hollow mast and natural objects. Keeping still to the animal world, we find the quill feathers of the flying birds to supply examples of the combination of great strength with great lightness and very little expenditure of material. Their wing bones, too, are hollow, communicating with the lungs, and are consequently light as well as strong.

Passing to the vegetable world, we find a familiar example of this structure in the common Wheat Straw. The ripe ear is so heavy, when compared with the amount of material which can be spared to carry it, that if the stalk were solid it would give way under the mere weight of the ear. Moreover, the full-grown corn has to endure much additional weight when wetted with rain, and to resist much additional force when bowed by the wind, so that a slight and solid stalk would be quite inadequate to the task of supporting the ear.

The material of the stalk is therefore utilised in a different manner, being formed into a hollow cylinder, the exterior of which is coated with a very thin shell of flint, or "silex" as it is scientifically termed. The result of this structure is that the stem possesses strength, lightness, and elasticity, so as to be equal to the burden which is laid upon it.

Then there is the common Bamboo, which is little more than a magnified straw, being constructed in much the same manner, and possessing almost the same constituents of vegetable matter and silex.

Perhaps the most extraordinary of the tubal system is to be found in the remarkable plant of Guiana called by the natives Ourah, and scientifically known by the name of *Arundinaria Schomburgkii*. Like the bamboo, it grows in clusters, and has a feathery top, which waves about in the breeze. But, instead of decreasing gradually in size from the base upwards, the Ourah, although it runs to some fifty feet in height, is nowhere more than half an inch in diameter. The first joint is about sixteen feet in length, and uniform in diameter throughout.

It is scarcely thicker than ordinary pasteboard, and yet so strong and elastic is it, that it can sustain with ease the weight and strain of its feathery top as it blows about in the breeze. The natives of certain parts of Guiana use this reed as a blow-gun, and I have a specimen, presented to me by the late Mr. Waterton, which is eleven feet in length.

So the reader will see that when engineers found that hollow iron beams were not only lighter, but stronger than solid beams, they were simply copying the hollow beams formed by Nature thousands of years ago.

ANOTHER great improvement in ship-building now comes before us.

We have already seen that the earliest boats were merely hollowed logs, just as Robinson Crusoe is represented to have made. But these had many disadvantages. They were always too heavy. They were liable to split, on account of flaws in the wood, and if a large vessel were needed, it was difficult to find a tree sufficiently large, or to get it down to the water when finished.

So the next idea was to build a skeleton, so to speak, of light wooden beams, and to surround it with an outer clothing, or skin, if it may be so termed. As far as I know, the two original types of this structure are the Coracle of the ancient Briton, and the birch-bark Canoe of the North American Indian, and it is not a little remarkable that both exist to the present day, with scarcely any modification.

The Coracle has been already represented on page 22. It is, perhaps, or was in its original form, the simplest boat in existence, next to the "dug-out." In the times of the very ancient Britons, who were content with blue paint by way of dress, and lived by hunting and fishing, the Coracle was a basin-shaped basket of wicker-work, rather longer than wide, and covered with the skin of a wild ox. This was sufficiently light to be carried by one man, and sufficiently buoyant to bear him down rapids, if he were a skilful paddler, and, of course, formed a considerable step in civilisation.

The modern Coracle is identical in form, and almost in material. The frame is still oval and basin-shaped, and made of wicker, but the outer covering is not the same. An ox-hide is an expensive article in these days, and, especially when wetted, is very heavy. So the modern Coracle builder covers the wicker skin with a piece of tarpaulin, which is much cheaper than the ox-hide, much lighter, is equally water-tight, and has the great advantage of not absorbing moisture, so that it is as light after use as before.

The Esquimaux make a boat on very similar principles. It is simply hideous in form, resembling a huge washing-tub in shape, but, as it is only intended for the inferior beings called women, this does not signify.

Best, most perfect, and most graceful of all such boats is the Birch-bark Canoe of the North American Indians, whose shape has evidently been borrowed from that of a fish. I have seen many of these canoes, and have now before me several models which are exactly like the originals, except in point of size. Instead of being mere elongated bowls, like the coracle, they are long and slender, swelling out considerably in the middle, and coming to an almost knife-like edge at each end. Both stem and stern are alike, so that the canoe can be paddled in either direction, and, as one of the paddlers always acts as steersman, no rudder is needed.

The mode of construction is perfectly simple. The labour is

divided between the sexes: the women cut large sheets of bark from the birch-trees, scrape and smooth them, and then sew them together, so as to form the outer skin, or "cloak" as it is called, of the canoe. Meanwhile the men are making the skeleton of strips of white cedar-wood, and binding them into shape with thongs made of the inner bark of the same tree, just like the "bass" of our gardeners. The "cloak" is then gradually worked over the skeleton, sewn into its place, and the canoe is finished. A figure of this canoe, as completed, is given in the same illustration as that which represents various forms of boat, page 7.

The last improvement is that which was caused by the necessity for large vessels, when planks or iron plates were fastened over the skeleton. But, in all these cases, the vessel is built on the principle of the thorax of a vertebrate animal, that of the whale or a fish being an admirable example. It only needs to take the skeleton of a whale, turn it on its back, and the ribs will be seen to form an almost exact reproduction of those of any ship being built in the nearest dockyard.

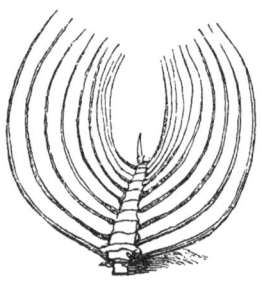
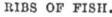
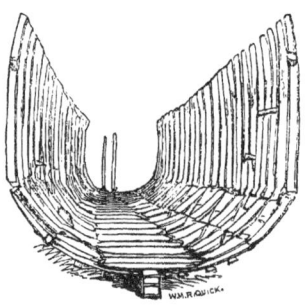

RIBS OF FISH.   RIBS OF SHIP.

I have now before me the spine and ribs of a herring. The fish was over-boiled, and the flesh fell off the bones as it was being lifted out of the dish, leaving most of the ribs in their places. When held with the spine downwards, and viewed from one end, the resemblance to the framework of a ship is absolutely startling, the ribs representing the beams, and the spine taking the place of the keel. I have also before me a sketch representing a section of a Fijian canoe, and it is remarkable that even the very curve of the ribs of the herring is reproduced in those of the canoe.

Whether the Fijians derived this peculiar and beautiful curve from the ribs of a fish I cannot say, but think it very likely.

A STILL greater improvement in ship-building now comes before us, and this also has been anticipated both in the animal and vegetable kingdoms.   There are so many examples of this anticipation that I can only give one or two.

The improvement to which I refer is that which is now almost universally employed in the construction of iron ships, namely, the making the outer shell double instead of single, and dividing it into a number of separate compartments.   Putting aside the advantage that if the vessel were stove, only one compartment would fill, we have the fact that the ship is at the same time

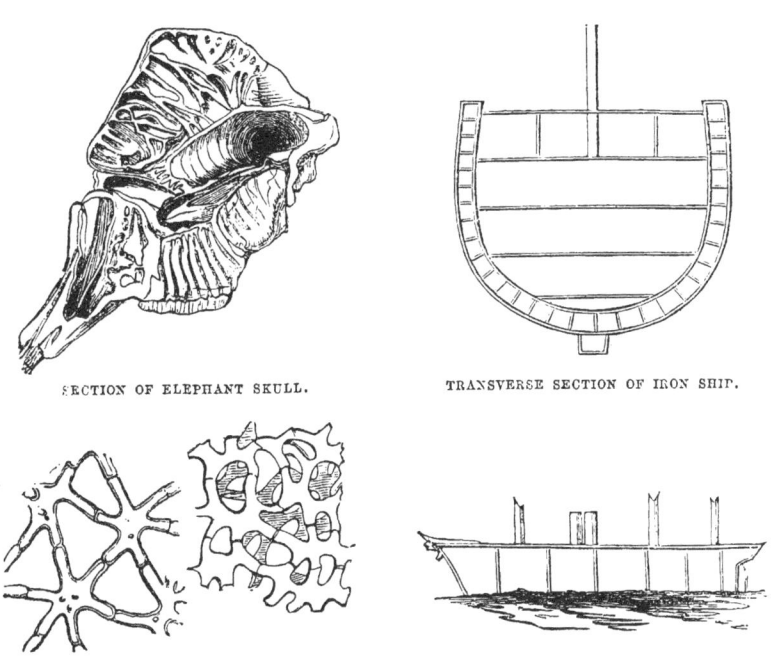

SECTION OF ELEPHANT SKULL.   TRANSVERSE SECTION OF IRON SHIP.

STELLATE TISSUES.   LONGITUDINAL SECTION OF IRON SHIP.

enormously strengthened and very light in proportion to her bulk.

Perhaps the best, and certainly the most obvious, example of this principle in the animal world is to be found in the skull of the Elephant.   The enormous tusks, with their powerful leverage, the massive teeth, and the large and weighty proboscis, require a corresponding supply of muscles, and consequently a large surface of bone for the attachments of these muscles. Now, were the skull solid in proportion to its requisite size, its weight would be too much for the neck to endure, however short and sturdy it might be. The mode of attaining expanse of surface, together with lightness of structure, is singularly beautiful.

Perhaps some of my readers may not be aware that the bone of the skull consists of an outer and inner plate, with a variable arrangement of cells between them. In many animals, such, for example, as man, where the jaws are comparatively feeble, and the teeth small and light, the size of the skull is practically that of the brain, to which it affords a covering. The same structure may be observed in the skull of the common sparrow, where, as in man, the two bony plates are set almost in contact.

But in the elephant these external and internal plates are set widely apart, and the space between them is filled with bony cells, much resembling those of a honeycomb. They are, in fact, just the same cells as those which exist in the skull of man and sparrow, but they are very much enlarged, and in consequence give a large surface, accompanied with united strength and lightness.

There are many other examples in the animal kingdom, but our limited space will not allow them to be even mentioned.

As to the vegetable examples of this principle, they are so multitudinous that only a very slight description can be given of them.

I suppose that most boys have seen a " cane " (whether they have felt it or not is not to the purpose), and some boys have made sham cigars from pieces of cane. In either case they must have noticed that the cane is not solid, but is pierced with a vast number of holes, passing longitudinally through it, and is, in fact, a collection of little tubes connected and bound together by a common envelope.

The Sugar-cane, if cut across, is seen also to consist of multitudinous cells, which, however, are not hollow, but filled with the sweet liquid from which sugar is obtained by boiling. Then there are many of our common English plants, like the ordinary rush or reed, which are very slight in diameter in comparison with their length, and in which the cells are still further strengthened and lightened by the projection of their sides into a number of points which meet each other, and leave interstices between them. This modification of the cellular system is called "Stellate" (or star-like) Tissue, and two examples of it are given in the illustration, one being taken from the common rush, and the other from the seed-coat of the privet. A very good specimen of stellate tissue may be obtained by cutting a thin section of the white inner peel of the orange.

NAUTICAL.

CHAPTER IV.

SUBSIDIARY APPLIANCES.—Part II.

The Cable and its Variations.—Material of Cables.—Hempen and Iron Cables, and Elasticity of the latter.—Natural Cables.—The "Byssus" of the Pinna and the common Mussel.—The Water-snail and its Cable.—A similar Cable produced by the common White Slug.—The Principle of Elasticity.—Elastic Cable of the Garden Spider.—Tendrilous Cables of the Pea and the Bryony.—The Vallisneria, and its Development through the Elastic Cable.—Proposed Submarine Telegraph Cable.—The Anchor, Grapnel, and their Varieties.—Natural Anchors.—Spicule of Synapta.—The Grapnel, natural and artificial.—Ice-anchor and Walrus Tusks.—The Mushroom Kedge.—The Flesh-hook.—Eagle-claw.—The Grapple-plant of South Africa.—The Drag.

AMONG the most important accessories to a ship are the Cable, by which she can be anchored to the bed of the sea, and the ropes called "warps," by which she can be fastened to the land.

Perhaps my readers may not know the old riddle—"How many ropes are there on board a man-of-war?" The non-nautical individual cannot answer, but the initiated replies that there are only three, namely, the man-rope, the tiller-rope, and the rope's-end, all the others being "tacks," "sheets," "haulyards," "stays," "braces," &c.

Formerly cables were always made of hemp, enormously thick, and most carefully twisted by hand. Now, even in small vessels, the hempen cable has been superseded by the iron chain, and this for several reasons.

In the first place, it is much smaller in bulk, and therefore does not occupy so much room. In the next place, it is even lighter than the hempen cable of corresponding strength; and, in the third, its specific gravity—*i.e.* its weight when compared with an equal bulk of water—is so great, that when submerged, it falls into a sort of arch-like form, and so attains an elasticity which takes off much of the strain on the anchor, and protects it from dragging.

WE will now look to Nature for Cables.

The natural cable which will first suggest itself is evidently that of the Pinna Shell (*Pinna pectinata*), which fixes its shell

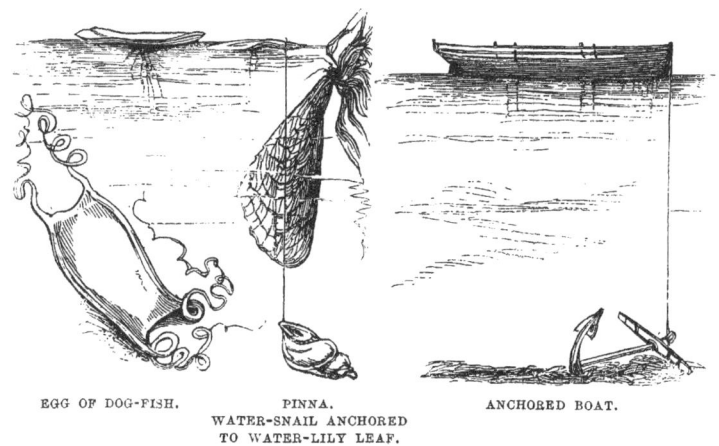

EGG OF DOG-FISH.   PINNA.   ANCHORED BOAT.
                    WATER-SNAIL ANCHORED
                    TO WATER-LILY LEAF.

to some rock or stone with a number of silk-like threads, spun by itself, and protruding from the base, just as a vessel on a lee shore throws out a number of cables. The threads which compose the "byssus," as it is called, are only a few inches in length, and apparently slight. They are, however, really strong, and by acting in unison enable the shell, though sometimes two feet in length, to be held firmly to the rock. I may here mention that they have been occasionally woven into gloves, and other articles of apparel, to which their natural soft grey-brown hue gives a very pleasing appearance.

A still more familiar instance of a natural marine cable is given by the common Mussel, which can be found in thousands on almost every solid substance which affords it a hold. Even copper-bottomed ships are often covered with Mussels, all clinging by their natural cables, and it is thought that the cases which sometimes occur of being poisoned by eating Mussels, or "musselled," as the malady is called by the sea-faring population, are due to the fact that the Mussels have anchored themselves to copper, and have in consequence imbibed the verdigris.

PASSING from salt to fresh water, we come to a natural cable which is very common, and yet, on account of its practical invisibility, is almost unknown, except by naturalists. I refer to the curious cable which is constructed by the common Water-snail (*Limnæa stagnalis*), which has already been mentioned in its capacity of a boat.

This creature has a way of attaching itself to some fixed object, such as a water-lily leaf, by means of a gelatinous thread, which it can elongate at pleasure, and by means of which it can retain its position in a stream, or in still water can sink itself to the bottom, and ascend to the same spot. This cable seems to be made of the same glairy secretion as that which surrounds the egg-masses which are found so plentifully on leaves and stones in our fresh waters, and, like that substance, is all but invisible in the water, so that an inexperienced eye would not be able to see it, even if it were pointed out

Slight, gelatinous, and almost invisible in the water as is this thread, its strength is very much greater than might be supposed. Not only can a mollusc be safely moored in the water by such a cable, but it can be actually suspended in the air, as may be seen from a letter in Hardwicke's *Science Gossip* for 1875, p. 190 :—

"Last summer (September 29) I met with the following unusual fact. In a green-house, from a vine-leaf which was within a few inches of the glass . . . a slug was hanging by a thread, which was more than four feet in length, not unlike a spider-web, but evidently much stronger.

"The slug was descending by means of this thread, and, as the glutinous matter from the under part of the body was drawn out by the weight of the creature, it was consolidated into a compact thread by the slug twisting itself in the direction of the hands of a clock, the power of twisting being given by the head, and the part of the body nearest the head being turned in the direction of the twist. There was no tendency to turn in the contrary direction. Evidently the thread became hard as soon as it was drawn away from the body.

"By wetting the sides of slips of glass, I secured two specimens of the thread. In one of these, part was stretched, and part quite loose, the latter appearing flat when seen through a microscope. The thread, which was highly elastic, was increased about three inches in a minute. The slug was white, and about an inch and a half in length."

Now we come to the elastic system of the Chain Cable, and find it anticipated in Nature in various ways.

One curious example was that of a Spider, which found its wheel-like net in danger from a tempestuous wind. The Spider descended to the ground, a depth of about seven feet, and, instead of attaching its thread to a stone or plant, fastened it to a piece of loose stick, hauled it up a few feet clear of the ground, and then went back to its web. The piece of stick thus left suspended acted in a most admirable manner, giving strength and support, and at the same time yielding partly to the wind.

By accident the thread became broken, and the stick, which was about as thick as an ordinary pencil, and not quite three inches in length, fell to the ground. The Spider immediately descended, attached another thread, and hauled it up as before. In a day or two, when the tempestuous weather had ceased, the Spider voluntarily cut the thread, and allowed the then useless stick to drop.

A curious example of the elastic cable is seen in the egg-case of the Dog-fish, which is given on page 35. The egg-case is formed like that of the common skate, and has a projection from each of its angles. But the projections, instead of being mere flattened horns, are lengthened into long elastic strings, tapering towards the ends, and twisted spirally, like the tendrils of a grape-vine.

These tendril-like appendages twist themselves round sea-weeds and other objects, and, on account of their spiral form, can hardly ever be torn from their attachments. Sometimes after a storm the egg is thrown on the shore, still clinging to the seaweed, but to find an egg detached is very rarely done.

I have already mentioned the tendrils of the vine, and their great strength. The reader may remember the corresponding cases of the Pea and the Bryony, the latter being a most remarkable example of the strength gained by the spiral form.

It clambers about hedges, is exposed to the fiercest winds, has large and broad leaves, and yet such a thing as a Bryony being blown off a hedge is scarcely, if ever, seen. I never saw an example myself, though I have had long experience in hedges.

ANOTHER excellent example of this principle is found in the Vallisnéria plant, which of late years has become tolerably familiar to us through the means of fresh-water aquaria, though it is not indigenous to this country.

In this plant the elastic power of the spiral cable is beautifully developed. It is an aquatic plant, mostly found in running waters, and has a most singular mode of development. It is diœcious—*i.e.* the male, or stamen-bearing, and the female, or pistil-bearing flowers, grow upon separate plants.

It has to deposit its seeds in the bed of the stream, and yet it is necessary that both sets of flowers should be exposed to the air and sun before they become able to perform their several duties. Add to this the fact that the male flower is quite as small in proportion to the female as is the case with the lac and scale insects, and the problem of their reaching each other becomes apparently intricate, though it is solved in a beautifully simple manner.

Fertilisation cannot be conducted by means of insects, as is the case with so many diœcious terrestrial plants, and it is absolutely necessary that actual contact should take place between them. This difficult process is effected as follows :—

The female flowers are attached to a very long spiral and closely coiled footstalk, and, when they are sufficiently developed, the footstalk elongates itself until the flower rests on the surface of the water, where it is safely anchored by its spiral cable, the coils yielding to the wavelets, and keeping the flower in its place.

Meanwhile the tiny male flowers are being developed at the bottom of the river, and are attached to very short footstalks. When they are quite ripe they disengage themselves from their footstalks, and rise to the surface of the river. Being carried along by the stream, they are sure to come in contact with the anchored female flowers. This having been done, and the seeds beginning to be developed, the spiral footstalk again coils itself tightly, and brings the seeds close to the bed of the stream, where they can take root.

There are other numerous examples, of which any reader, even slightly skilled in botany, need not be reminded, most of them being, in one form or another, modifications of the leaf or the petal, which, after all, are much the same thing. The vine and passion-flower are, however, partial exceptions.

I may here mention that soon after the failure of the first Atlantic telegraph cable, an invention was patented of a very much lighter cable, enclosed in a tube of india-rubber, and being coiled spirally at certain distances, so that the coils might give the elasticity which constitutes strength. The cable was never made, its manufacture proving to be too costly; but the idea of lightness and elasticity, having been evidently taken from the spiral tendrils of the bryony, was certainly a good one, and I should have wished to see it tried on a smaller scale than the Atlantic requires.

As a natural consequence, after the cable comes the Anchor, which in almost every form has been anticipated by Nature, whether it be called by the name of anchor, kedge, drag, or grapnel.

On the accompanying illustrations are shown a number of corresponding forms of the Anchor, together with a few others, which, although they may not necessarily be used in the water, are nevertheless constructed on the same principle—*i.e.* for the purpose of grappling.

ONE of the most startling parallels may be seen on the right

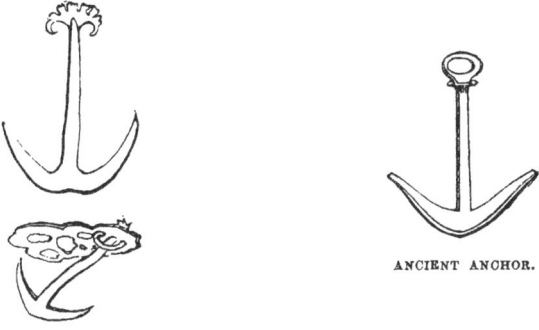

SPICULES OF SYNAPTA.

ANCIENT ANCHOR.

hand of the illustration, the figure having been drawn from an old Roman coin. On the other side of the same illustration may be seen an anchor so exactly similar in form, that the outline of the one would almost answer for that of the other. This object is a much-magnified representation of a spicule which is found on the skin of the Synapta, one of the so-called Sea-slugs, which are so extensively sold under the name of Bêche de Mer. It forms one of the curious group called the Holothuridæ.

Each of these anchors is affixed to a sort of open-worked shield, as shown above, and on the left hand; and it is a curious fact that in the various species of Synapta the anchor is rather different in form, and the shield very different in pattern. They are lovely objects, and I recommend any of my readers who possess a microscope to procure one. They need a power of at least 150 diameters to show their full beauties.

An ordinary Grapnel is here shown, and in the corresponding position on the opposite side is an almost exactly similar object, except that it is double, having the grapnel at both ends of the stem. This is a spicule of a species of sponge, and is one of the vast numbers of which the sponge principally consists.

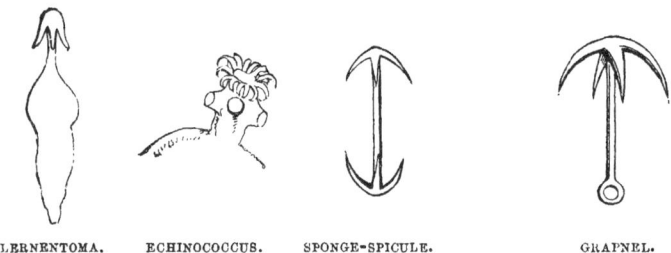

LERNENTOMA.   ECHINOCOCCUS.   SPONGE-SPICULE.   GRAPNEL.

Next to the sponge-spicule is a still more perfect example of a natural Grapnel. This is the head of an internal parasite called Echinococcus, which holds itself in its position by means of the circle of hooks with which the head is surrounded. These hooks are easily detached, and have a curious resemblance to the claw of the lion or tiger.

On the left-hand side is a representation of a parasitic crustacean animal called Lernentoma, which adheres to various fishes, and is mostly found upon the sprat, clinging to the gills by means of its grapnel-shaped head.

On the right hand of the accompanying illustration is an ice-anchor, copied from one of those which were taken out in the

Arctic expedition of 1875. Opposite is the skull of the Walrus, the tusks of which are said to be used for exactly the same

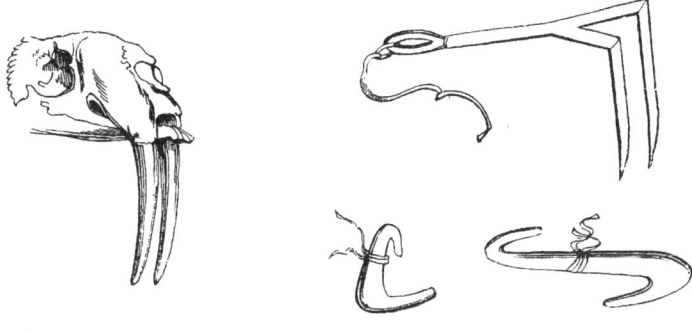

TUSKS OF WALRUS.   ICE-ANCHOR AND ICE-HOOKS.

purpose. Below are ice-hooks, also used for the same expedition.

The next illustration exhibits a butcher's hook and a common porter's hook, by which he lifts sacks on his back ; and oppo-

SPONGE-SPICULES.   BUTCHER'S HOOK.   PORTER'S HOOK.

site them are some sponge-spicules, the similarity of which in form is so remarkable that the former might have been copied from the latter.

OUR next sketch shows a remarkable example of similitude in form. There are certain small anchors called Kedges, which are very useful for mooring a boat where no great power of

MUSHROOM.   MUSHROOM KEDGE.

resistance has to be overcome, and a large anchor would be cumbersome. One of these is called, from its shape, the "Mushroom Kedge," and is very useful, as, however it may be dropped, some part of the edge is sure to take the ground. This Kedge is shown on the right hand of the illustration, and the Mushroom, from which it was borrowed, is seen on the left.

WE now come to some more examples of the principle of the Grapnel, some of which are applied to nautical, and others to terrestrial objects.

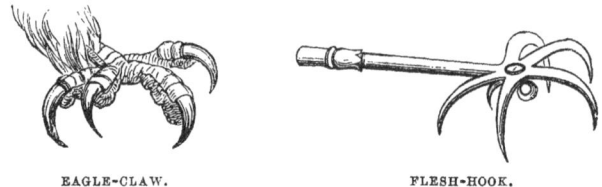

EAGLE-CLAW.   FLESH-HOOK.

The right-hand upper figure represents the "Flesh-hook," used for taking boiled meat out of the caldron, so familiar to us by the reference to it in Exodus xxvii. 3, and the still better-known allusion to its office in 1 Samuel ii. 13, 14. In the former passage, even the material, brass, which was really what we now call bronze, is mentioned, and it is a curious fact that all the specimens in the British Museum, from one of which the drawing was taken, are made of bronze. I need hardly state that the hollow handle is meant to receive a wooden staff.

On comparing this figure with that of the Eagle's foot on the opposite side, the reader cannot but be struck with the exact resemblance between the two. Indeed, there is very little doubt that the flesh-hook was intentionally copied from the foot of some bird of prey. Perhaps the Osprey would have furnished even a better example than the Eagle, the claws being sharper and more boldly curved, so as to hold their slippery prey the better.

ON the left hand of the next illustration is a figure of the seed-vessel of the Grapple-plant of Southern Africa, drawn from a specimen in my collection. The seed-vessel is several inches in length, and the traveller who is caught by a single hook had better wait for assistance than try to release himself. The stems of the plant are so slender, and the armed seed-vessels so numerous, that in attempting to rescue one portion of the dress, another portion becomes entangled, and the traveller gets hopelessly captured. Besides the hooks of the seed-vessels, the branches themselves are armed with long thorns, set in pairs. The scientific name of this plant is *Uncinaria procumbens*, the former word signifying "a hook," and the latter "trailing." It is also known by the popular name of Hook-plant.

GRAPPLE-PLANT.   DRAG.

In the late Kafir wars the natives made great use of this and other plants with similar properties, their own naked, dark, and oiled bodies slipping through them easily and unseen, while the scarlet coats of the soldiers were quickly entangled, and made them an easy mark for the Kafir's spear. In this way many more of our soldiers were killed by the spears than by the bullets of their enemies.

Opposite to the Grapple-plant is shown the common Drag, which is utilised for so many purposes. Generally it is employed for recovering objects that have sunk to the bottom of the water, and its use by the officers of the Humane Society is perfectly well known, the Drag being sometimes affixed to the end of a long pole, like the flesh-hook already described, and sometimes tied to a rope.

It can also be used as an anchor, after the manner of a kedge, and has been often employed in naval engagements for the purpose of drawing two ships together, and preventing the escape of the vessel which is being worsted. My relative, the late Admiral Sir J. Harvey, K.B., used drags in this manner, and secured two French ships, one on either side, namely, *L'Achille* and *Le Vengeur*. The first was sunk, and the second captured.

NAUTICAL.

CHAPTER V.

SUBSIDIARY APPLIANCES.

PART III.—THE BOAT-HOOK AND PUNT-POLE.—THE LIFE-BUOY AND PONTOON-RAFT.

The Boat-hook and its varied Uses.—The Earth-worm and the Serpula.—Microscopic Boat-hooks.—The Life-belt.—Life-boats and their Structure.—Uses of Cork.—Wine Corks made serviceable.—The Life-collar.—Portuguese Man-of-war.—Captain Boyton's Life-dress.—The Life-raft.—Victualling a Yacht and Boat.—The Janthina and its Air-vessels.—Cask-pontoon —Pottery-raft and its Uses.

AS all rowing men know, an indispensable appliance to the boat is the Boat-hook, which can be used either as a pole, wherewith to push the boat along, or as a grapnel, by which it can be drawn towards the shore or a ship. As the latter portion has been discussed at the close of the preceding chapter, we may proceed to the former.

Every one knows how a boat may be propelled by a pole pressed against the bank or the bottom of the water, and that there are certain boats, called punts, which are propelled in no other way.

Now, the punt-poles and boat-hooks, of which some examples are given in the accompanying illustration, have long been anticipated in Nature, there being many creatures which have no other mode of progression; such, for example, as the common Earth-worm, which pushes itself along by certain bristles which project from the rings of which the body is composed, and which have the power of extension and contraction to a wonderful extent. As, however, I shall advert to these in another part of the work, I will content myself at present with a single example, namely, the beautiful marine worm known as the Serpula.

This worm lives in a shelly tube, which is lined with a delicate membrane, up and down which it passes with ease, ascending slowly, but generally descending with such wonderful rapidity that the eye cannot follow its movements. The latter movement will be explained in a subsequent part of the book, and we will at present only treat of the former.

If the creature be removed from the tube, and carefully examined, a number of projections will be seen, in each of which is a perforation. If the animal be pressed, a slight glass-like bristle passes through the perforation, and can easily

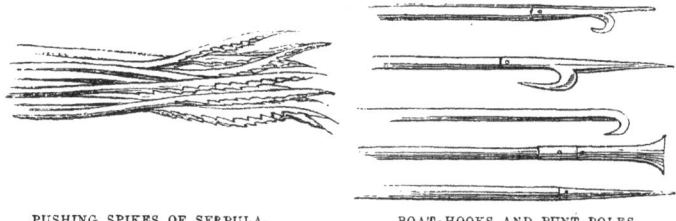

PUSHING SPIKES OF SERPULA.   BOAT-HOOKS AND PUNT-POLES.

be removed. If properly treated, and placed under a high power of the microscope, the tiny bristle resolves itself into the remarkable object which is shown on the left hand of the illustration.

It consists of a number of spear-like rods, each having a straight shaft, and a curved and pointed tip, deeply barbed on the inner portion of the curve. These curious bundles of spicules can be protruded or retracted at pleasure, and, as they are all directed backwards, it is evident that when they are pushed against the sides of the tube, either the points or the barbs must catch against the membrane which lines the tube, and so propel the animal upwards. When it wishes to descend, it uses another set of implements, and withdraws the first within their sheaths.

This is exactly analogous to the mode of progression employed by punters, who, after they have placed the pole against the bed of the stream, and run along the punt so as to push it as fast as possible, immediately withdraw the pole, and take it to the head of the punt, ready for another push. This, as the reader will see, is exactly the plan pursued by the Serpula in lengthening itself when it wishes to advance, and so to press its spicules against the sides of its tube, and in shortening itself and withdrawing the spicules ready for another push.

ANOTHER needful accessory of vessels now comes before us, namely, the capability of forming rafts or life-belts, which will float under any circumstances. Here, again, every human invention of which I know has been anticipated by Nature. Take, for example, the familiar instance of the cork life-belt and the cork edgings of the life-boat. Both are constructed on the same principle, *i.e.* the maintenance of cells which are filled by air instead of water, and are impervious to the latter.

The material most used for this purpose is cork, and life-belts constructed of it have long been in well-deserved use, the cork-bark having the property of holding much air and excluding water. Many of our life-boats are furnished with a broad and thick streak of cork, so that even if the boat be filled with water and upset, she will right herself and swim. I regret to say that many of the so-called " life-belts " which are offered for sale ought rather to be called " death-belts," they having been found to be filled with hay and straw, with only a few shavings of cork just under the covering of the belt.

Indeed, so buoyant is this substance that a very efficient belt can be made by stringing together three or four rows of ordinary wine corks, and tying them round the neck like a collar. Under these circumstances it is simply impossible to sink, and though any one may collapse from exhaustion, drowning is almost out of the question. The now well-known cork mattress, which is used in many ships, is another example of the same principle.

Lately there has been invented a " life-collar," which possesses similar advantages, but occupies less space when not wanted. It is nothing more than a tube of caoutchouc, which can be inflated at pleasure, and tied round the neck. The ordinary life-belt goes round the waist, and needs much more material without obtaining a better result, which is simply the keeping of the mouth and nostrils out of the water.

Perhaps the most buoyant of living beings is the Portuguese Man-of-war (*Physalis pelagicus*), which floats on the surface of the ocean like a bubble. It can at pleasure distend itself with air and float, or discharge the air and sink.

Now, there is a very remarkable swimming dress, which,

though not entirely invented, was at least perfected by Captain
Boyton, and which, as it enabled the wearer to cross from
France to England under rather unfavourable circumstances,
is clearly a most valuable invention.

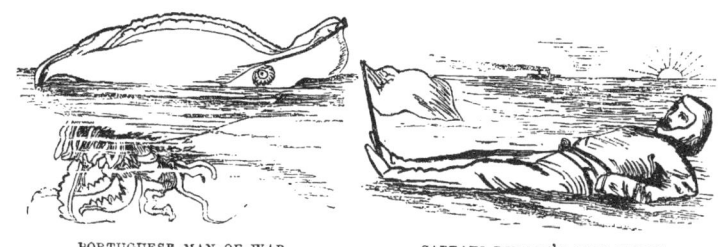

PORTUGUESE MAN-OF-WAR.   CAPTAIN BOYTON'S LIFE-DRESS.

Whether the inventor knew it or not I cannot say, but the
Boyton life-dress is simply a modification of the Physalis,
being capable of dilatation with air at will.

So much for the individual life-belt, and we will now pass to
those which are intended to sustain more than one individual.
It has almost invariably been found that when a ship has been
wrecked on a rock, or stove in by the sea, that, although there
may be plenty of boats, there is great difficulty in getting them
into the water rightly.

Now, if parts of the ship itself could be made of materials
which could not be sunk except by enormous pressure, and
which might be released by a touch if the vessel were sinking,
it is evident that many lives would be saved which have now
been lost.

And if such movable parts of the vessel were supplied with
water and provisions in air-tight cases, there is no doubt that
the number of "missing" ships would be very greatly diminished. I remember an instance where a yacht was "hung
up" on a mud-bank, whence there was no escape, for twenty-
four hours, and there was one sandwich on board to be divided
among the owner, two men, and a boy. Of course the boy had
the sandwich, and the men sustained themselves as well as they
could with tea, of which there was, fortunately, a canister on
board. As it was, they were some thirty-six hours without food.

After such an experience he had special lockers made in the
yacht and her boat, containing biscuit, potted meats, water,

wine, spirits, tobacco, tea, an "etna" for heating the water, and matches. Of course these were on a smaller scale in the boat; but several thick rugs were also stowed away, in case of being separated from the yacht at night. It so happened that they were never needed; but the sense of security which they imparted was worth ten times the expense and trouble, which included a careful inspection of all the stores before each voyage.

In Nature there is just such a raft as is needed, capable of carrying a heavy freight, and which cannot be upset. And it is rather remarkable that it has been unconsciously imitated in various parts of the world.

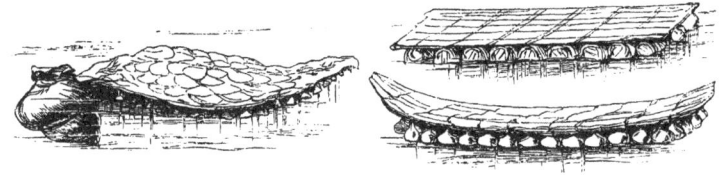

JANTHINA AND AIR-RAFT.

CASK-PONTOON.
POTTERY-RAFT OF THE NILE.

This is the singular apparatus attached to the Violet Snail (*Janthina communis*), which is common enough in the Atlantic, and derives its name of Violet-shell from its beautiful colour. The chief interest, however, centres in the apparatus which is popularly called the "raft," and which sustains the shell and eggs. It is made of a great number of air-vessels, affixed closely to each other, and by the curious property of bearing its cargo slung beneath it instead of being laid upon it.

Beneath the raft are the eggs, or rather, the capsules which contain the eggs, and at one end is the beautiful violet shell itself. The floating power of the raft is really astonishing, and even in severe tempests, when it is broken away from the animal, the raft continues to float on the surface of the waves, bearing its cargo with it.

On the opposite side of the illustration are two examples of rafts constructed so exactly on the same principle as that of the Violet Snail, that they both might have been borrowed from it.

The upper is the kind of raft which has often been constructed by sailors when trying to escape from a sinking ship, or by soldiers when wishing to convey troops across a river, and having no regular "pontoons" at hand. It is made simply by lashing a number of empty casks to a flooring of beams and planks.

The amount of weight which such a structure will support is really astonishing, as long as the casks remain whole, and to upset it is almost impossible. Even cannon can be taken across wide expanses of water in perfect safety, and there is hardly anything more awkward of conveyance than a cannon, with its own enormous and concentrated weight, and all the needful paraphernalia of limber, ammunition (which may not be wetted, and of immense weight), horses, and men.

Yet even this heterogeneous mass of living and lifeless weight can be carried on the cask-raft, which is an exact imitation of the living raft of the Violet Snail.

BENEATH the cask-pontoon is to be seen a sketch of a very curious vessel which is in use on the Nile, and I rather think on the Ganges also, though I am not quite sure. It is formed in the following manner :—

In both countries there are whole families who from generation to generation have lived in little villages up the river, and gained their living by making pottery, mostly of a simple though artistic form, the vessel having a rather long and slender neck, and a more or less globular body.

When a man has made a sufficient number of these vessels, he lashes them together with their mouths uppermost, and then fixes upon them a simple platform of reeds. The papyrus was once largely used for this purpose, but it seems to be gradually abandoned.

He thus forms a pontoon exactly similar in principle with the cask-pontoon which has just been described. Then, taking his place on his buoyant raft, he floats down the river until he comes to some populous town, takes his raft to pieces, sells the pots and reeds, and makes his way home again by land.

E

WAR AND HUNTING.

CHAPTER I.

THE PITFALL, THE CLUB, THE SWORD, THE SPEAR AND DAGGER.

Analogy between War and Hunting.—The Pitfall as used for both Purposes.—
African Pitfalls for large Game, and their Armature for preventing the Escape
of Prey.—Its Use in this Country on a miniature scale.—Mr. Waterton's
Mouse-trap.—Pitfall of the Ant-lion, and its Armature for preventing the
Escape of Prey.—The Club and its Origin.—Gradual Development of the
Weapon.—The "Pine-apple" Club of Fiji.—The Game of Pallone and the
"Bracciale."—The Irish Shillelagh.—Clubs and Maces of Wood, Metal,
or mixed.—The Morgenstern.—Ominous Jesting.—Natural Clubs.—The
Durian, the Diodon, and the Horse-chestnut.—The Sword, or flattened and
sharpened Club.—Natural and artificial Armature of the Edge.—The Sword-
grass, Leech, and Saw-fish.—Spears and Swords armed with Bones and
Stones.—The Spear and Dagger, and their Analogies.—Structure of the
Spear.—The Bamboo as a Weapon of War or Hunting.—Singular Combat,
and its Results.

THE two subjects which are here mentioned are practically
one, the warfare being in the one case carried on against
mankind, and in the other against the lower animals, the
means employed being often the same in both cases.

THE PITFALL.

ONE of the simplest examples of this double use is afforded
by the PITFALL, which is employed in almost every part of the
world, and, although mostly used for hunting, still keeps its
place in warfare.

On the right hand of the accompanying illustration is shown
a section of the Pitfall which is so commonly used in Africa for
the capture of large game. It is, as may be seen, a conical
hole, the bottom of which is armed with a pointed stake.
Should a large animal fall into the pit, the shape of the sides

forces it upon the stake, by which it is transfixed. Even elephants of the largest size often fall victims to this simple trap. It is only large enough to receive the fore-legs and chest, but that is quite sufficient to cause the death of the animal, the stake penetrating to the heart.

Many a hunter has fallen into these traps, and found great difficulty in escaping, while some have not escaped at all. Indeed, in many parts of Southern Africa, when part of one

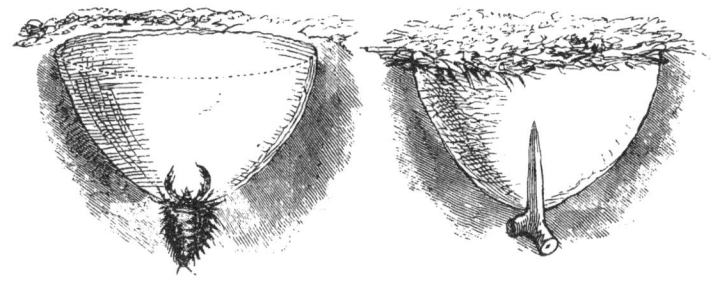

PITFALL OF ANT-LION FOR CATCHING INSECTS.   AFRICAN PITFALL FOR CATCHING LARGE GAME.

tribe is about to visit another, the pitfalls are always unmasked, lest the intended guests should fall into them.

Even without the spike, the elephant would scarcely be able to save itself, owing to its enormous weight, unless helped out by its comrades before the hunters came up. Indeed, many pitfalls are intentionally made for this purpose, and are of a different shape, i.e. about eight feet in length and four in breadth.

In those which are made for the capture of the giraffe, the pit is very deep, and the place of the stake is occupied by a transverse wall, which prevents the feet of the captive from touching the ground, and keeps it suspended until the hunters can come and kill it at leisure.

Even in Belgium and our own country the pitfall is in use. When the field-mice were devastating the districts about Liege some years ago, their ravages were effectually checked by pitfalls, in which they were caught by bushels, the pitfalls being simple holes some two feet deep, and made wider below than above.

The late Mr. Waterton contrived to rid his garden of field-mice by pitfalls constructed on the same principle, though more permanent. Finding that the little animals made great havoc among his peas just as they were starting out of the ground, he buried between the rows a number of earthen pickle-jars, sinking them to the level of the ground. He then rubbed the inside of the neck with bacon, and left them. The mice stooped down to lick off the bacon, fell into the jars, and, the neck being narrow and the sides slippery, they could not get out again.

ON the left hand of the illustration is the section of a pitfall made by the well-known Ant-lion (*Myrmeleo*), of which there are several species. The history of this wonderful insect is so familiar to us that it need not be repeated at length. Suffice it to say that it digs conical pitfalls in loose sandy soil, and that it places itself at the bottom of the pit, securing the insect victims with its jaws just as the larger animals are secured by the stake of the human hunter.

It makes no false cover, as does the human hunter, but it always chooses soil so loose that if an insect approach the edge, the sand gives way, and it goes sliding down into the pit, whence its chance of escape is very small, even were there no deadly jaws at the bottom ready to receive it.

THE CLUB.

THE simplest of all offensive weapons is necessarily the CLUB. At first, this was but a simple stick, such as any savage might form from a branch of a tree by knocking off the small boughs with a stone or another stick. Such clubs are still used in Australia, and I have several in my collection.

Then the inventive genius of man improved their destructive power by various means. The most obvious plan was to add to the force of its blow by simply making one end much thicker and heavier than the other. This is done in the "Knob-kerry" of Southern Africa, and it is worthy of remark that in Fiji a weapon exists so exactly like the short knob-kerry of Africa, that an inexperienced eye would scarcely be able to distinguish between them.

The next plan was to arm the enlarged head with projecting pieces or spikes, sometimes cut out of the solid wood, and sometimes artificially inserted. The "Shillelagh" of Ireland is a simple example of this kind of club. One of the best and most elaborate examples of this sort of weapon is the "Pine-apple" Club of Fiji, a figure of which may be seen in the illustration, drawn from a specimen in my collection.

It is made in the most ingenious manner from a tree which is trained for the purpose. There are certain trees belonging to the palm tribe which possess "aërial" roots, *i.e.* subsidiary roots, which surround the trunk at some distance from the ground, and assist in supporting it. Some trees have no central root, and are entirely upborne by the aërial roots, while others have both.

One of these latter is selected, and when it is very young is bent over and fastened to the ground almost at right angles, as shown in the illustration. When it has grown to a sufficient age it is cut to the requisite length, the central root is sharpened to a point, and the aërial roots are also cut down in such

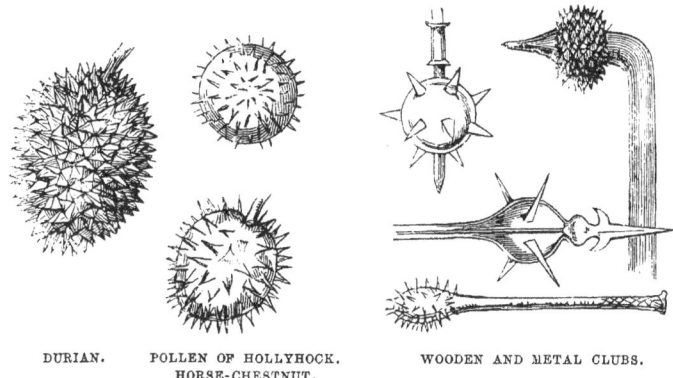

DURIAN.   POLLEN OF HOLLYHOCK.   WOODEN AND METAL CLUBS.
HORSE-CHESTNUT.

a way that they radiate very much like the projections on a pine-apple. This is really an ingenious weapon, for if the long and sharpened end should miss its aim, the projections would be tolerably sure to inflict painful if not immediately dangerous injuries.

As the pine-apple is so well known, I have given in the opposite side of the illustration a figure of the Durian, a large Bornean fruit, which is covered with projections almost identical in appearance with those of the pine-apple club, and almost equally hard and heavy.

Perhaps some of my readers may have heard of the grand Italian game of Pallone, the "game of giants," as it has been called.   The ball, which is a large and rather heavy one, weighing more than twice as much as a cricket-ball, is struck with a wooden gauntlet reaching nearly half-way up the fore-arm.   The original gauntlet was cut entirely out of the solid wood, and exactly resembled the exterior of the Durian.   The modern gauntlet, however, has the spikes fixed separately into a wooden frame, so that they can be replaced if broken in the course of the game.   The principle, however, is identical in all three cases.   The technical name of this gauntlet is Bracciale.

The next improvement was to add still further to the destructive powers of the club by arming it with stones, so as to make it harder and heavier.   Sometimes a stone is perforated, and the end of the club forced into it.   Sometimes the stone is lashed to the club, and sometimes a hole is bored in the club, and the stone driven into it.   This kind of club, made of a sort of rosewood, may be found among some of the tribes inhabiting the district of the Essequibo.

The next improvement was to make the weapon entirely of metal, and such clubs are plentiful in every good collection of arms.   There was, for example, the common mace, which was used for the purpose of stunning an adversary clothed in armour which the sword could not penetrate.   As this, how-ever, was nothing more than an ordinary wooden club executed in iron, we need not produce examples.

Other and more complicated forms were soon made, and were wonderfully valuable until the rapidly improving fire-arms kept combatants at a distance, and rendered a hand-to-hand fight almost impossible.

Three examples of such clubs are given in the illustration, and are taken from Demmin's valuable work called " Weapons of War."

The upper left-hand specimen is called Morgenstern, *i.e.* Morning Star.   It is a large, heavy wooden ball studded with steel spikes, and affixed to a handle usually some six or seven feet, but sometimes exceeding eleven feet, in length.   It was chiefly used by infantry when attacking cavalry, the long shaft enabling the foot-soldier to be tolerably sure of dealing the cavalier or his horse a severe blow, while himself out of reach of the latter's sword.

Behind it is another Morgenstern in which there is an

improvement, the armed ball being furnished at the end with a spike, so that it could be used either as a mace or a spear.

The commonest form of the Morning Star is shown below, and is thus described by Demmin :—

" This mace had generally a long handle, and its head bristled with wooden or iron points. It was common among the ancients, for many museums possess several fragments of these weapons belonging to the age of bronze.

" The Morning Star was very well known and much used in Germany and Switzerland. It received its name from the ominous jest of wishing the enemy ' good morning' with the Morning Star when they had been surprised in camp or city.

" This weapon became very popular on account of the facility and quickness with which it could be manufactured. The peasants made it easily with the trunk of a small shrub and a handful of large nails. It was also in great request during the wars of the peasantry which have devastated Germany at different times, and the Swiss arsenals possess great numbers of them."

One of these primitive weapons may be seen in the lower figure of the illustration.

Sometimes the spiked ball was attached to a chain, and fastened to the end of a handle varying greatly in length, measuring from two to ten feet. One of these weapons may be seen in the Guildhall of London, being held by one of the celebrated giants.

If the reader will now turn to the illustration on page 53, he will see that on the right of the Durian there are two spherical objects covered with spikes. The upper is the pollen of the Hollyhock, and the lower the common Horse-chestnut. The reader will see that these are precisely similar in form to the spiked balls of the Morgenstern, whether they be used at the end of a staff or slung to a chain. There are many similar examples in the vegetable kingdom which will doubtless suggest themselves to the reader, but these are amply sufficient for this purpose.

Then, in the animal world, the curious Diodons, sometimes called Urchin-fishes, or Prickly Globe-fishes, are good examples. These fishes are covered with sharp spines, and, as they have the power of swelling their bodies into a globular form, the spikes project on all sides just like those of the pollen or chestnut. There is a specimen in my collection, which, if the tail and fins were removed, and a cast taken in metal, would make a very good Morgenstern ball.

THE SWORD.

THE next improvement on the club was evidently to flatten it, and sharpen one or both edges, so as to make it a cutting as well as a stunning implement—in fact, the club was changed into a SWORD.

A GOOD example of this weapon in its simplest form is the wooden sword of Australia, now an exceedingly rare weapon.

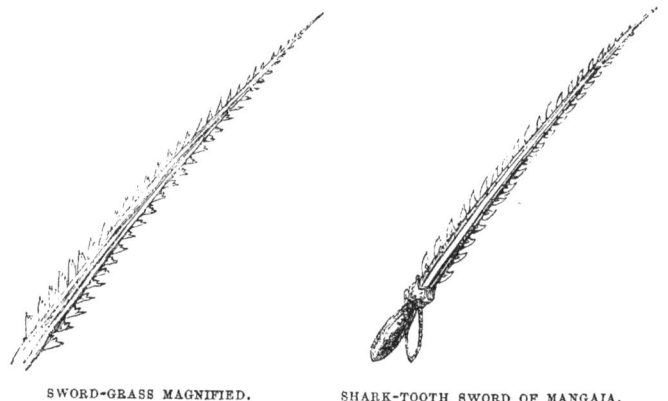

SWORD-GRASS MAGNIFIED.   SHARK-TOOTH SWORD OF MANGAIA.

It looks like a very large boomerang, but is nearly straight, and is made from the hard, tough wood of the gum-tree. Travellers say that the natives can cut off a man's head with this very simple weapon.

I just missed obtaining one of these swords from a man-of-war, but, unfortunately, a few hours before my arrival the zealous first lieutenant had ordered a large collection of savage weapons to be thrown overboard, among which were several Australian swords.

Finding that the edges were not sufficiently sharp, and were liable to break, the maker next turned his attention to arming them with some substance harder than wood. Various materials were used for this purpose, some of which will be mentioned.

One of these is given in the illustration, and is taken from a specimen in my collection. It is made of wood, rather more than two feet in length, and would in itself be an insignificant weapon but for its armature.

This consists of a number of sharks' teeth, which are fixed along either side, and are a most formidable apparatus, each tooth cutting like a lancet-blade, and not only being very sharp, but having their edges finely notched like the teeth of a saw. I have a series of these weapons in my collection, some being curved, some straight, and one very remarkable weapon having four blades, one straight and long blade in the centre, and three curved and short blades springing from the handle towards the point.

Opposite the shark-tooth sword is an object which might almost be taken for a similar weapon, but is, in fact, nothing but a common grass-blade, such as may be found in any of our lanes. I suppose that most of my readers must at some time have cut their fingers with grass, and the reason why is shown in the illustration, which represents a much-magnified blade of grass. The edges of the leaf are armed with sharp teeth of flint, set exactly like those of the sword, with their points directed towards the tip of the blade. The whole of the under surface of the blade is thickly set with similar but smaller teeth, arranged in the same manner. I have just brought a blade of grass from a lane near my house, and when it was placed under the half-inch power of the microscope, the resemblance to the sword was absolutely startling to some spectators who came to look at it.

As if to make the resemblance closer, many savage weapons are edged with flat stones, flint chips, or pieces of obsidian, so that the flint teeth of the grass are exactly copied by the flint edgings of the sword. The old Mexican swords were nearly all edged with obsidian, as is seen in the lower right-hand figure of the next illustration. I possess a number of obsidian flakes which were intended for that purpose, but do not appear to have been used.

The second figure from the top represents the head of a spear similarly armed, and I possess a small Australian implement in which the flakes of obsidian are set only on one side, so that the instrument can be used as a rude saw.

Between these two weapons is a spear-head armed with shark-teeth. I have a very remarkable weapon of this kind, made in Mangaia. It is eleven feet in length, and, besides being armed with a double row of sharks' teeth nearly to the handle, it has three curved blades similarly armed, set at dis-

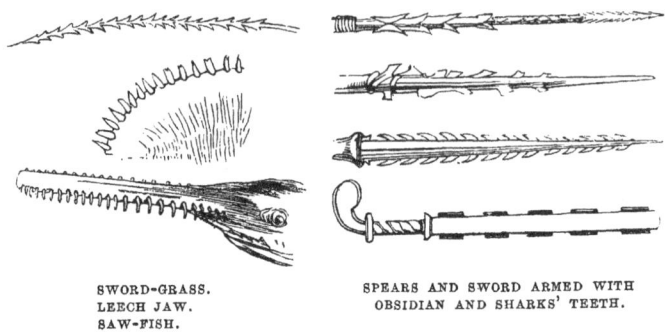

SWORD-GRASS.
LEECH JAW.
SAW-FISH.

SPEARS AND SWORD ARMED WITH
OBSIDIAN AND SHARKS' TEETH.

tances of about two feet, and projecting at right angles. Thus, if the foe were missed with the point of the spear, he would probably be wounded by one of the blades.

The upper figure represents a weapon where the natural bone of the sting-ray has been used as the point.

On the opposite side are seen three natural objects similarly armed. The uppermost is another species of sword-grass, like that which has already been described.

Next comes a magnified view of one of the three cutting instruments of the leech, showing the serrated teeth set along its edge, by means of which it produces the sharply-cut wounds through which it sucks the blood.

The last figure represents the head of the common Saw-fish, in which a vast number of flat and sharply-edged teeth are set upon the blade-like head. The fish has been observed to use this weapon just as the Mangaian uses his sword-spear. It dashes among a shoal of fish, sweeps its head violently backwards and forwards, and then, after they have dispersed, picks up at its leisure the dead and disabled.

THE SPEAR AND THE DAGGER.

IT is tolerably evident that the invention of the spear and dagger must have been nearly, if not quite, contemporaneous with that of the club. I place these weapons together because there is great difficulty in assigning to either of them the precedence, the spear being but a more or less elongated dagger, and the dagger a shortened spear.

As a good example of this fact, I have in my collection a number of spears and daggers belonging to the Fan tribe of Western Africa. In every case the weapons correspond so closely with each other, that if the daggers were attached to shafts they would exactly resemble the spears, and if the spears were cut off within a few inches of the head, they would be taken for daggers.

I may here mention that as this part of the subject merely involves the employment of a pointed or thrusting weapon, instead of the club or sword, both of which are used for striking, the question of poison, barbs, and sheaths will be treated on another page.

The primary origin of the Spear is probably the thorn, as a savage who had been wounded by a thorn would easily pass to the conclusion that a thorn of larger size would enable him to kill an enemy in war, or an animal in hunting. Anything of sufficient dimensions, which either possessed a natural point or could be sharpened into a point, would be available for the purpose of the hunter or warrior.

Accordingly we find that such objects as the beak of the heron or stork, the sharp hind-claw of the kangaroo, the bone of the sting-ray, the beak of the sword-fish, and many similar objects, are employed for the heads of spears, or used simply as daggers.

As to artificial spears, nothing is easier than to scrape a stick to a point, and then, if needful, to harden it in the fire. This is, indeed, one of the commonest forms of primitive spears, and I have in my collection many examples of such weapons. Another simple form of this weapon is that which is made by cutting a stick or similar object diagonally.

Hollow rods—such, for example, as the bamboo—are the best for this purpose. I have now before me a cast of a most interesting weapon discovered by Colonel Lane Fox. It is the head of a spear, and is formed from part of the leg-bone of a sheep. At one end there is a simple round hole, which acted as a socket for the reception of the shaft, and the other end is cut away diagonally, so as to leave a tolerably sharp point.

As to the bamboo, it has a great advantage in the thinness of its walls, and the coating of flinty substance with which it is surrounded, and which gives its edges a knife-like sharpness. Indeed, so very sharp is the silex, that splinters of bamboo are still used as knives, and with them a skilful operator can cut up a large hog as expeditiously as one of our pork-butchers could do with the best knife that Sheffield produces.

I possess several of these weapons, and formidable arms of offence they are. If the reader can imagine to himself a tooth-pick, a foot or more in length, made from bamboo instead of quill, and having its edges nearly as sharp as a razor, he can realise the force of even so simple a weapon. In the case of the bamboo, too, celerity of manufacture has its value, for any one can make a couple of spears in less than as many minutes. All he has to do is to cut down a joint of bamboo transversely, and then with a diagonal blow of his knife at the other end to form the point.

The force of such a weapon may be inferred from a remarkable combat that took place some sixty years ago, when the roads were not so safe as they are at present.

A gentleman, who happened to be a consummate master of the sword, was going along the highway at night, and was attacked by two footpads, he having no weapon but a bamboo cane.

One of them he temporarily disabled by a severe kick, and then turned to the other, whom he found to be pretty well as good a swordsman as himself, and to possess a good stick instead of a slight cane. The footpad soon discovered the dis-crepancy of weapons, and with a sharp blow smashed the cane to pieces, leaving only about eighteen inches in his antagonist's hand.

Almost instinctively Baron —— sprang under the man's guard, and dashed the broken cane in his face. The footpad staggered with a groan, put his hands to his face, and ran away, followed by his companion, who did not desire another encounter with such an antagonist. When the victor reached his destination, he found that the footpad's face must have been torn to pieces, for the clefts of the split bamboo were full of scraps of skin, flesh, and whisker hair.

It is worthy of notice that the combination of the club and

the dagger is common to savage and civilised life, as may be seen by reference to the illustration in page 53, where the wooden club of savage warfare and the metal club and maces of civilisation are alike armed with a piercing as well as a bruising apparatus. Mostly the dagger is on the head of the mace or battle-axe, but, in some cases, the end of the handle acts as the dagger, and the head as the axe or mace.

A very good example of this formation is found in the wooden battle-axe, or "Patoo," of New Zealand, a weapon which has been long superseded by modern fire-arms. A specimen in my possession is rather more than five feet in length. The head is just like that of an ordinary axe, while the handle tapers gradually to the end, where it terminates in a sharp spike. In actual combat the point was used much more than the axe.

WAR AND HUNTING.

CHAPTER II.

POISON, ANIMAL AND VEGETABLE.—PRINCIPLE OF THE BARB.

Poison as applied to Weapons.—Its limited Use. —Animal and Vegetable Poisons.—Animal Poisons.—The Malayan Dagger, or Kris, and two Modes of poisoning it.—The Bosjesmans and their Arrows.—Snake Poison and its Preparation.—The Pseudo-barb.—The Poison-grub, or N'gwa.—Simple Mode of Preparation, and its terrible Effects.—Vegetable Poisons.—The Upas of Malacca.—The Wourali Poison of Tropical America.—Mode of preparing the various Arrows.—The Fan Tribe of West Africa, and their poisoned Arrows.—Subcutaneous Injection.—Examples in Nature.—The Poison-fang of the Serpent.—Sting of the Bee.—Tail of the Scorpion.—Fang of the Spider.—Sting of the Nettle.—Exotic Nettles and their Effects.—The Barb and its Developments.—The "Bunday" of Java.—Reversed Barbs of Western Africa.—Tongans and their Spears.—The Harpoon and Lernentoma, or Sprat-sucker.—The Main Gauche, or Brise-épée.

ANOTHER advance, if it may so be called, lay in increasing the deadly effect of the weapons by arming them with poison.

Without the poison, it was necessary to inflict wounds which in themselves were mortal; but with it a comparatively slight wound would suffice for death, providing only that the poison mixes with the blood. It is worthy of notice that cutting weapons, such as swords and axes, seldom, if ever, have been envenomed, the poison being reserved for piercing weapons, such as the dagger, the spear, and the arrow.

ANIMAL POISONS.

PERHAPS the most diabolical invention of this kind was the Venetian stiletto, made of glass. It came to a very sharp point, and was hollow, the tube containing a liquid poison. When the dagger was used, it was driven into the body of the victim, and then snapped off in the wound, so that the poison was able to have its full effect.

Such poisons are of different kinds, and invariably animal or vegetable in their origin. Taking the animal poisons first, we come to the curious mode of poisoning the Malayan dagger, or "Kris." The blade of the weapon is not smooth, but is forged from very fibrous steel, and then laid in strong acid until it is covered with multitudinous grooves, some of them being often so deep that the acid has eaten its way completely through the blade.

Among some tribes the kris is poisoned by being thrust into a putrefying human body, and allowed to remain there until the grooves are filled with the decaying matter. It is also said that if the kris be similarly plunged into the thick stem that grows just at the base of the pine-apple, the result is nearly the same.

As a rule, however, the Arrow is generally the weapon which is poisoned, and a few examples will be mentioned of each kind of poisoning.

The two most formidable animal poisons are those which are made by the Bosjesmans of Southern Africa. Their bows are but toys, and their arrows only slender reeds. But they arm these apparently insignificant weapons with poison so potent, that even the brave and bellicose Kafir warrior does not like to fight a Bosjesman, though he be protected by his enormous shield.

There are two kinds of animal poison used by the Bosjesmans. The first is made from the secretion of the poison-glands of the cobra, puff-adder, and cerastes. Knowing the sluggish nature of snakes in general, the Bosjesman kills them in a very simple manner. He steals cautiously towards the serpent, boldly sets his foot upon its neck, and cuts off its head. The body makes a dainty feast for him, and the head is soon opened, and the poison-glands removed.

By itself, the poison would not adhere to the point of the weapon, and so it is mixed with the gummy juice of certain euphorbias, until it attains a pitch-like consistency. It is then laid thickly upon the bone point of the arrow, and a little strip of quill is stuck into it like a barb. The object of the quill is, that if a man, or even an animal, be wounded, and the arrow torn away, the quill remains in the wound, retaining sufficient poison to insure death. I have a quiverful of such arrows in my collection.

That arrows so armed should be very terrible weapons is
easily to be imagined, but there is another kind of poison which
is even more to be dreaded. This is procured from the innocent-
looking, but most venomous, Poison-grub. It is called N'gwa by
the Bosjesmans, and is the larval state of a small beetle. When
the arrow is to be poisoned, the grub is broken in half, and the
juices squeezed upon the arrow in small spots.

Both Livingstone and Baines give full and graphic accounts
of the horrible effect produced by this dread poison, which, as
soon as it mixes with the blood, drives the victim into raging
madness. A lion wounded by one of these arrows has been
known nearly to tear himself to pieces in his agonies. M.
Baines was good enough to present me with the N'gwa grub in
its different stages, together with an arrow which has been
poisoned with its juices.

The Bosjesmans are themselves so afraid of the weapon, that
they always carry the arrows with the points reversed, the
poisoned end being thrust into the hollow reed which forms
the shaft of the arrow. Not until the arrow is to be discharged
does its owner place the tip with its point uncovered.

Vegetable Poisons.

WE now come to the Vegetable Poisons, the two best known
of which are the Upas poison of Borneo, and the Wourali of
South America. It is rather remarkable that in both these
cases the arrows are very small, and are blown through a hollow
tube, after the manner of the well-known " Puff-and-dart " toy
of the present day.

The Upas poison is simply the juice of the tree, and it does
not retain its strength for more than a few hours after it has
been placed on the arrow-points. A supply of the same liquid
is therefore kept in an air-tight vessel made of bamboo, the
opening being closed by a large lump of wax kneaded over it at
the mouth. One of these little flasks, taken from a specimen in
my collection, is seen on the extreme right of the illustration.

The Wourali poison owes all its power to its vegetable element, though certain animal substances are generally mixed
with it. The principal ingredient is the juice of one of the
strychnine vines, which is extracted by boiling, and then carefully inspissated until it is about the consistency of treacle.

This poison differs from the Upas in the fact that it retains its potency after very many years, if only kept dry. I have a number of arrows poisoned with the Wourali. They were given to me by the late Mr. Waterton, who procured them in 1812, and even in the present year (1875) they are as deadly as when they were first made.

A bundle of these tiny arrows, surmounted by the little wheel which is used to guard the hand from being pricked, is seen next to the Bornean poison-flask.

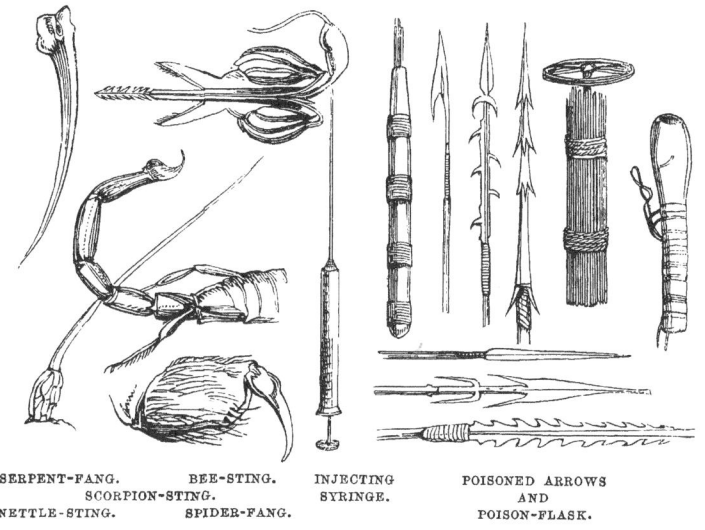

SERPENT-FANG.   BEE-STING.   INJECTING   POISONED ARROWS
SCORPION-STING.   SYRINGE.   AND
NETTLE-STING.   SPIDER-FANG.   POISON-FLASK.

Beside these little arrows, which are only about ten inches in length, very much larger arrows are used both for war and hunting, and are propelled by the bow, and not with the breath. Many of these arrows are nearly six feet in length. In all, the head is movable fitting quite loosely into a socket, so that when an animal is struck and springs forward, the shaft is shaken off, to be picked up by the hunter, and fitted with another point, while the poisoned head remains in the wound.

Another kind of poison, also of a vegetable origin, is used by the Fan tribe. The arrows are mere little slips of bamboo, and are propelled by a slight crossbow. But the poison is so potent, that even these tiny weapons produce a fatal effect.

Nearly in the centre of the illustration is seen a rather curiously formed syringe, with an extremely long and slender tip. This is a recently invented instrument, used for the purpose of subcutaneous injection—*i.e.* of injecting any liquid under the skin. It is mostly employed for injecting opium and other drugs of similar qualities, for the purpose of obtaining relief from local pain. The slender spike-like point is hollow, and ends in a sharp tip, formed like the head of a lance. Just below the head there is a little hole, communicating with the interior of the tube.

The mode of operating is simple enough. The syringe is filled with the drug, and the point introduced under the skin at any given spot. Pressure on the piston then forces out the liquid, and causes it to mix with the blood.

NATURAL ANIMAL POISONS.

Now, both in the animal and vegetable worlds may be found several examples of an apparatus which acts in exactly the same manner.

The first is the poison-fang of the Serpent, a specimen of which is given on the left hand of the illustration. This fang answers in every respect to the syringe above mentioned. The long and slender fang is hollow, and answers to the pipe of the syringe. It communicates at the base with a reservoir of liquid poison, which answers to the body of the syringe, and there is a little hole, or rather slit, just above the point, which allows the poison to escape.

When the serpent makes its stroke, the base of the fang is driven against the reservoir, so that the liquid is urged through the hollow tube, and forced into the wound. Even in large serpents these fangs are very small. I have now before me some fangs of the cobra, puff-adder, rattlesnake, and viper, and it is astonishing how small and slender are these most deadly weapons. The figure in the illustration is much magnified, in order to show the aperture at the base, where communication is made with the interior of the fang. As the exit hole is on the upper curve of the fang, it is not visible in the figure.

Next to the serpent's fang is a representation of the Bee-sting, the poisonous reservoir being seen at the base, and having attached to it the tiny thread-like gland by which the poison is secreted.

In the centre is seen the tail of a Scorpion, with its hooked sting. The last joint is formed just like the serpent's fang, being hollow, having a sharp point with a slit near the end, and a poison reservoir in the rounded base. When the scorpion attacks an enemy, it strikes violently with the tail, and the force of the blow drives out the poison just as is done with the serpent's fang.

At the bottom of the illustration is shown the poison-fang of a Spider, which, as the reader may see, is formed just on the principle of the scorpion-sting.

NATURAL VEGETABLE POISONS.

So much for animal poisons. We will now pass to the vegetable world.

Of the vegetable sting-bearers none are more familiar to us than the Nettle, three species of which inhabit this country. The two commonest are the Great Nettle (*Urtica diœcea*) and the Small Nettle (*Urtica urens*), and both of them are armed with venomous stings, which cause the plants to be so much dreaded.

The structure of these stings is very simple, and can be made out with an ordinary microscope, or even a good pocket lens. Each of these stings is, in fact, a rather elaborately constructed hair, hollow throughout its length, coming to a point at the tip, and having the base swollen into a receptacle containing the poisonous juice. When any object—such, for example, as the human hand—touches a nettle, the points of the stings slightly penetrate the skin, and the hair is pressed downwards against the base, so that the poison is forced through the hole.

One of these hairs is shown in the left-hand bottom corner of the illustration.

Even the tiny stings of our English nettles are sufficiently venomous to cause considerable pain, and, in some cases, even to affect the whole nervous system. But some of the exotic nettles are infinitely more formidable, and are, indeed, so dangerous that, when they are grown in a botanical garden, a fence is placed round them, so as to prevent visitors even from touching a single leaf.

The two most dreaded species are called *Urtica heterophylla* and *Urtica crenulata*. The former is thought to be the more dangerous of the two, and a good idea of its venomous qualities may be gathered from an account of an adventure with *Urtica crenulata*. The narrator is M. L. de la Tour.

"One of the leaves slightly touched the first three fingers of my left hand; at the time I only perceived a slight pricking, to which I paid no attention. This was at seven in the morning. The pain continued to increase, and in an hour it became intolerable; it seemed as if some one were rubbing my fingers with a hot iron. Nevertheless, there was no remarkable appearance, neither swelling, nor pustules, nor inflammation.

"The pain spread rapidly along the arm as far as the arm-pit. I was then seized with frequent sneezing, and with a copious running at the nose, as if I had caught a violent cold in the head. About noon I experienced a painful attack of cramp at the back of the jaws, which made me fear an attack of tetanus. I then went to bed, hoping that repose would alleviate my suffering, but it did not abate. On the contrary, it continued nearly the whole of the following night; but I lost the contraction of the jaws about seven in the evening.

"The next morning the pain began to leave me, and I fell asleep. I continued to suffer for two days, and the pain returned in full force when I put my hand into water. I did not finally lose it for nine days."

There is another of these formidable nettles, called in the East by a name which signifies "Devil's Leaf," and which is sufficiently venomous to cause death. There is but little doubt, however, that in the present instance, if a larger portion of the body—say the whole arm—instead of three fingers, had been stung, death would have ensued from the injury.

The Barb.

WE now come to another improvement, or rather addition, in the various piercing weapons. Sometimes, as in the case of the dagger or the hand-spear, it was necessary that when a blow had been struck the weapon should be easily withdrawn from the wound, so as not to disarm the assailant, and to enable him to repeat the stroke if needful. But in the case of a missile weapon, such as a javelin or an arrow, it was often useful, both in war and hunting, to form the head in such a way that when it had once entered it could scarcely be withdrawn For this purpose the Barb was invented, taking different forms, according to the object of the weapon and the nationality of the maker.

As in this work I prefer to show the gradual development of human inventions, I shall take my examples of barbs entirely from the weapons of uncivilised nations, six examples of which are given in the accompanying illustration, and five of them being drawn from specimens in my collection.

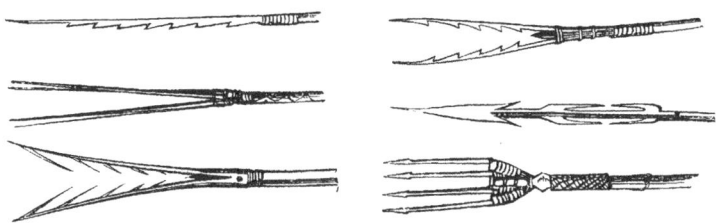

BARBED WEAPONS.

The upper left-hand figure is rather a curious one, the position of the barbs being nearly reversed, so that they serve to tear the flesh rather than adhere to it. The opposite figure represents an arrow with a doubly barbed point. It is chiefly used for shooting fish as they lie dozing on or near the surface of the water, but it is an effective weapon for ordinary hunting purposes, and, as the shaft is fully five feet in length, is quite formidable enough for war.

The left-hand bottom figure represents a very remarkable instrument, for it can hardly be called a weapon, and is, in fact, the head of a policeman's staff. It is peculiar to Java, and is called by the name of " Bunday." As may be seen by reference to the illustration, the head of the Bunday is formed of two diverging slips of wood. To each of these is lashed a row of long and sharp thorns, all pointing inwards, and the whole is attached to a tolerably long shaft.

When a prisoner is brought before the chief, a policeman stands behind him, armed with the Bunday, and, if the man should try to escape, he is immediately arrested by thrusting the weapon at him, so as to catch him by the waist, neck, or arm, or a leg. Escape is impossible, especially as in Java the prisoner wears nothing but his waist-cloth.

A weapon formed on exactly the same principle was used in the fifteenth and sixteenth centuries, and was employed for

dragging knights off their horses. It was of steel instead of
wood, and the place of the thorns was taken by two movable
barbs, working on hinges, and kept open by springs. When
a thrust was made at the knight's neck the barbs gave way, so
as to allow the prongs to envelop the throat, and they then
sprang back again, preventing the horseman from disengaging
himself. This weapon is technically named a "catchpoll."

An illustration of one of these weapons will be given on
another page.

The right-hand central figure is an arrow from Western
Africa. In a previous illustration (page 65) a head of one of
these arrows is given on rather a larger scale, so as to show
the very peculiar barbs. These are of such a nature that when
they have well sunk into the body they cannot be withdrawn,
but must be pushed through, and drawn out on the opposite
side. This is drawn from one of my own specimens.

In some cases, with an almost diabolical ingenuity, the
native arrow-maker has set on a couple of similar barbs,
directed towards the point, so that the weapon can neither be
pushed through nor drawn back. One of these arrows is shown
in the illustration, but, for want of space, the artist has placed
the opposing barbs too near each other.

In some parts of Southern Africa a similar weapon was
used for securing a prisoner, the barbed point being thrust
down his throat and left there. If it were pushed through the
neck it killed him on the spot, and if it remained in the
wound the man could not eat nor drink, and the best thing
for him was to die as soon as he could.

With similar ingenuity, the Tongans and Samoans made
their war-spears with eight or nine barbs, and, before going
into action, used to cut the wood almost through between
each barb, so that when the body was pierced, the head, with
several of the barbs, was sure to break off and leave a large
portion in the wound. In Mariner's well-known book there is
an admirable account of the mode employed by a native
surgeon for extracting one of these spear-heads. So common
was this weapon that every Tongan gentleman carried a many-
barbed spear about five feet long, and used it either as a
walking-stick or a weapon. It is needless to say that this
spear is almost an exact copy of the tail-bone of the Sting-

ray. A dagger made of this bone was used in the Pelew Islands in 1780, but seemed to be rather scarce.

The left-hand central figure is a Fijian fish-spear of four points, and the last figure on the right hand represents a large four-pronged spear of Borneo. Both these weapons are in my collection.

ANOTHER example of a weapon where a large and powerful barb is needful is the Harpoon. As the harpoon is used in capturing the whale, the largest and most powerful of living mammalia, it is evident that a barb which will hold such a prey must be rather peculiarly made. The head and part of the shaft of the harpoon are shown in the right-hand figure of the accompanying illustration.

The left-hand figure represents a curious parasitic crustacean, popularly called the Sprat-sucker, because it is usually found on sprats. It affixes itself mostly to the eye, the deeply barbed head being introduced between the eye and the socket. In

LERNENTOMA.   HARPOON.

some seasons this remarkable parasite is quite plentiful, while in others scarcely a specimen can be found. Its total length is slightly under an inch, and its scientific name is *Lernentoma Spratti.*

The following graphic account of some prototypic weapons belonging to a marine worm is given by Mr. Rymer Jones, and is well worthy of perusal, not only for the vividness of the description, but for its exact accuracy:—

"Here is a Polynoe, a curious genus, very common under stones at low water on our rocky shores.

"It is remarkable on several accounts. All down the back we discover a set of oval or kidney-shaped plates, which are called the back-plates (*dorsal elytra*); these are flat, and are planted upon the back by little footstalks, set on near the margin of the under surface: they are arranged in two rows, overlapping each other at the edge. These kidney-shaped shields, which can be detached with slight violence, are studded over with little transparent oval bodies, set on short footstalks, which are, perhaps, delicate organs of touch. The intermediate antennæ, the tentacles, and the cirrhi or filaments of the feet, are similarly fringed with these little appendages, which resemble the glands of certain plants, and have a most singular appearance.

" If we remove the shields, we discover, on each side of the body, a row of wart-like feet, from each of which project two bundles of spines of exquisite structure. The bundles, expanding on all sides, resemble so many sheaves of wheat, or you may more appropriately fancy you behold the armoury of some belligerent sea-fairy, with stacks of arms enough to accoutre a numerous host.

"But, if you look closely at the weapons themselves, they rather resemble those which we are accustomed to wonder at in missionary museums,—the arms of some ingenious but barbarous people from the South Sea Islands,—than such as are used in civilised warfare. Here are long lances, made like scythe-blades, set on a staff, with a hook on the tip, as if to capture the fleeing foe, and bring him within reach of the blade. Among them are others of similar shape, but with the edge cut into delicate slanting notches, which run along the sides of the blade like those on the edge of our reaping-hooks.

" These are chiefly the weapons of the lower bundle ; those of the upper are still more imposing. The outermost are short curved clubs, armed with a row of shark's teeth to make them more fatal ; these surround a cluster of spears, the long heads of which are furnished with a double row of the same appendages, and lengthened scimitars, the curved edges of which are cut into teeth like a saw.

" Though a stranger might think I had drawn copiously on my fancy for this description, I am sure, with your eye upon what is on the stage of the microscope at this moment, you will acknowledge that the resemblances are not at all forced or unnatural. To add to the effect, imagine that all these weapons are forged out of the clearest glass instead of steel ; that the larger bundles may contain about fifty, and the smaller half as many each ; that there are four bundles upon every segment, and that the body is composed of twenty-five such segments, and you will have a tolerable idea of the garniture and armature of this little worm, which grubs about in the mud at low-water mark."

Somewhere between the fifteenth and sixteenth centuries a sort of anomalous weapon was in use, namely, a dagger, with a

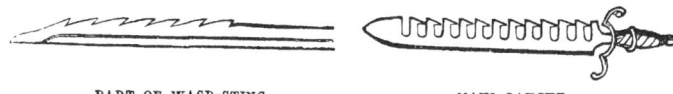

PART OF WASP-STING.   MAIN GAUCHE.

number of very deep and bold barbs. It was not, however, employed for offence, but for defence, and was used in the "rapier and dagger" mode of fighting, when the dagger, which was held in the left hand, was employed to parry the thrusts of the rapier, which was held in the right. From the mode of holding it, the weapon was called "Main Gauche."

Sometimes the blade was quite plain, and, indeed, an ordinary dagger answered the purpose. But in most cases the Main Gauche was made for this special purpose, and was furnished either with strong diverging projections, or with a series of deep notches, so that the sword of the enemy might be caught in them and broken. In consequence of this use these notched or guarded weapons were also called by the name of *Brise-épée*, or Sword-breaker.

The resemblance between this weapon and the blade of a wasp's sting can be seen at a glance. There is another form of the *Brise-épée* which is so strangely like the cutting apparatus of one of the saw-flies, that an outline sketch of the one would answer very well for the other.

WAR AND HUNTING.

CHAPTER III.

PROJECTILE WEAPONS AND THE SHEATH.

Propulsive Power.—The Pea-shooter and its Powers.—An Attack repulsed.—Clay
Bullets.—Puff and Dart.—The Sumpitan of Borneo, and its Arrows.—The
Zarabatana or Pucunha of South America, and its Arrows.—The Air-gun.—
Modern Firearms.—The Chœtodon, or Archer-fish.—The Pneumatic Rail-
way.—The Throwing-stick and its Powers.—Australians, Esquimaux, and
New Caledonians.—Principle of the Sheath.—Waganda Spears.—Sheathed
Piercing Apparatus of the Gnat, Flea, and Bombylius.—Indian Tulwar and
Cat's Claw.—The Surgeon's Lancet, and Piercing Apparatus of the Gad-fly
and Mosquito.

WE will now take some of the analogies between Projectile
Weapons of Art and Nature, selecting those in which the
propulsive power is air or gases within a tube. Whether the
weapon be a blow-gun, an air-gun, or a firearm of any description, the principle is the same. We will take them in succession, choosing first those of the simplest and most primitive
character.

Taking ourselves as examples, and looking upon the toys of
children as precursors of more important inventions, we find
that the simplest and most primitive of projectiles is the
Pea-shooter, so familiar to all boys.

Insignificant as is the little tin tube, and small as are the
missiles which are propelled through it, the blow which can be
struck by a pea properly shot is no trifle. At college I have
seen a night attack upon an undergraduate's rooms successfully
repelled by a pea-shooter made for the nonce of a glass tube,
the owner of the rooms having a taste for chemicals, and
possessing a fair stock of the usual apparatus. Though the
assaulted rooms were on the top set, and the assailants began
their storming approaches below, the peas were too much

for the stones, taking stinging effect on the hands and faces, and preventing any good aim being taken at the windows. Only two panes of glass were broken through a siege that lasted for several hours.

There is another toy which is a development of the pea-shooter, and carries a small clay bullet instead of a pea. When the tube is quite straight and the balls fit well, the force of this missile is very great, as it can be used for killing small birds. Indeed, such an instrument is largely employed by the native hunters in procuring humming-birds for the European market. These weapons are generally lined with metal in this country, but a simple bamboo tube is sufficient for the native hunters.

A still further improvement occurs where the place of the bullet is taken by a small dart or arrow, which is usually made to fit the bore by having a tuft of wool, or some similar substance, at the butt. The arrow is aimed at a target, and the toy is popularly known as "Puff and Dart."

With us this apparatus is only a toy, but in several parts of the world it becomes a deadly weapon, namely, in Borneo and over a large part of tropical America. In both cases the arrows are poisoned, as has already been mentioned when treating of poisoned weapons.

THE first and best known of these weapons is the dreaded Sumpitan, or Blow-gun, of Borneo, the arrows of which are poisoned with the deadly juice of the upas-tree. Here I may as well mention that the scientific name of the upas-tree is *Antiaris toxicaria*. It belongs to a large group of plants, all of which have an abundance of milk-like and sometimes poisonous juice. We are most of us familiar with the old story of the upas-tree and its deadly power, and how the tree stood in a valley, in which nothing else could live, and that condemned criminals might compound for their inevitable fate by venturing into the valley of death and bringing back a flask of the dread poison. Even birds were supposed to be unable to fly over the valley, but to fall into it, being poisoned by the exhalations of the tree.

Now, there is a saying that there is no smoke without fire, and though this account is evidently incredible, it is not altogether without foundation. In Java, as in many other

parts of the world, there are low-lying places where carbonic
acid gas exudes from the earth, and no living creature can
exist in them. Even in this country scarcely a year passes
without several' deaths occurring from inhalation of the same
fatal gas, which has collected in some disused excavation.
That there is, therefore, a deadly valley in Java may be true
enough, and it is also true that the juice of the upas-tree is
poisonous when it mixes with the blood. But the two have no
connection with each other, and, so far from the upas-tree
poisoning the valley by its exhalations, it could not exist in
such an atmosphere.

Now for the Sumpitan and the arrows. The former is a tube,
some seven feet in length, with a bore of about half an inch in
diameter, and often elaborately inlaid with metal. I have one
in which the whole of the mouthpiece is brass, and the other
end of the weapon has been fitted with a large spear-head,
exactly on the principle of the bayonet.

The arrows are very slight, and, in order to make them fit
the tube, are furnished at their bases with a conical piece of
soft wood. In themselves they would be almost useless as
weapons, but when the poison with which their points are
armed is fresh, these tiny arrows, of which sixty or seventy
are but an ordinary handful, carry death in their points.
Though they have no great range, they are projected with
much force, and with such rapidity that they cannot be
avoided, their slender shafts being almost invisible as they pass
through the air.

THE second weapon is the still more dangerous blow-gun of
tropical America, called Zarabatana, or Pucunha, according to
the locality. Some of these tubes measure more than eleven
feet in length, and through them the arrow can be propelled
with wonderful force. I have often sent an arrow to a distance
of a hundred yards, and with a good aim.

A native, however, can send it much farther, knack, and not
mere capacity of lung, supplying the propelling power, just
as it is with the pea-shooter. When the arrow is properly
blown through the zarabatana a sharp "pop" ought to be
heard, like the sound produced by a finger forced into a
thimble and quickly withdrawn, or a cork drawn from a bottle.

As to seeing the diminutive arrow in its flight, it is out of the question, and no agility can be of the least use in avoiding it. One of my friends, a peculiarly sharp-sighted officer of artillery, has often tested this point, and although there was but one arrow to watch, and it was blown in the open air, he could not see it until it either struck or passed him (of course the poisoned end was cut off). What, then, would be the result of a number of these deadly missiles hurled out of a dense bush may easily be imagined.

An account of the poison with which these arrows are armed will be found on p. 64.

THE reader will please to remember that in all these cases the missile is propelled by air which is compressed by the aid of the lungs, and forced into the tube behind the bullet or arrow. Now, the AIR-GUN, which really can be made a formidable weapon, is constructed on exactly the same principle as the pea-shooter and the blow-guns, except that the air is compressed by the human arm instead of the human lungs. There are various modifications of this weapon, but in all of them air is driven into a strong chamber by means of a forcing syringe, and is released by the pull of the trigger, so as to drive out the missile which has been placed in the barrel.

It is worthy of notice that the term "noiselessly destructive" weapon, as applied to the air-gun, is entirely false. I have already mentioned that with the blow-gun of tropical America a definite explosion accompanies the flight of each arrow. The same result occurs with the air-gun, the loudness of the report being in exact proportion to the force of the air, each successive report becoming slighter and the propulsive power weaker until a new supply of air is forced into the chamber.

HOWEVER dissimilar in appearance may be the cannon, rifle, pistol, or any other firearm, to the pea-shooter and its kin, the principle is exactly the same in all. It has been already mentioned that in the blow-guns the air is compressed by the exertion of human lungs, and in the air-gun the compression is achieved by human hands.

But with the firearm a vast volume of expansible gas is kept locked up in the form of gunpowder, gun-cotton, fulminating silver, or other explosive compound, and is let loose, when wanted, by the aid of fire.

IN the illustration are represented on the right hand the blow-guns of America and Borneo, and below them is the cannon as at present made. On the left hand of the same

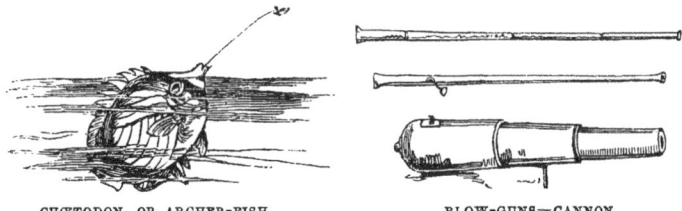

CHŒTODON, OR ARCHER-FISH.   BLOW-GUNS—CANNON.

illustration is seen a representation of a natural gun which has existed for thousands of years before gunpowder was invented, and very long before the savage of Borneo or America discovered the blow-gun.

It is the ARCHER-FISH (*Chœtodon*), which possesses the curious power of feeding itself by shooting drops of water at flies, and very seldom failing to secure its prey.

There are several species of this very curious fish spread over the warmer parts of the world, and their remarkable mode of obtaining prey is very well known in all. There is, indeed, scarcely any phenomenon in Nature more remarkable than the fact of a fish being able to shoot a fly with a drop of water projected through its tubular beak, if we may use that expression for so curiously modified a mouth.

Indeed, so certain is the fish of its aim, that in Japan it is kept as a pet in glass vases, just as we keep gold fish in England, and is fed by holding flies or other insects to it on the end of a rod a few inches above the surface of the water. The fish is sure to see the insect, and equally sure to bring it down with a drop of water propelled through its beak.

It is worthy of remark that the same principle was once, though unsuccessfully, employed in the propulsion of carriages, under the name of the Pneumatic Railway. Some of my readers may remember the railway itself, or at all events the disused tubes which lay for so many years along the Croydon

Railway. Speed was obtained, as I can testify from personal experience, but the expense of air-pumps and air-tight tubing was too great to be covered by the income, especially as the rats ate the oiled leather which covered the valves.

I FIND some little difficulty in arranging the subject which comes next in order. It might very properly be ranked among the Levers, which will be treated of in another chapter; or it might be placed among the examples of centrifugal force, together with the sling, the " governor " of the steam-engine, &c., all of which will be more fully described in their places. However, as we are on the subject of Projectiles, we may as well take it in the present place.

It is the THROWING-STICK, by which the power of the human arm is enormously increased, when a spear is to be hurled. Perhaps the most expert spear-throwers in the world are to be found among the Kafir tribes of Southern Africa, and yet the most experienced among them could not make sure of hitting a man at any distance above thirty or forty yards. But the throwing-stick gives nearly double the range, and I have seen the comparatively slight and feeble Australian hurl a spear to a distance of a hundred yards, and with an aim as perfect as that of a Kafir at one-fourth of the distance.

The mode in which this feat is performed is shown in the accompanying diagram. Instead of holding the spear itself, the native furnishes himself with a "Throwing-stick." This weapon varies greatly in shape and size, but a very good idea of its form, and the manner of using it, may be obtained from the accompanying illustration, which was drawn from the actual specimen as held by an Australian native.

The throwing-stick is armed at the tip with a short spike, which fits into a little hole in the but of the spear. The stick and spear being then held as shown in the illustration, it is evident that a powerful leverage is obtained, varying according to the length of the stick. I possess several of these instruments, no two of which are alike.

It is rather remarkable that among the Esquimaux a throwing-stick is also used, exactly similar in principle, but differing slightly in structure, the but of the spear fitting into a hole at the end of the throwing-stick. Wood being scarce among

the Esquimaux, these instruments are mostly made of bone.
I possess one, however, which is made of wood, beautifully
polished, and adorned with a large blue stone, something like
a turquoise, set almost in its middle.   One of the most curious
points in the formation of the Esquimaux weapon is, that the

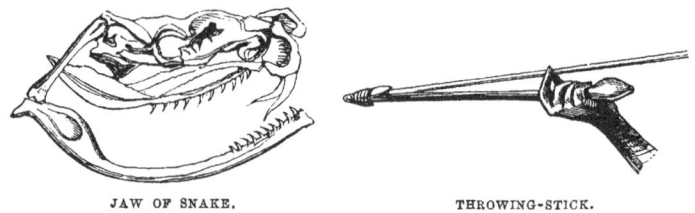

JAW OF SNAKE.   THROWING-STICK.

but is grooved and channelled so as to admit the fingers and
thumb of the right hand.   The average length of this instrument is twenty inches.

In New Caledonia the natives use a contrivance for increasing the power of the spear, which is based on exactly identical
principles, though the mode of carrying them out is different.
A thong or cord of some eighteen inches in length is kept in
the right hand, one end being looped over the forefinger, and
the other, which is terminated by a button, being twisted round
the shaft of the spear.   When the weapon is thrown, the
additional leverage gives it great power ; and it is a note-
worthy fact that the sling-spear of New Caledonia has enabled
us to understand the otherwise unintelligible "amentum" of
the ancient classic writers.

Passing from Art to Nature, we have in the jaw of the
serpent an exact type of the peculiar leverage by which the
spear is thrown.   If the reader will refer to the illustration,
he will see that the lower jaw of the snake, instead of being
set directly on the upper jaw, is attached to an elongated bone,
which gives the additional leverage which is needful in the
act of swallowing prey, after the manner of serpents.

In War and in Peace we have been long accustomed to
shield the edges and points of our sharp weapons with sheaths,
and even the very savages have been driven to this device.

I have in my collection a number of sheathed weapons from nearly all parts of the world, and it is a remarkable fact that the Fan tribe, who are themselves absolutely naked, sheathe their daggers and axes as carefully as we sheathe our swords and bayonets. In some points, indeed, they go beyond us; for the most ignorant Fan savage would never think of blunting the edge of his weapon by sheathing it in a metal scabbard. Their sheaths are beautifully made of two flat pieces of wood, just sufficiently hollowed to allow the blade to lie between them, and bound together with various substances. For example, the sheaths of one or two daggers in my possession are made of wood covered with snake-skin, while others are simply wood bound with a sort of rattan. Even the curious missile-axe which the Fan warrior uses with such power is covered with a sheath when not in actual use.

The figure on the right hand of the illustration represents the heads of two spears of Waganda warriors. When they

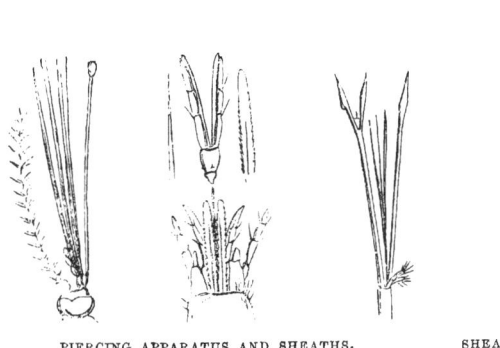 

PIERCING APPARATUS AND SHEATHS.   SHEATHED SPEARS OF WAGANDA.
GNAT.   FLEA.   BOMBYLIUS.

present themselves before their king, the warriors must not appear without their weapons, and it would be contrary to all etiquette to show a bare blade except in action. The sheath can be slipped off in a moment, but there it is, and any man who dared to appear before his sovereign without his weapon, or with an unsheathed spear, would lose his life on the spot, so exact is the code of etiquette among these savages.

The sheathed spears of Nature are shown in the same illustration. On the left is a side view of the piercing apparatus of the common Gnat.

In the middle is the compound piercing apparatus of the common Flea, with which we are sometimes too well acquainted, the upper figure showing the lancets and sheaths together, and the lower exhibiting them when separated.

On the right is shown the group of mouth-lancets belonging to one of the Humble-bee flies (*Bombylius*). These flies do not suck blood like the Mosquito, the Flea, and the Gad-fly, but they use the long proboscis for sucking the sweet juices out of flowers, and in consequence it is nearly of the same form as if it were meant for sucking blood. Indeed, there are some insects which do not seem to care very much whether the juice which they suck is animal or vegetable.

On the right hand of the illustration is seen an Indian sword, or "Tulwar," drawn from one of my own specimens. I have selected this example on account of the structure of the sheath. It is evident, from the form of the blade, that the sword cannot be sheathed point foremost, and that therefore

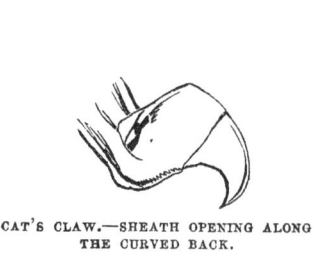

CAT'S CLAW.—SHEATH OPENING ALONG THE CURVED BACK.

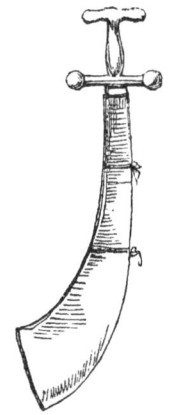

INDIAN TULWAR.—SHEATH OPENING ALONG THE CURVED BACK.

some other plan must be used. In this weapon the sheath is left open on one side, the two portions being held together by the straps which are shown in the figure. Of course there is loss of time in sheathing and drawing such a sword, but the peculiar shape of the blade entails a necessity for a special scabbard.

On the other side is shown one of the fore-claws of a cat, which, as we all know, can be drawn back into its simple sheath between the toes, when it is not in use. This sheath is exactly the same in principle as that of the Indian tulwar, and any one can examine it by looking at the foot of a good-tempered cat. I have done so even with a chetah, which is not a subject that would generally be chosen for such a purpose.

On the next illustration is shown an ordinary Lancet, in which the blade is guarded between a double sheath, the two halves and the blade itself working upon a common pivot. As for the ordinary sword and dagger sheaths, it is not worth while to figure them.

TURNING to the opposite side of the illustration, we shall see a few of the innumerable examples in which the principle of the sheath was carried out in Nature long before man came on the earth.

The reader should compare this figure with the side view of the Gnat's lancets given on p. 81.

They represent the cutting and piercing instruments of several insects, all of which are very complicated, and are

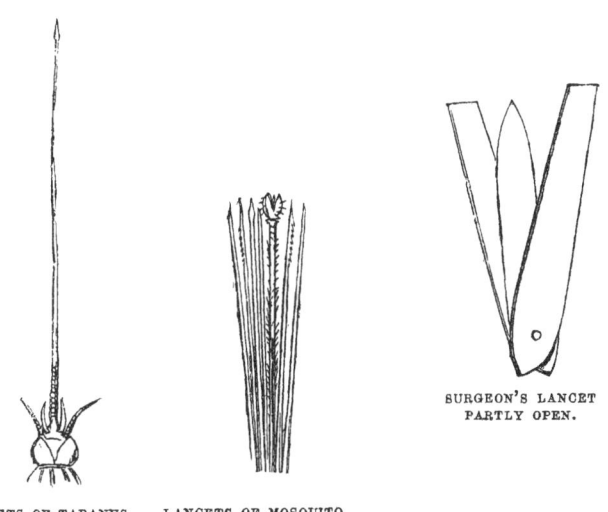

LANCETS OF TABANUS CLOSED.

LANCETS OF MOSQUITO PARTLY OPEN.

SURGEON'S LANCET PARTLY OPEN.

sheathed after the manner of the lancet. Indeed, they are popularly known as "mouth-lancets," and with reason, as the reader may see by reference to the illustration.

On the extreme left are shown the head and closed lancets of a foreign Gad-fly, the lancets being all in their sheaths, and showing the character of the weapon which enables a small fly to be master, or rather mistress, of the forest. I say mistress, because in all these cases it is the female alone that possesses these instruments of torture.

Next it is a magnified representation of the lancets of the common Mosquito, as seen from above, both lancets being removed from their sheaths and separated.

WAR AND HUNTING.

CHAPTER IV.

THE NET.

ALTHOUGH the Net is but seldom employed for the purposes of general warfare, it was once largely used in
individual combats, of which we will presently treat. In
hunting, however, especially in fishing, the Net has been in
constant use, and is equally valued by savages and the most
civilised nations.

To begin with the fisheries. Even among ourselves there
are so many varieties of fishing-nets that even to enumerate
them would be a work of time. However, they are all based
on one of two principles, *i.e.* the nets which are set and the
nets which are thrown.

We will begin with the first.

ON the right hand of the illustration, and at the bottom,
may be seen a common Seine-net being " shot " in the sea.
This form of net is very long in proportion to its width, some
of these nets being several miles long. The upper edge

of the net is furnished with a series of cork bungs, which maintain it on the surface, while the lower edge has a corresponding set of weights, which keep the net extended like a wall of meshes. Any fish which come against this wall are, of course, arrested, and are generally caught by the gill-covers in their vain attempts to force themselves through the meshes.

We may see representations of fishing with the seine-net in the sculptures and paintings of Egypt and Assyria; and in the Berlin Museum there is a part of an Egyptian seine-net with the leads still upon the lower edge, and the upper edge bearing a number of large pieces of wood, which acted as buoys, and served the same purpose as our corks.

In hunting, this plan has been adopted for many centuries, the upper edge of the net being supported on poles, and the lower fastened to the ground in such a manner as to leave the net hanging in loose folds. While this part of the business

SPIDER-WEB.   HUNTING-NET.   THE SEINE-NET.

is being completed by the servants, the hunters are forming a large semicircle, in which they enclose a number of wild beasts, which they drive into the nets or "toils" by gradually contracting the semicircle. The ancient sculptures give us accounts of nets used in exactly this manner. There are represented the nets rolled up ready for use, and being carried on the shoulders of several attendants, who are bearing them to the field. Then there are the nets set up on their poles, and having enclosed within them a number of wild animals, such as boars and deer.

In various parts of India, hunting with the net is one of the chief amusements of their principal men, and the variety of game driven into the toils is really surprising, and affords a magnificent sight to those who view it for the first time. Even the tiger himself cannot leap over the nets because they are so high, nor force his way through them, because their folds hang so lightly that they offer no resistance to his efforts.

A very simple net on similar principles is used for catching elephants. It is formed of the long creeping plants that fling themselves in tangled masses from tree to tree. These creepers are carefully twisted into a net-like form, without being removed from the trees, and when a sufficient space has been enclosed the elephants are driven into it. Not even their gigantic strength and tons of weight are capable of breaking through a barrier which, apparently slight, is as strong as if it were built of the tree-trunks on which the creepers are hung.

This net is seldom used for military purposes, though I have seen one, which I believe still exists, and would do good service. In one of our largest fortresses there is a subterranean corridor, through which it is desirous that the enemy should not penetrate. One mode of defence consists of a large net made of steel hanging loosely across it. The meshes are about ten inches square, so that the defenders can fire from their loopholes through the meshes, while the assailants, even if they knew of its position, would find that nothing smaller than a field-gun would have any effect on this formidable net.

THE natural analogy of the fixed net is evidently the web of the common Garden Spider, or Cross Spider (*Epeira diadema*), whose beautiful nets we all must have admired, especially when we are wise enough to get up sufficiently early in the morning to see the webs with the dewdrops glittering on them.

Last year there was a wonderful sight. Within a mile of my house there is a long iron fence, which in one night had been covered with the webs of the garden spider. The following morning, though bright, was chilly, so that the dewdrops were untouched. I happened to pass by the fence soon after sunrise, and was greatly struck with the astonishing effects which could be produced with such simple materials as water and web. The dewdrops were set at regular intervals upon the web, so as to produce a definite and beautiful pattern, the whole line of fence looking as if it had been woven in fine lace.

Then, as the fence runs north and south, and the path is on the westward of it, every passenger saw the rays of the rising sun dart through these tiny globules, and convert every one of them into a jewel of ever-changing colours. It seemed a pity that such beauty could but last for an hour or so, or that these exquisite webs should only be used for catching flies.

NEXT comes the Casting-net in its various forms. This net is mostly circular, and is loaded round the edge with small leaden plummets. It is evident that, if such a net could be laid quite flat upon the water, it would assume a dome-like shape, in consequence of the circumference being heavier than the centre, and would sink to the bottom, enclosing anything which came within its scope.

The difficulty is to place the net in such a manner, and this is accomplished by throwing it in a very peculiar way. The net is gathered in folds upon the shoulder, which it partially envelops. By a sudden jerk the thrower causes it to fly open with a sort of spinning movement, and when well cast it will fall on the water perfectly flat.

After allowing it to sink to the bottom, the fisherman draws it very gently by a cord attached to its middle. As he raises it the weights of the leaded circumference are drawn nearer and nearer together by their own weight, and finally form it into a bag, within which are all the living creatures which it has enclosed.

Though the Casting-net has never been used in warfare, it was one of the favourite implements in gladiatorial combats among the Romans. Two men were opposed to each other; one, called the Retiarius or Netsman, being quite naked, except sometimes a slight covering round the waist, and armed with nothing but a Casting-net and a slight trident, which could not inflict a deadly wound. The other, called the Secutor or Follower, from his mode of fighting, was armed with a visored helmet, a broad metal belt, and armour for the legs and arms. He also carried a shield large enough to protect the upper part of the body, and a sword. It will be seen, therefore, how great was the power of the Casting-net, when it enabled its naked bearer to face such odds of offensive and defensive armour.

When the two met in combat, the Retiarius tried to fling his net over his adversary, and if he succeeded, the fate of the latter was sealed. Entangled in the loose meshes, he could scarcely move his limbs, while the sharp prongs of the long-shafted trident came darting in at every exposed point, and exhausting the man with pain and loss of blood. The trident

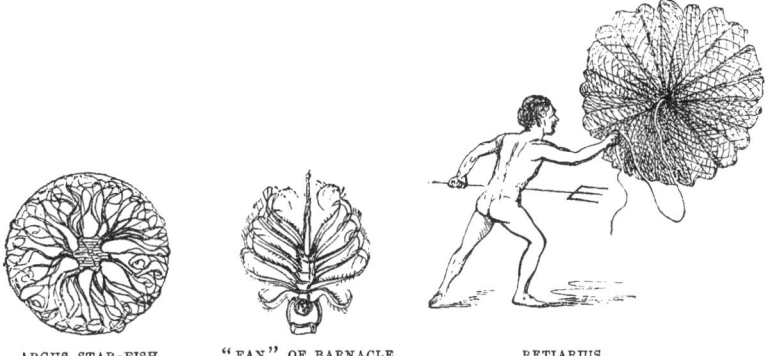

ARGUS STAR-FISH.   "FAN" OF BARNACLE.   RETIARIUS.

was in itself so feeble a weapon, that if the Secutor were vanquished and condemned to death by the spectators, his antagonist could not kill him, but had to call another Secutor to act as executioner with his sword.

Should he fail in his cast, the Retiarius drew back his net by the central cord, and took to flight, followed by the Secutor, who tried to wound him before he could re-fold his net upon his shoulder, ready for another cast. It is worthy of notice that in these singular combats the netsman seems generally to have been the victor. A Retiarius with his net is shown in the illustration.

I may mention that our ordinary bird-catchers' nets, and even the entomologist's insect-net, are only modifications of the Casting-net.

Now for Nature's Casting-nets, two examples of which are figured, though there are many more. These two have been selected because they are familiar to all naturalists.

The first is the Argus Star-fish, Basket-urchin, or Sea-basket.

The innumerable rays and their subdivisions, amounting to some eighty thousand in number, act as the meshes of the net. All the rays are flexible and under control. When the creature wishes to catch any animal for prey, it throws its tentacles over it, just like the meshes of a net. It then draws the tips of the rays together, just as is done by the circumference of the casting-net, and so encloses its prey effectually.

THE next specimen is the net-like apparatus of the common Acorn Barnacles, with which our marine rocks are nearly covered. These curious beings belong to the Crustacea, and the apparatus which is figured on page 89, and popularly called the " fan," is, in fact, a combination of the legs and their appendages of bristles, &c. When the creature is living and covered with water, the fan is thrust out of the top of the shell, expanded as far as possible, swept through the water, closed, and then drawn back again. With these natural casting-nets the Barnacles feed themselves, for, being fixed to the rock, they could not in any other way supply themselves with food. There are many similar examples in Nature, but these will suffice.

THE ROD AND LINE.

THAT both terrestrial and aquatic nets should have their parallels in Nature is clear enough to all who have ever seen a spider's web, or watched the " fan " of the barnacle. But that the rod and baited line, as well as the net, should have existed in Nature long before man came on earth, is not so well known. Yet, as we shall presently see, not only is the bait represented in Nature, but even our inventions for " playing " a powerful fish are actually surpassed.

We will begin with the Bait.

In nearly all traps a bait of some kind is required, in order to attract the prey, and when we come from land to attract the dwellers in water to our hooks, it is needful that bait of some kind should be used, were it only to deceive the eye, though not the nostrils or palate, of the fish.

A notable example of the deception is given in the common artificial baits of the present day, which are made to imitate almost any British insect which a fish might be disposed to eat.

Perhaps the best instance of this deception is that which is practised by sundry Polynesian tribes. They have seen that the Coryphene or Dorado, and other similar fish, are in the habit of preying upon the flying-fish, and springing at them when they are tolerably high in the air. So these ingenious semi-savages dress up a hook made of bone, ormer-shell, and other materials, making the body of it into a rudely designed form of a fish. A hole is bored transversely through it at the shoulders, and a bunch of stiff fibres is inserted to represent the wings. Another bunch does duty for the tail.

The imitation bait being thus complete, it is hung to a long and slender bamboo rod, which projects well beyond the stern of a canoe, and is so arranged that the hook is about two feet or so from the surface. The Coryphene, seeing this object skimming along, takes it for a flying-fish, leaps at it, and is caught by the hook. There are in several collections specimens of these ingenious hooks, and I possess one which is made on similar principles, but intended for use in the water, and not in the air. It is, in fact, a " spoon-bait."

One point of ingenuity must be mentioned, as it really belongs to the principle of the bait. These same savages, having noticed that large sea-birds are in the habit of hovering over the flying-fish, and would probably be seen by the Coryphenes, rig up a very long bamboo rod, tie to its end a large bundle of leaves and fibres, and then fix it in the stern of the boat, the sham bird being hung some twenty feet above the sham fish. There is a refinement of deception here, for which we should scarcely give such savages their due credit.

In Art, then, we bait our hooks either with real or false food, and so attract the fish.

In Nature we have a most accomplished master of the art of baiting, who has the wonderful power of never needing a renewal of his bait. A glance at the left-hand figure of the next illustration will show that I allude to the Angler-fish, sometimes called the Fishing-frog (*Lophius piscatorius*). This remarkable creature has a most enormous mouth, and comparatively small body. On the top of its head are some curious bones, set just like a ring and staple, so as to move freely in every direction. A figure of this piece of mechanism will be given in a future page. At the end of these bones are little fleshy appendages, which must be very tempting to most fish, which are always looking out for something to eat. As they are being waved about, they look as if they were alive. The fish darts at the supposed morsel, and is at once engulfed in the huge jaws of the Angler-fish, which, but for this remarkable apparatus, would be scarcely able to support existence, as it is

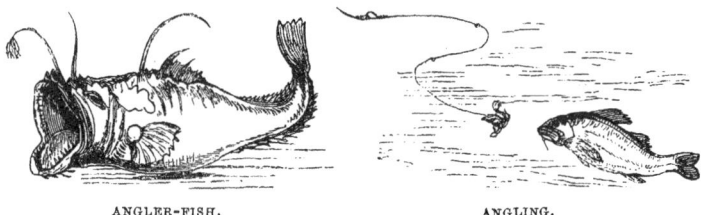

ANGLER-FISH.   ANGLING.

but a sluggish swimmer, and yet needs a large supply of food. The illustration, representing on the right hand a fish attracted to a bait, and on the left, the Angler-fish, with its bait-like appendage to the head, speaks for itself.

PASSING to the art of Angling with a rod and line, we now arrive at another development.

Supposing a fish to have taken the bait, and to have been firmly hooked, how is it to be landed ? The simplest plan is, of course, to have a very thick and strong line which will not break with the weight of any ordinary fish.

This is very well in sea-fishing, where a line made of whip-cord will answer the purpose in most cases. But, in river fishing, we have the fact that the fish are so shy that a linen thread would scare them, and so strong and active, that even whip-cord would not prevent them from breaking the line, or tearing the hook out of their mouths. So the modern angler sets himself to the task of combating both these conditions. In the first place, he makes the last yard or two of his line of "silkworm-gut"—a curious substance made from the silk-vessels of silkworms, and nearly invisible in the water. In the next place, he has a very elastic rod ; and, in the third, he has forty or more yards of line, though perhaps only twenty feet are in actual use until the fish is hooked. The remainder of the line is wound upon a winch fixed to the handle of the rod.

Thus, when a powerful fish is hooked and tries to escape, the line is gradually let loose, so as to yield to its efforts. When it becomes tired by the gradual strain, the line is again wound in, and in this way a fish which would at the first effort smash rod and line of a novice will, in the hands of an experienced fisherman, be landed as surely as if it were no bigger than a gudgeon.

NATURE has in this case also anticipated Art, and surpassed all her powers.

There is a wonderful worm, common on our southern coasts, and bearing, as far as I know, no popular name. It is known to

NEMERTES. "PLAYING" A FISH.

the scientific world as *Nemertes Borlasii*. It possesses the power of extension and contraction more than any known creature, and uses those powers for the purpose of capturing prey. The fishermen say that this worm can extend itself to a length of ninety feet, and as Mr. Davis found one to measure twenty-two feet, after being immersed in spirits of wine, it is likely that their account may be true, especially as the spirit greatly contracted the animal in point of length.

A most vivid description of this worm is given by C. Kingsley, in his " Glaucus," and was written before he knew its name.

" Whether we were intruding or not, in turning this stone, we must pay a fine for having done so ; for there lies an animal as foul and monstrous to the eye as ' hydra, gorgon, or chimæra dire,' and yet so wondrously fitted to its work that we must needs endure for our own instruction to handle and to look at it. Its name I know not (though it lurks here under every stone), and should be glad to know. It seems some very 'low' Ascarid or Planarian worm.

"You see it? That black, shiny, knotted lump among the gravel, small enough to be taken up in a dessert spoon. Look now, as it is raised and its coils drawn out. Three feet, six, nine at least; with a capability of seemingly endless expansion; a slimy tape of living caoutchouc, some eighth of an inch in diameter, a dark chocolate black, with paler longitudinal lines.

"Is it alive? It hangs helpless and motionless, a mere velvet string, across the hand. Ask the neighbouring Annelids and the fry of the rock-fishes, or put it into a vase at home, and see. It lies motionless, trailing itself among the gravel; you cannot tell where it begins or ends; it may be a dead strip of seaweed, *Himanthalia lorea*, perhaps, or *Chorda filum*, or even a tarred string.

"So thinks the little fish who plays over and over it, till he touches at last what is too surely a head. In an instant a bell-shaped sucker mouth has fastened to his side. In another instant, from one lip, a concave double proboscis, just like a tapir's (another instance of the repetition of forms), has clasped him like a finger; and now begins the struggle : but in vain. He is being 'played' with such a fishing-line as the skill of a Wilson or a Stoddart never could invent; a living line, with elasticity beyond that of the most delicate fly-rod, which follows every lunge, shortening and lengthening, slipping and twining round every piece of gravel and stem of seaweed, with a tiring drag such as no Highland wrist or step could ever bring to bear on salmon or on trout.

"The victim is tired now; and slowly, and yet dexterously, his blind assailant is feeling and shifting along his side, till he reaches one end of him; and then the black lips expand, and slowly and surely the curved finger begins packing him end foremost down into the gullet, where he sinks, inch by inch. till the swelling which marks his place is lost among the coils, and he is probably macerated to a pulp long before he has reached the opposite extremity of his cave of doom.

"Once safe down, the black murderer slowly contracts again into a knotted heap, and lies, like a boa with a stag inside him, motionless and blest."

The accuracy as well as the pictorial effect of this description cannot be surpassed. The "velvety" feel of the creature is most wonderful, as it slips and slides over and among the fingers, and makes the task of gathering it together appear quite hopeless.

This astonishing worm is drawn on the left hand of the illustration on page 93, so as to show the way in which the body is contracted or relaxed at will. On the other side of the illustration is an angler, armed with all the paraphernalia of his craft, and doing imperfectly that which the Nemertes does with absolute perfection.

A similar property belongs to the long, trailing tentacles of the Cydippe, which is described and figured on page 16. When they come in contact with suitable prey, all struggle is useless, the tentacles contracting or elongating to suit the circumstances, and at last lodging the prey within the body of the Cydippe.

THE SPRING-TRAP.

WE are all familiar with the common Spring-trap, or Gin, as it is sometimes called.

It varies much in form and size, sometimes being square and sometimes round ; sometimes small enough to be used as a rat-trap, and sometimes large enough to catch and hold human beings, in which case it was known by the name of man-trap. This latter form is now as illegal as the spring-gun, and though the advertisement "Man-traps and Spring-guns are set in these grounds" is still to be seen, neither one nor the other can be there.

They are all constructed on the same principle, namely, a couple of toothed jaws which are driven together by a spring, when the spring is not controlled by a catch. They are evidently borrowed from actual jaws, the same words being used to signify the movable portions and notches of the trap as are employed to designate the corresponding parts in the real jaw.

In both figures of the accompanying illustration we shall see how exact is the parallel. On the right hand is a

common rat-trap, or gin, such as is sold for eightpence, with
the jaws wide open, so as to show the teeth.   On the left is a

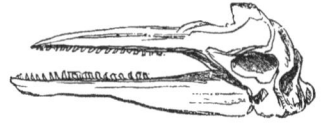

JAWS OF DOLPHIN (OPEN).

RAT-TRAP (OPEN).

sketch of the upper and lower jaws of the Dolphin, in which
an exactly analogous structure is to be seen.

The figure on the right hand of the lower illustration shows
a man-trap as it appears when closed, the teeth interlocking so
as exactly to fit between each other.   The same principle is
exhibited in the jaws of the Porpoise, which are seen on the left
of the illustration.   The jaws of an Alligator or Crocodile would
have answered the purpose quite as well, inasmuch as their

JAWS OF PORPOISE (CLOSED).

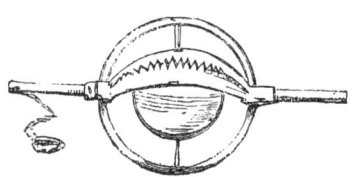

MAN-TRAP (CLOSED).

teeth interlock in a similar fashion, but I thought that it would
be better to give as examples the jaws of allied animals.   The
reason for this interlocking is evident.   All these creatures feed
principally on fish, and this mode of constructing the jaws
enables them to secure their prey when once seized.

Another example of such teeth is to be found in the fore-legs
of various species of Phasma and Mantis, as may be seen by

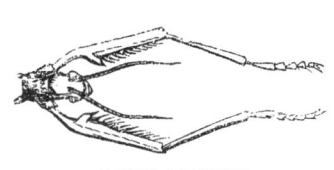

FORE-LEGS OF PHASMA.

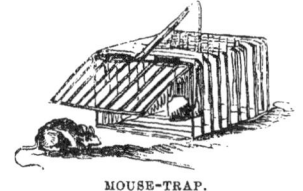

MOUSE-TRAP.

reference to the illustration.   The latter insects are wonderfully
fierce and pugnacious, fighting with each other on the least
provocation, and feeding mostly on other insects, which they

secure in their deeply-toothed fore-legs. They use these legs with wonderful force and rapidity, and it is said that a pair of these insects fighting remind the observer of a duel with sabres.

THE BAITED TRAP.

OUR space being valuable, we are not able to give many examples of Baited Traps, whether in Art or Nature.

The most familiar example of this trap is the common Mouse-trap, the most ordinary form of which is shown at the right hand of the illustration on page 96. In all the varieties of these traps, whether for mice or rats, the prey is induced to enter by means of some tempting food, and then is secured or killed by the action of the trap. Sometimes these traps are made of considerable size for catching large game, and in Africa are employed in the capture of the leopard, in India for taking both tigers and leopards, and in North America for killing bears.

We have already noticed one instance of a bait in the Angler-fish, described in page 92, but in this case the bait serves only for attraction, and the trap, or mouth, is not acted upon by the prey.

There are, however, many examples in the botanical world, where the plant is directly acted upon by the creature which is to be entrapped, such being known by the now familiar term "Carnivorous Plants." Of these there is a great variety, but under this head I only figure two of them.

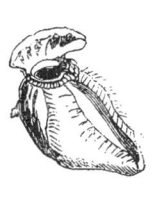

CEPHALOTUS.

DIONEA.

The plant on the right hand is the Venus Fly-trap (*Dionea muscipula*), which is common in the Carolinas. The leaves of this plant are singularly irritable, and when a fly or other insect

H

alights on the open leaf, it seems to touch a sort of spring, and the two sides of the leaf suddenly collapse and hold the insect in their grasp. The strange point about it is, that not only is the insect caught, but is held until it is quite digested, the process being almost exactly the same as if it had been placed in the stomach of some insect-eating animal.

So carnivorous, indeed, is the Dionea, that plants have been fed with chopped meat laid on the leaves, and have thriven wonderfully. Experiments have been tried with other substances, but the Dionea would have nothing to do with them. The natural irritability of the leaves caused them to contract, but they soon opened and rejected the spurious food.

On the left is the Cephalotus. This plant, instead of catching the insect by the folding of the leaf, secures it by means of a sort of trap-door at the upper end. The insect is attracted by the moisture in the cup, and, as soon as it enters, the trap-door shuts upon it, and confines it until it is digested, when the door opens in readiness to admit more prey.

BIRDLIME.

By a natural transition we pass to those traps which secure their prey by means of adhesive substances.

With us, the material called "birdlime" is usually employed. This is obtained from the bark of the holly, and is of the most singular tenacity. An inexperienced person who touches birdlime is sure to repent it. The horrid stuff clings to the fingers, and the more attempts are made to clear them, the more points of attachment are formed. The novice ought to have dipped his hands in water before he touched the birdlime, and then he might have manipulated it with impunity.

The most familiar mode of using the birdlime is by "pegging" for chaffinches.

In the spring, when the male birds are all in anxious rivalry to find mates, or, having found them, to defend them, the "peggers" go into the fields armed with a pot of birdlime and a stuffed chaffinch set on a peg of wood. At one end of this peg is a sharp iron spike. They also have a "call-bird," *i.e.* a chaffinch which has been trained to sing at a given signal.

When the "peggers" hear a chaffinch which is worth taking, they feel as sure of him as if he were in their cage. They take the peg, and stick it into the nearest tree-trunk. Round the decoy they place half-a-dozen twigs which have been smeared with birdlime, and arrange them so that no bird flying at the decoy can avoid touching one of them.

The next point is, to order the call-bird to sing. His song is taken as a personal insult by the chaffinch, which is always madly jealous at this time of year. Seeing the stuffed bird, he takes it for a rival, dashes at it, and touches one of the

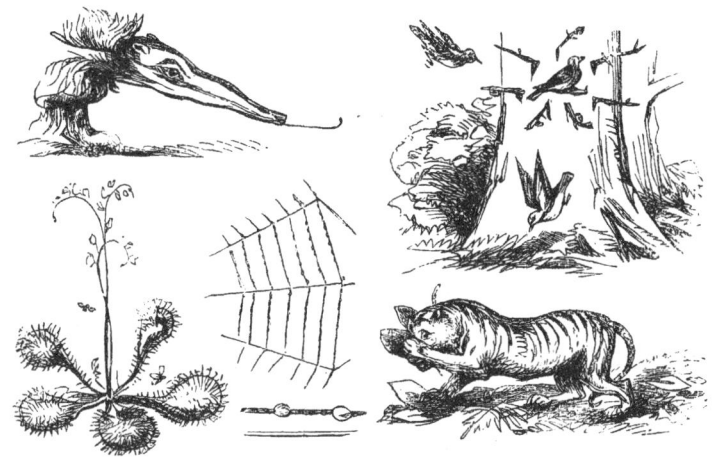

ANT-BEAR. DROSERA. SPIDER'S WEB. PEGGING CHAFFINCHES. TIGER AND LIMED LEAVES.

twigs. It is all over with him, for the more he struggles and flutters, the tighter is he bound by the tenacious cords of the birdlime, and is easily picked up by the "pegger."

EVEN the fierce and powerful tiger is taken with this simple, but terrible means of destruction. It is always known by what path a tiger will pass, and upon this path the native hunter lays a number of leaves smeared with birdlime. The tiger treads on one of them, and, cat-like, shakes his paw to rid himself of it. Finding that it will not come off, he rubs his paw on his head, transferring the leaf and lime to his face.

By this time he is in the middle of the leaves, and works himself into a paroxysm of rage and terror, finishing by blinding himself with the leaves that he has rubbed upon his head.   The hunters allow him to exhaust his strength by his struggles, and then kill him, or, if possible, capture him alive.

Both these scenes are represented on the right hand of the illustration.

On the left hand are several examples of natural birdlime, if we may use the term.   The upper represents the Ant-bear, or Great Ant-eater. This animal feeds in a very curious manner. It goes to an ant-hill, and tears it open with its powerful claws.   The ants, of course, rush about in wild confusion. Now, the Ant-eater is provided with a long, cylindrical tongue, which looks very like a huge earth-worm, and which is covered with a tenacious slimy secretion.   As the ants run to and fro, they adhere to the tongue, and are swept into the mouth of their destroyer.

Below the Ant-eater is the common Drosera, or Sundew, one of our British carnivorous plants.   It captures insects, just as has been narrated of the Dionea.   But, instead of the leaf closing upon the insect, it arrests its prey by means of little globules of viscous fluid, which exude from the tips of the hairs with which the surface of the leaf is covered.   As soon as the insect touches the hairs, they close over it, bind it down, and keep it there until it is digested.   Several species of Drosera are known in England, and are found in wet and marshy places.

Another plant, the Green-winged Meadow Orchis (*Orchis morio*), has been known to act the part of the Drosera.   A fly had contrived to push its head against the viscous fluid of the stigmatic surface, and, not being able to extricate itself, was found sticking there.

Next comes a portion of the web of the common Garden Spider (*Epeira diadema*).   We have already treated of this web as a net, and we will now see how it comes within the present category.

In the web of the spider there are at least two distinct kinds of threads.   Those which radiate from the centre to the cir- cumference are strong and smooth, while those which unite them are much slighter, and are covered with tiny globules set at regular intervals.   When the web is newly spun, these globules are found to be nearly as tenacious as birdlime, and it is by these means that an insect which falls into the web is arrested, and cannot extricate itself until the spider can seize it. After awhile the globules become dry, refuse to perform their office, and then the spider has to construct another web. So numerous are these globules that, according to Mr. Blackwall's calculations, an ordinary net contains between eighty and ninety thousand. Below the figure of the web itself are shown the two kinds of thread, the upper bearing the globules, and the lower representing one of the plain radiating threads.

WAR AND HUNTING.

CHAPTER V.

Reverted Spikes and their Modifications.—The Wire Mouse-trap.—George III. and the Trap.—Fate of a Royal Finger.—The Crab and Lobster Pot.—The Eel-pot.—Cocoon of the Emperor-moth and its Structure.—"Catchpoll" of the Middle Ages.—Deer-trap of India.—Jaws of Pike and Serpent.—The Grass-snake.—Jaws of Shark and their Power.—Spiked Defences.—The Park Fence, the Garden Wall, and the Chevaux-de-frise.—The "Square" of Infantry Manœuvres.—The Abattis, and its Structure and Power.—Ranjows and Caltrops.—Ancient Ranjows in Ireland.—Hedgehog.—Porcupine Echidna.—House-builder Caterpillar and its Home.—Repagula of Ascalaphus.—Tearing Weapons.—The "Wag-nuk" of India.—Armed Gauntlet of the Middle Ages.—Shark-tooth Gauntlet of Samoa, and the Uses to which it was put.—A terrible Warrior.—The Tiger's Claw.—Sport and Earnest.

REVERTED SPIKES.

I AM not quite satisfied with this title, but it is the best that I can find. By it I mean that mode of mechanism which, by means of an array of sharp spikes, permits an animal to enter a passage easily, and yet prevents it from emerging.

Whether or not this principle be now employed in warfare I cannot say, but it is at all events used extensively in a small

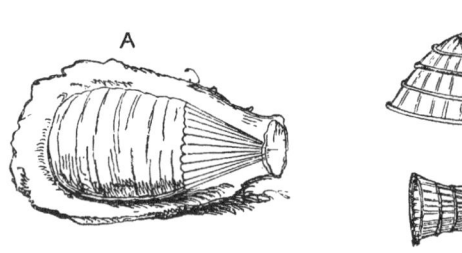
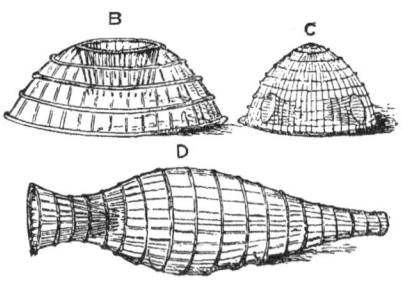

COCOON OF EMPEROR-MOTH.   CRAB-POT.   EEL-POT.   MOUSE-TRAP.

way of hunting, the best known of which is the wire Mouse-trap, one of which is shown at Fig. C on the illustration.   A glance at the figure will explain the trap, even to those who have never seen it. It is composed entirely of wire, and has several round holes just above its lower edge. Each of these holes is the entrance to a conical tunnel made of wires with sharpened ends.

The mouse, being attracted by a bait placed within the trap, tries to get at it. The doomed animal soon finds its way to one of the entrances, and with little difficulty pushes itself through the tunnel. Entering, however, is one thing, and returning is another. The wire yielded easily enough in one direction, but for the mouse to force itself against the converging points is an impossible task.

Readers of the last century literature may perhaps remember, in the pages of " Peter Pindar," a very clever and sarcastic account of the astonishment created in the mind of George III. by a mouse-trap seen accidentally in the house of a widow living at Salt Hill.

> "Eager did Solomon, so curious, clap
> His rare round optics on the widow's trap,
> That did the duty of a cat.
>   And, always fond of useful information,
>   Thus wisely spoke he with vociferation,—
> ' What's that? what? what?   Hæ, hæ? what's that?'
>
> To whom replied the mistress of the house,
> ' A trap, an't please you, sir, to catch a mouse.'
>
> ' Mouse—catch a mouse!' said Solomon with glee;
> ' Let's see, let's see—'tis comical—let's see—
> Mouse! mouse!'—then pleased his eyes began to roll—
> ' Where, where doth he go in?' he marvelling cried.
> ' There,' pointing to the hole, the dame replied.
> ' What! here?' cried Solomon, ' this hole? this hole?'
> Then in he pushed his finger 'midst the wire,
> That with such pains that finger did inspire,
>   He wished it out again with all his soul."

For my part I think that the King was quite right. If he did not know the philosophy of a mouse-trap he ought to have asked, and to have been rewarded, as in that case, by catching with a trap of his own baiting, six mice on six successive days.

At Fig. B on the same illustration is shown the simple apparatus by which crabs and lobsters are caught. The reader will see that the principle is exactly the same in both cases, the only difference being in material, the mouse-trap being made of wire, and the crab-pot of wicker.

At Fig. D is shown the common Eel-pot, or Eel-basket. In order to suit the peculiar shape of an eel, this basket is much longer in proportion to its diameter than either of the preceding traps, but it is formed on the same plan. An eel can easily pass into the basket through the conical tunnel, but it is next to impossible that it should find its way out again.

So much for Art, and now for Nature.

On the left hand of the illustration, at Fig. A, is the cocoon of the common Emperor-moth (*Saturnia pavonia minor*), the cocoon having been stripped of its outer envelope, so as to allow its structure to be better seen.

The reader will at once perceive that the entrance of the cocoon is guarded by an arrangement exactly like that of the above-mentioned traps, except that the cone is reversed, so as to allow of exit and to debar entrance. Guarded by this conical arrangement of stout bristly appendages, the pupa can remain in quiet during the time of its transformation, for nothing can force its way through such a defence, and yet the moth, when fully developed, can push its way out with perfect ease.

So admirably is this cocoon formed, that even after the moth has escaped, it is impossible to tell by mere sight whether or not it is within, the elastic wires closing on it after its passage.

ANOTHER modification of the same principle now comes before us. In the above-mentioned examples the arrangement of the reverted spikes is more or less conical, and they lead into a chamber. In the present instances, however, the mere reversion of the points is all that is needed.

The upper figure on the right hand represents the " Catch-poll " of the Middle Ages, an allusion to which has already been made. The reverted spikes turn on hinges, and are kept apart by springs. This beautifully formed head was attached to a long shaft, and was used for the purpose of dragging horsemen from the saddle. It was thrust at the neck of the rider, generally from behind. If a successful thrust were made, the spring-points gave way, sprang back again, and thus clasped the neck with a hold that was fatal to the rider.

BELOW it is the Deer-trap which is used in many parts of India, and to which allusion has already been made. The reader will see at once that if a deer should get its foot through the converging spikes, its doom is sealed, especially as there is a heavy log of wood attached to the trap by a rope.

On the left hand of the illustration are two examples of the same principle taken from Nature, one belonging to fresh and the other to salt water.

The upper figure represents the jaws of a Pike, with their terrible array of reverted teeth. The Pike, as every one knows,

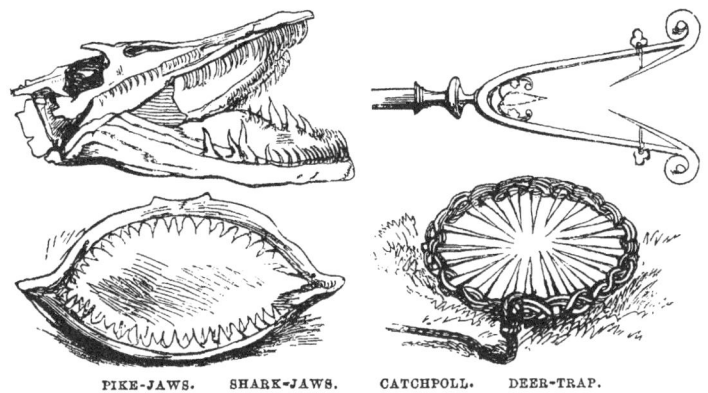

PIKE-JAWS.   SHARK-JAWS.   CATCHPOLL.   DEER-TRAP.

feeds upon other fish, and eats them in a curious manner. It darts at them furiously, and generally catches them in the middle of the body. After holding them for a time, for the purpose, as I imagine, of disabling them, it loosens its hold, makes another snap, seizes the fish by the head, and swallows it.

The Pike is so voracious that it will attack and eat fish not very much smaller than itself, for its digestion is so rapid that the head and shoulders of a swallowed fish have been found to be half digested, while the tail was sticking out of the Pike's mouth. Unless, therefore, the teeth of the Pike were so formed as to resist any retrograde movement on the part of the prey, the fish would starve; for, lank and lean as it is, the Pike is one of the most voracious creatures in existence, never seeming able to get enough to eat, and yet, as is often found in such cases, capable of sustaining a lengthened fast.

How well adapted is this arrangement of teeth for preventing the escape of prey, any one can tell who, in his early days of angling, caught a Pike, and, after killing it, tried to extract the hook without previously propping the jaws open. If once the hand be inserted between the jaws, to get it out again is almost impossible without assistance, and often has the spectacle been exhibited of a youthful angler returning disconsolately home, with his right hand in the mouth of a Pike, and supporting the weight of the fish with his left.

THE teeth of a serpent are set in a similar manner, as can be seen by reference to the illustration on page 80. An admirable example of the power of this arrangement may be seen in the jaws of our common Grass or Ringed Snake (*Coluber natrix*). The teeth are quite small, very short, and not thicker than fine needle-points. Yet, when once the snake has seized one of the hind-feet of a frog, all efforts to escape on the part of the latter are useless. The lower jaw is pushed forward, and then retracted, and at each movement the leg is drawn further into the snake's mouth, until it reaches the junction.

The snake then waits quietly until the frog tries to free itself by pushing with its other foot against the snake's mouth. That foot is then seized, the leg gradually following its companion, and in this way the whole frog is drawn into the interior of the snake. I have seen many frogs thus eaten, but never knew one to escape after it had been once seized by the snake. As these reptiles are perfectly harmless, it is easy to try the experiment by putting the finger into a snake's mouth, when it will be found that the assistance of the other hand will be needful in order to extricate it.

BELOW the head of the pike is a view of a Shark's jaws, as seen from the front.

Here, again, we have a similar arrangement of teeth, row after row of which lie with their points directed towards the throat of the fish. As, however, the pike and the snake swallow their prey whole, their teeth need be nothing but points. But, as the Shark is obliged to mangle its prey, and seldom swallows it whole, its teeth are formed on a different principle, each tooth being flat, wide, sharply pointed, and having a double edge, each of which cuts like a razor. So knife-like are they, indeed, that when a whale is killed, the sharks which surround it bite off huge mouthfuls of blubber, and, as they swarm by hundreds, cause no small loss to the whalers.

Many a man has lost a leg by a shark, the fish having bitten it completely through, bone and all, and there have been cases where a shark has actually severed a man's body, going off with one half, and leaving the other clinging to the rope by which he was trying to haul himself on board.

Spiked Defences.

This mode of defence is, perhaps, one of the most primitive in existence, and takes a wonderful variety of forms. The spiked railings of our parks and gardens, the broken glass on walls, and even the spiked collars for dogs, are all modifications of this principle.

On the illustrations are several examples of spikes used for military purposes. The first is known by the name of "Chevaux-de-frise," and is extensively used in forming an extemporised fence where no great strength is needed. The structure is perfectly simple, consisting of a number of iron bars with sharpened ends, and an iron tube some inches in diameter, which is pierced with a double set of holes. When not in use, the bars and tube can be packed in a small compass, but when they are wanted, the bars are thrust through the holes as shown in the illustration, and the fence is completed in a few minutes. The horizontal bars are linked together by chains, so as to prevent them from being shifted, and a defence such as this is generally used for surrounding parks of artillery and the like.

All who have the least acquaintance with military matters must be familiar with the "Square," and its uses in the days of old. I say in the days of old, because in the present day the rapid development of guns and rifles has entirely destroyed the old arrangement. So lately, for example, as the day of Waterloo, troops might manœuvre in safety when they were more than two hundred yards from the enemy. Now, a regiment that attempted to manœuvre in open ground would be cut to pieces by the rifles of the enemy at a thousand yards' distance.

In those days, however, the square was a tower of safety when rightly formed. It was formed in several rows. The outer line knelt, with the butts of their muskets on the ground, and the bayonet pointing upwards at an angle of forty-five.

The others directed their muskets towards the enemy in such
a manner that nothing was presented to him but the points of
bayonets and the muzzles of loaded muskets. In all probability the battle of Waterloo would have been lost but for the
use of the "square," against which the French cuirassiers
dashed themselves repeatedly, but in vain.

However admirable may be the organization of the square,
whether it be hollow, or whether it be solid, like the "rallying
square," the principle is the same as that of the chevaux-de-
frise.

In the next illustration is shown the "Abattis," one of the
most important elements of extemporised fortifications, and as
simple as it is important.

In any wooded country an abattis can be made in a very
short time by practised hands. All that is required is to cut
down the requisite number of trees, strip off the leaves and
twigs, and then cut off the smaller branches with sloping

TREE-CADDIS.

CHEVAUX-DE-FRISE.

blows of the axe, so as to leave a tolerably sharp point on each.
The trees are then laid side by side, with the ends of the
branches towards the enemy, and, the trunks being chained
together, a wonderfully effective defence is constructed.

Not only is it almost impossible for the bravest and strongest
man to force his way through the branches, even if the abattis
were undefended, but the tree-trunks afford shelter for swarms
of riflemen, who can pick off their assailants by aiming between
the branches, themselves being almost unseen, and entirely
covered.

In Southern Africa, during the late wars, the abattis was found to afford the best defence against the Kafirs, and that when the waggons and abattis were united so as to form a fortress, not even the naked Kafir, with all his daring courage,

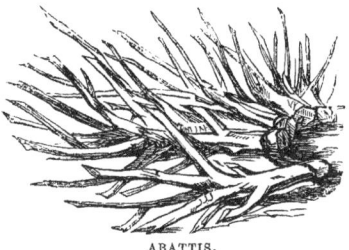

CALTROPS.   RANJOWS.   ABATTIS.

could force his way through them. Even artillery has but little power against the abattis, which allows the shot to pass between the branches, and is very little the worse for it. Accordingly, it is in great use for defending roads, especially those which are bounded by high banks, and makes a formidable obstacle in front of gates.

THE two figures on the left of the same illustration represent two modes of carrying out the same principle, the one showing it as used in European warfare, and the other as a weapon of defence which has been employed from time immemorial, and is now in full use in many parts of the world.

Both these weapons are intended either to obstruct the approach of an enemy, or to cover the flight of a retreating force. The most simple and most ancient is the Ranjow, which is shown on the right hand of the illustration. The ranjow is nothing but a wooden stick varying in length from eighteen inches to nearly three feet, and sharply pointed at each end. In Borneo, China, &c., the ranjows are almost invariably made of bamboo, as that plant can be cut to a sharp point by a single stroke of a knife. (See page 59.)

When they are to be used, each soldier carries about a dozen or so of them, and sticks one end of them into the ground, taking care to make the upper end lean towards the enemy. Simple as are these weapons, they are extremely formidable, for it is necessary to pull up every ranjow before the troops can advance. Sometimes it has happened that a body of soldiers are driven over their own ranjows, and then the slaughter is terrible.

Some years ago a number of sketches were taken on the spot from scenes in the Chinese war. Among them was one that was absolutely terrible in its grotesqueness. It represented a piece of ground thickly planted with ranjows, over which the Chinese who had fixed them had been driven. They were simply hung with human bodies in all imaginable and unimaginable attitudes, some transfixed on a single ranjow, and others hanging on three or four, the body and limbs being alike pierced by them.

That ranjows were once used in Great Britain is evident from a discovery made by Col. Lane Fox. He had been excavating the soil around an old Irish fort, and deep beneath the bog he found a vast quantity of ranjows still set as the ancient warriors had left them. They were evidently used to defend a passage leading to the fort, and all of them were carefully set with their points outwards. Col. L. Fox was good enough to present me with several of these ancient weapons, which are now in my collection.

On the left is seen a piece of ground strewed with Caltrops, or Crow's-feet, as they are sometimes called. These very unpleasant implements are made of iron, and have four sharp points, all radiating from one centre, so that no matter how they may be thrown, one point must be uppermost. They are used chiefly for the purpose of impeding cavalry, but I should think, judging from the specimens which I have seen, that infantry would find them very awkward impediments.

As for natural ranjows, they are so numerous that only a very few examples can be given.

The most perfect and most familiar example is, perhaps, the common Hedgehog, which, when rolled up, displays an array of sharp points so judiciously disposed, that it fears but very few foes. The same may be said of the Australian Echidna, or Porcupine Ant-eater, and the Porcupine itself. Whether the radiating bristles of the larva of the Tiger-moth, commonly called the Woolly Bear, come under the same category, I cannot say, but think it very likely.

Among vegetables the analogues are multitudinous. See, for example, the spikes of the Spanish and Horse Chestnuts, and especially the hair-like but formidable bristles which defend the common Prickly Pear. Indeed, all that tribe of plants is furnished so abundantly with natural ranjows, that a hedge of prickly pear forms the best defence which a house and garden can have.

Another example of natural ranjows is seen in the Tree-caddis, one of which is shown in the illustration on page 108, as it appears when suspended from a twig. It is the work of one of the House-builder Moths of the West Indies, and forms a sort of house in which the caterpillar can rest securely. It is built of bits of twigs and thorns, the latter being disposed so that their points are outwards, much after the fashion of a hedgehog's spines.

I possess many specimens of Tree-caddis, evidently belonging to several species, and in all of them the principle is the same, *i.e.* a number of spikes set with their ends outwards in order to defend a central position.

Sometimes these spikes are left exposed, as shown in the illustration, and sometimes they are covered with a slight but strong web. The principle, however, is the same in all.

Now I shall have to use two very long words, and much against my will. I very much fear that, if most of my readers were to hear any one speak of the " repagula of Ascalaphus," they would not be much the wiser. And yet there are no other words that can be used.

In the first place, Ascalaphus is a name belonging to a genus of Ant-lions, remarkable for having straight, knobbed antennæ, very much like those of a butterfly. This insect deposits its eggs in a double row on twigs, and then defends them with a series of natural ranjows, set in circular rows, and supposed to be without analogies in the animal creation. They are transparent, reddish, and "are expelled by the female with as much care as though they were real eggs, and are so placed that nothing can approach the brood ; nor can the young ramble abroad until they have acquired strength to resist the ants and other insect enemies."

The word "repagulum," by the way, signifies a bar or barrier.

A turnpike gate when closed would be a repagulum, and so would a chevaux-de-frise.

TEARING WEAPONS.

WE have already had examples of weapons, like the Club, which bruise ; of weapons, like the Spear and Dagger, which pierce ; and of weapons, like the Sword, which cut. We now come to a totally distinct set of weapons, those which wound by tearing, and not by any of the preceding modes.

In civilised warfare we have long abandoned such weapons, as belonging to a barbarous age, but they are even yet employed in some parts of the world.

The accompanying illustration shows three examples of such weapons. One is the celebrated Tiger-claw of India, known by the native name of Wag-nuk. It is about two inches and a half in length, and is made to fit on the hand. The first and fourth fingers are passed through the rings, and the curved claws are then within the hand, and hidden by the fingers. The mode of employing this treacherous weapon was by

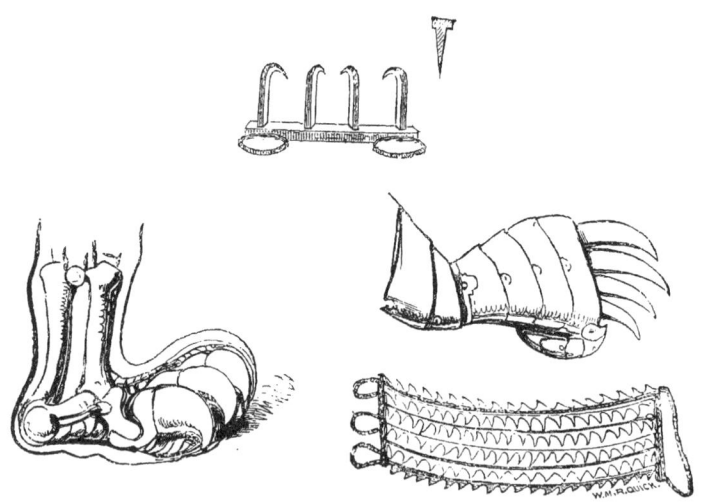

WAG-NUK OF INDIA.

HIND-CLAWS OF TIGER.

CLAWED GAUNTLET.
SHARK-TOOTH GAUNTLET.

engaging a foe in conversation, pretending to be very friendly, and then ripping up his stomach with an upward blow of the right hand.

It is comparatively a modern weapon, having been invented about two hundred years ago. A Hindoo, named Sewaja, was the inventor, and by means of the Wag-nuk he committed many murders unsuspected, the wounds being exactly like those which are made by the claw of the tiger. Sometimes there were four claws instead of three, as is the case with a specimen one in the Meyrick collection.

Perhaps the reader may be aware that the Transatlantic " knuckle-duster " is fitted on the hand in the same manner, only its object is to strike a heavy blow, and not to tear. History repeats itself, and the large and clumsy "cestus" of the ancient athlete is reproduced in the small but scarcely less formidable " knuckle-duster" of the modern rowdy.

The figures are remarkable, one representing the remaining epoch of chivalry, and the other that of barbarism. The upper figure shows a curious Gauntlet of the Middle Ages, in which the hand is not only defended by steel plates, but is also rendered an offensive weapon by the addition of four sharp spikes set just at the junction of the fingers with the hand. As long as the fingers are extended the spikes lie parallel with them, and are as harmless as a cat's claws in their sheaths. But when the fingers are closed, as shown in the illustration, the spikes come into use, and can be made into a formidable weapon of offence, just as are the cat's claws when protruded.

BELOW the gauntlet of civilised warfare is one of savage war, which has for many years been discontinued, partly on account of the introduction of firearms, and partly owing to the superficial coating of civilisation which is so easily adopted by the singular varieties of the human race which populate the isles where this remarkable weapon was once worn. The figure is taken from a specimen in the United Service Museum.

It is a Gauntlet, having at one end a band through which the whole hand is passed, and at the other three loops for the fingers, just like those of the Wag-nuk, which has already been described. The body of the weapon is made of cocoa-nut fibre, and upon it are strung six rows of sharks' teeth, the tips all pointing backwards. It is a Samoan weapon, some of the most renowned warriors never using club nor spear, but trusting entirely to their terrible gauntlets. With these they struck right and left, dashing beneath the clubs and spears of their enemies, and always trying to rip up their stomachs, just as is done with the Wag-nuk. In order to guard against this weapon, the Samoan warrior wears a belt of cocoa-nut fibre some eight inches wide, and thick enough to defy the best gauntlet that could be made.

One celebrated Samoan warrior, a man of gigantic stature and strength, was addicted to the amusement of seizing his enemies with the shark-tooth gauntlets, breaking their backs across his knee, throwing them down, and going' off after another victim.

ON the left hand of the illustration is seen the hind-foot of the Tiger. I have chosen the hind-foot for two reasons: firstly, because the fore-foot has already been figured; and secondly, because the hind-foot is used for tearing open the abdomen of the prey. Any one who has played with a kitten has noticed how the animal throws itself on its back, clasps the wrist with its fore-paws, and kicks vigorously with its hind-legs. It does not mean to hurt its playfellow, but the hand does not easily escape without sundry scratches.

Child's play though it may be in the kitten, it is no play at all with the tiger, or even the leopard, for either of these animals, when hard pressed, will throw itself on its back, clasp the foe in its fore-paws, and with the talons of the hind-feet tear him to pieces.

WAR AND HUNTING.

CHAPTER VI.

THE HOOK.—DEFENSIVE ARMOUR.—THE FORT.

Anglers and their Hooks.—Single and double Hooks.—Hook of British Columbia. —Seed of Galium, or Goose-grass, and its Armature of Hooks.—Seed of the Burdock, and its Annoyance to Sheep.—Hooked Sponge-spicules.—"Snatching" Fish.—The Fish-rake of British Columbia.—The "Gaff" and its Uses. —The Jaguar as a Fisher.—Defensive Armour and its Varieties.—Plate and Chain Mail.—The Shield.—Australian and West African Shields.—Fibre Armour.—Seal's-tooth Cuirass.—Joints of Armour.—"Tassets."—Scale Armour in Art and Nature.—The Manis and the Fish.—Feather Armour.—"Madoc in Aztlan."—Quilted Armour of Silk or Cotton.—Terrible Results from the latter.—Mr. Justice Maulstatute.—Natural Quilt Armour.—The Rhinoceros and the Whale.—The Testudo of the ancient Romans, and its Uses.—The common Tortoise.—The Fort.—Curious Transitions in Fort-building; first Earth, then Stone, then Earth again.—Advantage of Earthen Mounds.—Natural Snow-fort made by the Elk, and its Defensive Powers against the Wolf.

THE HOOK.

HAVING now seen that the rod and line of anglers have their prototypes in Nature, we will proceed to the hook, by which the fish are secured.

The two figures on the right hand of the accompanying illustration represent hooks which are familiar to every angler. The lower is the ordinary fish-hook, which can be used in so many ways. Generally it is employed singly, being fastened to the end of a line, and armed with a bait, either real or artificial. Sometimes, however, these hooks are whipped together, back to back, three or even four being so employed, and thus forming a combination of the hook and grapnel, and rendering the escape of a fish almost impossible.

Above it is a double hook, such as is used in "trolling" for pike, and with the use of which many of my readers are probably acquainted.

The third is a singularly ingenious hook made by the natives of British Columbia. It is almost entirely made of wood, with the exception of the barb, which is of bone. This, as the reader will see, is fixed, not to the point of the hook, as with us, but to its base, the point being directed towards the central portion of the curve.

At first sight this seems to be a singular arrangement, but it is a very effective one, as any one may see by placing the

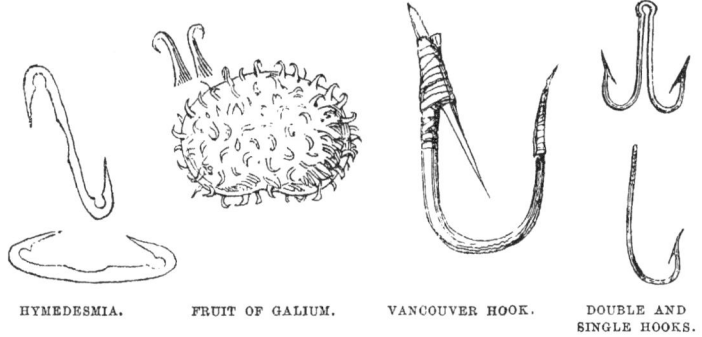

HYMEDESMIA.   FRUIT OF GALIUM.   VANCOUVER HOOK.   DOUBLE AND
                                                                       SINGLE HOOKS.

point between the fingers and pushing it through them. It will be found impossible to force it back again, the sharp point of the bone-barb coming against them and retaining them.

It has also another advantage. Very large fish, for which this hook is intended, are apt in their struggles to reverse the hook, and so to weaken its hold. In this hook, however, such a proceeding is impossible; for, even should the hook be reversed, it still retains its hold, the barb becoming the point, and the point keeping the lip of the fish against the tip of the barb. The figure is drawn from a specimen in my collection.

IF the reader will look at the illustration, he will see a globular object covered with little hooks. This is a magnified representation of the seed-vessel of the common Goose-grass (*Galium*), which is so luxuriant in our hedges, and often intrudes itself into our gardens. Its long, trailing stems, with their tightly-clinging leaves, are familiar to all, and there are few who have not, while children, pelted each other with the little round green seed-vessels during the time that the fruit is in season. That they clung so tightly as not to be removed without difficulty, we all knew, but we did not all know the cause. The magnifying-glass, however, reveals the secret at once. The whole of the surface is covered with little sharp prickles, curved like hooks, and turned in all directions, so that, however it may be thrown, some of them are sure to catch.

So readily do these hooks hold to anything which they touch, that if a lady only sweeps her dress against a plant of Goose-grass, she is sure to carry off a considerable number of the seed-vessels, and to waste much time afterwards in picking them off.

The seed-vessel of the common Burdock, known popularly by the name of Bur, is armed in a similar manner, but, as it is much larger, it is easily avoided. Sheep suffer greatly from burs, which twist themselves among the wool so firmly that it is hardly possible to remove them without cutting away bur and wool together. As to a Skye terrier, when once he gets among burs, his life is a misery to him (I was going to say, a burden to him, but it would have looked like a pun).

Below, and on the left of the Galium-seed, are some spicules of the Hymedesmia, a sponge which is found on the coast of Madeira. The following account of it occurs in the *Intellectual Observer*, vol. ii. p. 312 :—

"FISH-HOOK SPICULÆ.—We have received from Mr. Baker, of Holborn, a slide containing spicules of the *Hymedesmia Johnsonii*, which are stated to be rare objects in this country. They have the form of a double fish-hook, and on the inner surface of each hook is an extremely sharp knife-edge projection, corresponding with a similar and equally sharp projection from the inside of the shank.

"These minute knife-blades are so arranged that in addition to their cutting properties, they would act as barbs, obstructing the withdrawal of the hook. The two hooks attached to one shank are not in the same place, but nearly at right angles with one another, so that when one is horizontal the other is vertical, or nearly so. A magnification of four or five hundred linear does not in any way detract from the sharp appearance of the knife-edges, and they may take their place with the anchors of the Synapta as curious illustrations of the occurrence in living organisms of forms which man was apt to fancy were exclusively the products of his own contrivance and skill.

"We presume that these hooks of the Hymedesmia answer the usual purpose of spiculæ in strengthening the soft tissue, but they must likewise render the sponge an awkward article for the Madeira sea-slugs to eat."

For an account and figures of the Synapta anchor-spicules see page 39.

WE now come to another modification of the hook. I presume that many of my readers have heard of the practice called "snatching" fish, though I hope that they have never been unsportsmanlike enough to follow it.

This plan, which is only worthy of poachers, consists in taking several flights of treble or quadruple hooks, dropping them gently by the side of the fish, and then, with a sudden jerk, driving them into any part of its body which they may happen to strike. Most anglers have snatched fish accidentally, but to do so intentionally is ranked among the worst of an angler's crimes, and is equivalent to cheating at cards, or playing with false dice.

In some parts of the world, however, there are certain small fish which are never taken in any other way, and, indeed, are raked out of the water just as a gardener rakes dead leaves off the path or beds.

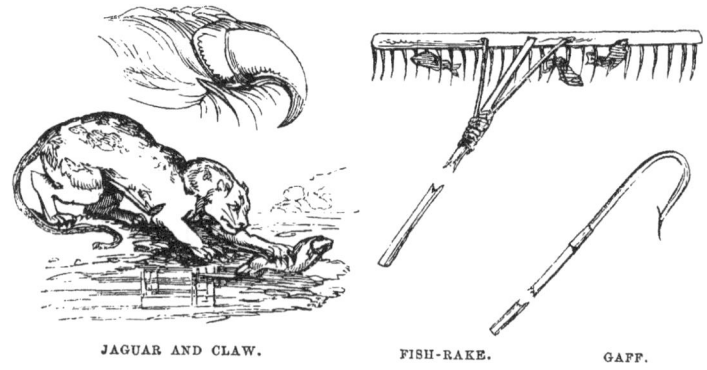

JAGUAR AND CLAW.   FISH-RAKE.   GAFF.

In British Columbia there are certain lakes tenanted largely with small fish which form a considerable portion of the natives' diet. They swim in vast shoals close to the surface of the water, and are captured by veritable rakes, one of which is shown in the illustration. The points of the rake are slightly curved, and very sharp, and so numerous are the fish that when the native has struck his rake among the shoal, and drawn it into the boat, he generally finds a fish on every tooth, while it often happens that two or three are transfixed by the same tooth. A sharp knock against the side of the boat shakes off the prey, and the fisherman again strikes his rake into the shoal. By this simple mode of fishing a couple of men will, in a few hours, load a canoe with small but valuable fish.

Below the rake is the "Gaff," an instrument, not to say a weapon, which is indispensable when salmon or other large fish are to be caught. For ordinary-sized fish a landing-net is sufficient, but no landing-net could either receive or retain a salmon of any size.

Recourse is then had to the Gaff, which is simply a huge hook at the end of a handle. The fish being "played" until it can be drawn within reach, the gaff is slipped under it, struck into the side of the salmon, and by its aid the fish is easily lifted out of the water.

On the left hand of the illustration are two figures showing how the principle of the fish-rake and gaff has been anticipated in Nature.

It is a well-known fact that the Jaguar feeds largely on fish, which it catches for itself. It goes down to the river-side as close to the water as possible, and waits patiently for its prey. As soon as a fish comes within reach, the Jaguar stretches out its paw to the fullest extent, and, with a stroke of the curved claws, hooks the fish on shore, just as the Vancouver Islander does with his fish-rake, or the English angler with his gaff.

Many persons have practically experienced the gaff-like powers of the feline claw by the loss of their gold-fish. It is seldom safe to leave a globe of gold-fish within reach of a cat. Nearly all cats are madly fond of fish, and, in spite of their instinctive hatred of water, will hook out the fish with their claws, and eat them. Indeed, there are several instances on record where a cat has regularly caught fish, and brought them home to its owner. Mr. F. Buckland gives an account of a fisherman's cat, which used to go out with her master, jump into the sea, secure a fish, and then be lifted on board with her prey.

Above the Jaguar is drawn a single claw, so as to show the form of the instrument by which the fish is captured.

ARMOUR.

WE will now take the subject of Defensive Armour, by which warriors are enabled to protect themselves against the offensive weapons of the enemy.

As many readers will probably know, armour reached its greatest development in the Middle Ages, when the knight was so completely cased in steel that no weapon then in use could penetrate his panoply.

The head, body, and limbs were covered with steel plates curiously articulated at the joints, so as to give freedom of motion, while guarding the wearer from any ordinary weapon. A warrior might be beaten from his horse by a mace, or struck down by a lance, or the horse itself might be killed under him.

In either of these cases the fallen knight was not much the worse, until a weapon called the " Misericorde," or dagger of

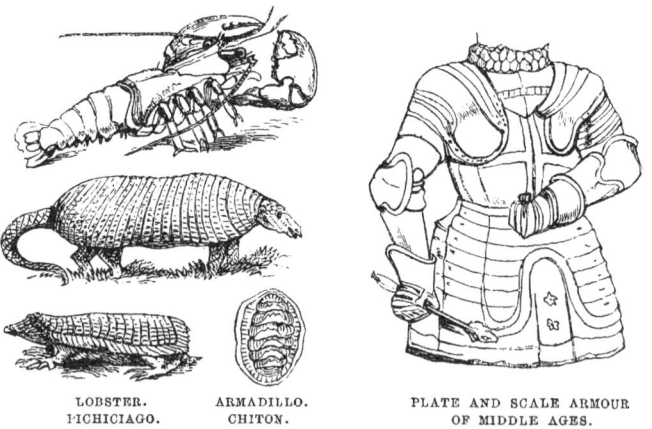

LOBSTER.   ARMADILLO.   PLATE AND SCALE ARMOUR
I'ICHICIAGO.   CHITON.   OF MIDDLE AGES.

mercy, was invented. This was a poniard with a very slender and very sharp blade, so constructed that it could be driven between the joints of the armour, and thus inflict a mortal wound. The Misericorde, however, was baffled by the use of chain or scale armour under the plate-mail, and then the only way of getting at the fallen knight was by breaking up the armour with hammers which were made for this express purpose.

Perhaps the reader may wonder that any one should lie quietly and allow himself to be so badly treated. The very strength of the armour, however, which rendered its wearer unassailable by ordinary weapons, involved so much weight, that when a knight had fallen, it was impossible for him to rise, much less to mount a horse, without help. Moreover, the first blow of a weighty hammer on the helmet would, although it could not kill the wearer, cause such a jar to his brain as partially, if not wholly, to stun him.

The rapidly increasing power of firearms soon caused armour to be laid aside, and now the only remains of it are to be found in the helmets and cuirasses worn by our dragoons.

THERE are few parts of the world where armour of some sort is not used. Putting aside civilised or semi-civilised nations, we find that in most cases, wherever there is war, there is armour of some kind. Sometimes it is movable, and in that case is called a shield.

The most singular shields that I know are those made by the Australians, which are so shaped that no one who did not know their use would take them for shields. They are about three feet long, four inches wide at the back, six inches or so thick in the middle, tapering towards the ends, and coming to an edge in front. They are held by the centre with one hand, so that they can be rapidly twisted from side to side, and so serve to parry the spear or stop the boomerang. The weight of the shield enables it to withstand the shock of the boomerang, which whirls through the air with terrific force.

Several warlike savage tribes have, however, no armour of any kind, such as the New Zealanders, the Samoans, and the Fijians.

Sometimes the armour is affixed to the body, and of such protection many examples are to be found in various museums, among which the Christy collection is pre-eminent.

Among the Polynesians cocoa-nut fibre was at one time employed as the material for armour. It was twisted into small cords, and with these a sort of armour was constructed, quite strong enough to resist any weapon that an enemy of their own kind could bring against them. Sometimes this armour was merely a belt wide enough to protect the abdomen, but sometimes the whole body was defended, from the neck to the hips.

In the United Service Museum there is a very remarkable cuirass, which is made of successive rows of seals' teeth, each row overlapping the other like the tiles of a house. It is very heavy, weighing quite as much as a steel cuirass, and was probably quite as effective against the primitive weapons which could be brought to bear upon it.

Now for Natural Armour.

There are so many examples of armour, as furnished by Nature, that I can only mention a few.

Any one who looks at a lobster, crayfish, prawn, or shrimp, must at once see that in it lies the prototype of plate armour. That portion of the lobster which is popularly called the head, and is scientifically known as the "carapace," is not jointed, and corresponds with the cuirass of ancient or modern armour. Then comes the part called the "tail," the joints of which are exactly like those employed in the shoulders, elbows, knees, and ankles of ancient armour. The lobster tail will again be mentioned in connection with another branch of human art.

As for the heavy, ungraceful armour which was used in tilting, we have an admirable example in the Trunk-fish of the tropical seas (*Ostracion*), the whole of which is enclosed in a bony case, the fins and tail protruding through openings in it. In fact, the scales, instead of being separate, are fused together so as to form a continuous covering. The Box-tortoise of South America is another good example, the creature being furnished with bony flaps with which it covers the apertures through which the head, legs, and tail are protruded, and so is as impervious as the knight of old.

In the later ages of armour, the thighs, instead of being enclosed in steel coverings with cuisses, were defended by a number of steel plates called " tassets." Now these tassets are exactly like the defensive armour of the Armadillo's back, and, though it is not likely that the inventor of tassets should have seen an Armadillo, the fact still remains, that Art has been anticipated by Nature.

Exactly the same principle is seen in that wonderful little animal, the Pichiciago of South America, which is shown in the lower left-hand figure of the illustration. This creature is not only furnished with bony rings on the body like those of the Armadillo, but has likewise a flap which comes over the hind-quarters, and effectually defends it against the attacks of any foe that might pursue it into its burrow.

In the lower right-hand corner of the illustration is seen a figure of a Chiton, several species of which are common on most of our coasts. This is one of the molluscs, which adheres to the rock just as limpets do. But, whereas the shell of the limpet is all in one piece and inflexible, that of the Chiton is composed of several pieces, which are arranged exactly like the tassets of armour, and enable the Chiton to accommodate itself to the inequalities of the rocks to which it is adhering.

The common Pill Millipede, which rolls itself up in a ball when alarmed, is a familiar instance of similar defensive armour, and much the same may be said of the Julus Millipede.

WE now come to Scale Armour, which is one of the earliest modes of protecting the body, and the idea of which was clearly taken from animal life. In Scale Armour, flat plates of metal, horn, or bone are sewn to a linen or leathern vest in such a

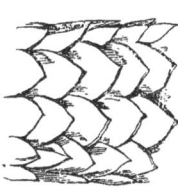

SCALES OF MANIS.

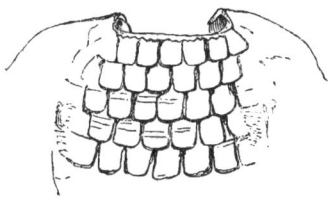

SCALE-MAIL.

way that the scales overlap each other, and so tend to throw off the blow of a weapon. One great advantage of this armour is its lightness and flexibility, the former quality allowing of more prolonged exertion than could be possible with the heavy plate armour, and the latter rendering that exertion less fatiguing to the limbs.

A glance at the preceding illustration will show how the scale armour of the human warrior has been anticipated by Nature.

On the right hand is an example of ordinary scale armour, while on the opposite side is a portion of a scaly surface. This figure represents some of the scales of a Manis. These scales are wonderfully hard, and scarcely to be penetrated. I have in my collection the skin of a Short-tailed Manis, which had been kept for some time in an Indian compound, but which made itself such a nuisance by its perpetual burrowing, that its owner was forced to condemn it to death.

So he took a Colt's revolver, and fired at it from a distance of a yard or two. The only result was to knock over the Manis, which rolled itself up, and appeared to be none the worse. A second and a third shot were fired with similar results, and the last bullet recoiled upon the firer. At last, the animal was killed by introducing the point of a dagger under the scales, and driving it in with a mallet. The Manis itself is given in the illustration on page 189.

SKIN OF SINGLETHORN.

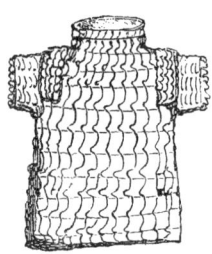
SCALE-MAIL.

Again, the scales of most fishes afford excellent examples of scale armour. I have selected one, the Japanese Singlethorn, on account of the strength of the scales, each of which is deeply ridged and furrowed. The reader will probably have noticed that the skin of the animal, into which are inserted the bases of the scales, is analogous to the linen or leathern foundation upon which the artificial scales are sewn.

Even feathers give a better protection than might be imagined from their individually fragile structure. This is well shown in the case of aquatic birds, whose feathers are very closely pressed together, each overlapping the next, and set in regular order. Not only is the plumage rendered water-tight, but it is able to resist a severe blow. This is well known by sportsmen, who do not fire at ducks or geese while they are approaching, knowing that their shot would only glide harmlessly from the feather-mail of the bird.

They wait until the birds have passed, and then find no difficulty in killing them, the shot penetrating under the feathers just as did the dagger under the scales of the manis. Even the diminutive puffin, or sea-parrot, as it is sometimes called, cares little for shot while it is sitting on the rocks with closed wings and feathers pressed together. When, however, it takes to flight, it can be killed without difficulty.

Perhaps some of my readers may be aware that the ancient Mexican warriors wore armour made of feathers, which I presume must have been arranged much after the fashion of those of a duck's breast.

This remarkable Feather-mail is mentioned by Southey in his poem, "Madoc in Aztlan." In canto xviii. is recounted the single combat between Madoc and Coanocotsin, the King of Aztlan. The contrasting armour and weapons of each are graphically described, and especial mention is made of the cuirass :—

> " Over the breast,
> And o'er the golden breastplate of the King,
> A feathery cuirass, beautiful to eye,
> Light as the robe of peace, yet strong to save ;
> For the sharp faulchion's baffled edge would glide
> From its smooth softness."

Then, in the course of the combat, when the King has been grappled in Madoc's arms and forced to drop his buckler and club, the narrative proceeds :—

> " Which when the Prince beheld,
> He thrust him off, and drawing back, resumed
> The sword that from his wrist suspended hung,
> And twice he smote the King. Twice from the quilt
> Of plumes the iron glides."

If such armour could in truth resist the weapons which have been discovered, it must have been a wonderfully strong garment, for the Mexican swords, though made of wood, are edged with flakes of obsidian, which cuts like a razor. I have a number of these flakes, which have evidently been intended for the edges of a sword, but have not been used.

THERE is another kind of armour which is still used in some parts of the world, and at one time was employed in this country. This is the Quilt Armour, which is made by enclosing a thick layer of some fibre, such as silk or cotton, between two pieces of fabric, and then sewing them across and across, so as to keep the lining or stuffing in its place.

The eider-down quilts are familiar examples of such fabrics, and so are the quilted petticoats, which are so comfortable in winter. Horsehair and flock mattresses are made in a similar manner.

Insufficient as it may appear to be, the quilt armour, when well made, is really proof against most weapons, even against firearms, as we shall presently see. Being very much lighter than steel, it was easier for the wearer, its chief drawback being that its extreme thickness gave it a very clumsy and awkward look. Those who wore it, however, cared more for their safety than their appearance, as was exemplified by James I., who lived in perpetual fear of assassination, but who had a nervous dislike to arms, whether offensive or defensive. He therefore wore a cuirass quilted with silk, which answered every purpose of defence, while it did not offend his nerves.

Perhaps the reader may remember that in "Peveril of the Peak" Sir Walter Scott gives a ludicrous picture of the timid justice, his fears of the Popish plot, his suit of quilted armour, and his "Protestant Flail" with which he hits himself on the head instead of striking his supposed enemy :—

"Some ingenious artist, belonging, we may presume, to the worshipful Mercers' Company, had contrived a species of armour of which neither the horse armoury in the Tower, nor Gwynnap's Gothic Hall, no, nor Dr. Meyrick's invaluable collection of ancient arms, has preserved any specimen.

"It was called Silk-armour, being composed of a doublet and breeches of quilted silk, so closely stitched, and of such thickness, as to be proof against either bullet or steel, while a thick bonnet of the same materials, with ear-flaps attached to it, and on the whole much resembling a nightcap, completed the equipment, and ascertained the security of the wearer from the head to the knee. Master Maulstatute, among other worthy citizens, had adopted this singular panoply, which had the advantage of being soft, and warm and flexible, as well as safe.

And he was sat in his judicial elbow-chair—a short, rotund figure, hung round, as it were, with cushions, for such was the appearance of the quilted garments—and with a nose protruded from under the silken casque, the size of which, together with the unwieldiness of the whole figure, gave his worship no indifferent resemblance to the sign of the Hog in Armour, which was considerably improved by the defensive garment being of a dusky orange colour, not altogether unlike the hue of those half-wild swine which are to be found in the forests of Hampshire."

Roger Nutt gives as a reason for the security of quilted armour, that it made the wearer look so ridiculous that no one could hit him for laughing. The reader will probably remember that the sign of the Hog in Armour was really a representation of the rhinoceros.

That such a cuirass is really impervious to ordinary weapons is shown by the following anecdote :—During one of the late Indian wars a trooper discharged his pistol close to the back of a fleeing horseman. The shot produced no apparent effect, and the man rode off. Presently, however, a thin cloud of smoke was seen to rise from his shoulders. The smoke thickened, then burst into flame, and after riding at desperate speed in hopes of overtaking his comrades, the unfortunate man fell from his horse, and was miserably burned to death.

The fact was that cotton being cheaper than silk, he had wadded his cuirass with cotton fibre. Had he chosen silk, he

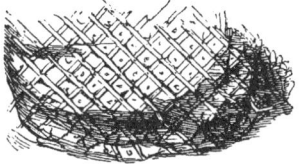

INDIAN RHINOCEROS.   QUILTED ARMOUR.

would have got off in safety. Among the Chinese this cotton mail is largely used. In consequence, many Chinese soldiers were found who had been burned to death in exactly the same way as the Indian warrior.

Towards the south-western parts of Africa there is a nation called the Begharmis. Their soldiers are mounted, and are all furnished with suits of quilted mail, which fall below the knee as the rider is seated on his horse.   Not only is the rider thus defended, but the horse also, which is covered with quilted armour like that of its rider, the appearance of both being exceedingly grotesque.

THERE are several examples of such armour in the animal world, the principal of which is the Indian Rhinoceros.   Any one who has seen this animal, or even a good portrait of it, will at once recognise the parallel between the heavy folds of its thick skin and the padded flaps of the quilted mail.   The blubber with which the whale is so thickly coated affords another example of the parallel between Nature and Art.

IN the days of ancient Rome there was a curious mili- tary manœuvre, by which the defensive armour of individual soldiers might be made collectively useful.   This manœuvre was called Forming a Tortoise (*testudinem facere*), and is thus described in Smith's " Dictionary of Greek and Roman Antiquities :"—

" The name of Testudo was also applied to the covering made by a close body of soldiers, who placed their shields over their heads to screen themselves against the darts of the enemy.   The shields fitted so closely together as to present one unbroken surface without any interstices between them, and were so firm that men could walk upon them, and even horses and chariots be driven over them.

" A Testudo was formed either in battle, to ward off the arrows and other missiles of the enemy, or, which was more

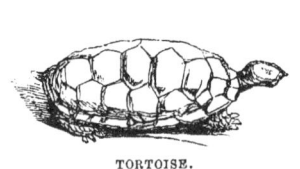

TORTOISE.

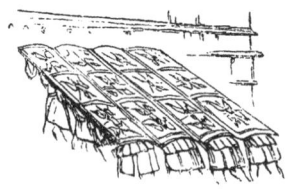

ROMAN TESTUDO.

frequently the case, to form a protection to the soldiers when they advanced to the walls or gates of a town for the purpose of attacking them.

" Sometimes the shields were disposed in such a way as to make the Testudo slope. The soldiers in the first line stood upright, those in the centre stooped a little, and each line successively was a little lower than the preceding, down to the last, where the soldiers rested on one knee. Such a disposition of the shields was called *Fastigata Testudo*, on account of their sloping like the roof of a building.

" The advantages of this plan were obvious. The stones and missiles thrown upon the shields rolled off them like water from a roof; besides which, other soldiers frequently advanced upon them to attack the enemy upon the walls. The Romans were accustomed to form this kind of Testudo as an exercise in the games of the Circus."

On the right hand of the illustration is shown a portion of a Testudo of three ranks, taken from the Antonine column. On the left is an ordinary Tortoise. Sometimes the Testudo was a covered machine on wheels, and guarded above with a supplementary roof of wet hides arranged in scale fashion, so as to prevent it from being set on fire by the besieged, and to throw off the heavy missiles which were dropped upon it. Under cover of this Testudo, the soldiers could either undermine the walls, or bring a battering-ram to bear upon them, while the men who worked it were safely under cover. As to the battering-ram itself, we shall presently treat of it.

THE FORT.

As we have treated of one of the modes by which Forts were assaulted, we will now come to the Fort itself.

The transitions in Fort-making are too curious to be omitted from the present book. As soon as war became organized, a Fort of some kind was necessary. The simplest mode of making a Fort was evidently to dig a deep trench, and throw up the earth on the inside, so as to form a wall. Let such a trench be square or circular, and there is a simple but powerful Fort, by means of which a comparatively small garrison could defend themselves against a superior force.

The Romans were great masters of this art, fighting as much with the spade as the sword. So strong and thorough was the old Roman work that many of their camps still remain, and will remain for centuries if man does not deface them. Such, for example, are Cæsar's camp, near Aldershot, and the fine camp at Lyddington, in Wiltshire, almost every detail of which is preserved. Roman camps are all constructed on the same model, the general's place, or Prætorium, being in the centre, whence he issued his orders, and the commanders under him occupying the corners. Thus, no matter how he might be shifted from one corps to another, every Roman soldier knew his way about the camp without needing to see it, and could tell at any moment where to find any officer.

ELK FORT.   MOUND FORT.

Other nations made their Forts circular, an example of which I lately saw a few miles from Bideford, while others consisted of nearly parallel lines, enclosures, and demi-lunes, like those wonderful dykes near Clovelly, which occupy more than thirty acres of land. One of the circular Forts is shown on the right hand of the illustration.

As time went on, stone took the place of earth, and the principal object of the builder was to give considerable thickness below, so as to resist the battering-ram, and great height both to walls and towers, so as to be comparatively out of the reach of the arrows and other missiles of the besiegers.

For awhile, such castles were impregnable, and the owners thereof were the irresponsible despots of the neighbourhood, recognising no law but their own will, robbing, torturing, and murdering at pleasure, and setting the king at open defiance. When, however, the tremendous powers of artillery became developed, the age of stone castles passed away. Height was found to be equivalent to weakness, as the strongest tower in existence could be knocked to pieces in an hour or two, and do infinite harm within the fortress by its falling fragments.

Fortification then returned to its original principles. Earth took the place of stone or brick; and at the present day, instead of erecting lofty walls and stately towers, the military engineer sinks his buildings as far as he can into the ground, and protects them with banks of simple earth, which is found to be the best defence against heavy shot. There is no masonry in existence that will endure the artillery fire of the present day, and even the solid rock can be knocked to pieces by it. But an earth-mound is a different business, and will absorb as many shot and shell as can be poured into it, without being much the worse for it. See, for example, the Proof-mound at Woolwich, which receives the shot of guns as they are being proved. Now, this mound has undergone perpetual battering for many years, and is as strong as ever. The same thing may be said of the celebrated Mamelon before Sebastopol.

So much for the Fort made by the hand of man. We now come to that which is formed by the feet of animals.

The Elk, or Moose, an inhabitant of Northern Europe, finds itself in great danger during the winter, the wolves being its chief enemies. At certain times of the year there comes a partial thaw during the day, followed by a frost at night. The result is, that a slight cake of ice forms on the surface of the snow, too slight to bear the weight of so heavy an animal, and strong enough to cut the legs of the elk as it ploughs its way along. Now, the wolves are sufficiently light to pass over the frozen surface without breaking it, and accordingly, they can easily run down and secure the elk.

In order, therefore, to counteract the wolves, a number of elks select a convenient spot where they can find food, and unite in trampling the snow down so as to sink themselves nearly to their own height below its surface. The wolves never dare attack an Elk-yard, as this enclosure is termed. In the first place, they are always haunted with suspicions of traps, and do not like the look of the yard; and in the next place, if some of the wolves did venture within the fort, the elks would soon demolish them with hoofs and horns. One of these Elk-yards is seen on the left hand of the illustration.

WAR AND HUNTING.

CHAPTER VII.

SCALING INSTRUMENTS.—DEFENCE OF FORT.—IMITATION.— THE FALL-TRAP.

Scaling-forks.—The Climbing-spur and its Use.—Larva of the Tiger-beetle.— Hooks of Serpula.—Mr. Gosse's Description.—Falling Stones.—A Stone roll- ing down a Precipice.—The Polar Bear and the Walrus.—Imitation.—The Polar Bear and the Seal.—The Esquimaux Hunter "Seal-talking."—En- ticing Mother by means of Young.—The Fall-trap and its Variations.—The Schoolboy's "Booby-trap."—Curious Mode of killing Elephants.—The Ele- phant-spear.—The Hippopotamus-trap of Southern Africa.—The Mangrove and its Seeds.—The Spring-gun and Spring-bow.

BEFORE dismissing the subject of the Fortress, we will glance at the Attack and Defence, as seen in Nature and Art.

SCALING INSTRUMENTS.

WE have already seen how the Battering-ram could be worked against the walls of a fort, or how the assailants could scale them by means of the Testudo. There must, however, be occasions when it would be impossible to bring together a sufficiently large body of men to form the Testudo, or even to place ladders, and in such instances it would be necessary that each soldier should be furnished with an instrument by which he could haul himself up the wall.

There are many examples still extant of such weapons, which were called "Scaling-forks," and their general appearance may be known by the two right-hand figures of the cut. The handles of these weapons were very long, and by them the soldier hauled himself to the top of the wall.   In some of these instruments the shafts were armed with projecting pegs, set at regular intervals, so that they acted as the steps of a ladder, and rendered the ascent comparatively easy.

Many of the long-handled partisans, such as the well-known Jedwood axe, were furnished with a hook upon the back of the blade, so that the weapon served the purpose of a scaling-fork as well as a battle-axe.

The Scaling-fork (German *Sturmgabel*), which is shown on the right hand of the illustration, was in use somewhere about A.D. 1500.   That which is shown next to it is about a hundred years later.

Demmin, from whose work these figures are taken, mentions that at the siege of Mons, in 1691, the grenadiers of the elder

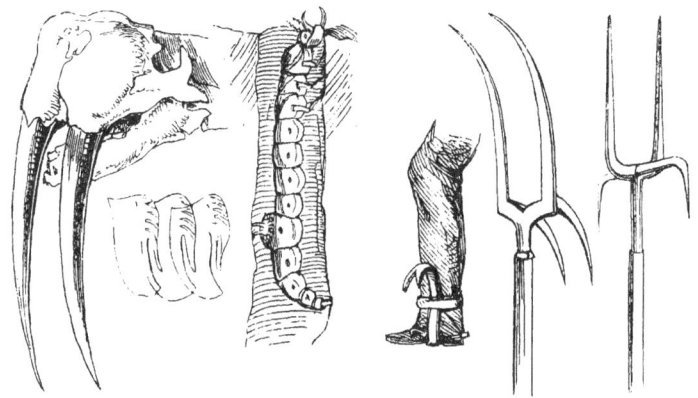

WALRUS TUSKS.   LARVA OF TIGER-BEETLE.   CLIMBING-SPUR.   SCALING-FORKS.
HOOKS OF SERPULA.

Dauphin's regiment stormed the walls under the command of Vauban, and, by means of the Scaling-fork, carried the breast-work, which they assaulted.   As a mark of honour to these gallant men, Louis XIV. ordered that the sergeants of the regiment should carry scaling-forks instead of halberds, which had been the peculiar weapon of the sergeant until compara-tively late days, just as the spontoon, or half-pike, was the weapon of the infantry officer from A.D. 1700 to A.D. 1800, or thereabouts.

The English student will remember that in the writings of Sterne, Fielding, and Smollett the half-pike is frequently mentioned as the weapon of a subaltern officer. Demmin states that the last spontoons used in France were carried by the French Guards in 1789.

PERHAPS the Climbing-spur may be familiar to some of my readers, and bring back a reminiscence of boyhood. There is nothing more tantalising to a boy than to see a hawk or magpie nest at the top of a tree which is too large to be climbed in the ordinary way, and which has no branches within many feet of the ground. However, boyish ingenuity has brought almost any tree within the power of a bird's-nester by the invention of the Climbing-irons.

These are made so as to pass under the foot like a stirrup, and can be secured to the leg by leathern straps, the hooks being, of course, on the inside of the leg. The cut represents the Climbing-iron of the right leg. By means of these instruments, a very large tree can be mounted, the irons being struck firmly into the bark, and the legs moved alternately, and not in the usual manner of climbing. Sometimes the hook of the Climbing-iron is terminated by a single instead of a double point, but the principle is the same in all.

WE will now look for similar examples in Nature.

On the right of the left-hand group is shown the larva or grub of the common Tiger-beetle, which is itself a curious creature.

It lives in perpendicular burrows, feeding upon those insects which come within its reach. Its usual position is at the upper part of the burrow, with its jaws widely extended, so as to snap up any insect that may venture too near.

When it has secured its prey, it seeks the bottom of its burrow, makes its meal in quiet, and reascends. How it does so we shall soon see. Towards the end of the body, one of the segments is much enlarged, and has a bold prominence upon the back. On the summit of this prominence there are two horn-like hooks, shaped as seen in the illustration. These hooks are used exactly like the boy's climbing-spurs, the alternate elongation and contraction of the body answering the same purpose as the movements of the boy's legs. When the larva has seized its prey and wishes to retreat, all that it has to do is to withdraw the hooks, straighten the body, and down it falls by its own weight.

In the nautical branch of this subject I have already treated of the curious pushing-poles by means of which the Serpula protrudes itself from its tube. As all must have noticed who have seen these creatures alive, the Serpula protrudes itself very slowly, but flies back into its tube with such velocity that the eye can scarcely follow its movements. Its difference of motion shows that there must be a difference in the means by which these movements are produced.

Referring to the illustration on page 45, the reader will see that the instruments with which the Serpula propels itself are used just after the fashion of punt-poles, and cannot act with any great swiftness. When, however, the creature wishes to withdraw itself, it employs a curious apparatus, consisting of many rows of little hooks. The points of these hooks readily catch against the lining of the tube, and by their aid the worm jerks itself back with wonderful celerity.

Three rows of these hooks are shown next to the Tiger-beetle larva.

The structure of these remarkable organs is elaborately described by Mr. Gosse in his "Evenings with the Microscope:"—

"If you look again at this Serpula recently extracted, you will find with a lens a pale yellow line running along the upper surface of each foot, transversely to the length of the body. This is the border of an exceedingly delicate membrane, and, on placing it under a high power (say six hundred diameters), you will be astonished at the elaborate provision here made for prehension.

"This yellow line, which cannot be appreciated by the unassisted eye, is a muscular ribbon, over which stand edgewise a multitude of what I will call combs, or rather subtriangular plates. These have a wide base, and the apex of the triangle is curved over into an abrupt hook, and then this cut into a number (from four to six) of sharp and long teeth.

"The plates stand side by side, parallel to each other, along the whole length of the ribbon, and there are muscular fibres seen affixed to the basal side of each plate, which doubtless give it independent motion.

" I have counted one hundred and thirty-six plates on one ribbon. There are two ribbons on each thoracic segment, and there are seven such segments. Hence, we may compute the total number of prehensile comb-like plates on this portion of the body to be about one thousand nine hundred, each of which is wielded by muscles at the will of the animal; while, as each plate carries on an average five teeth, there are nearly *ten thousand teeth* hooked into the lining membrane of the cell, when the animal chooses to descend.

" Even this, however, is far short of the total number, because long ribbons of hooks of a similar structure, but of smaller dimensions, run across the abdominal segments, which are more numerous than the thoracic. No wonder, with so many muscles wielding so many grappling-hooks, that the descent is so rapidly effected."

Lastly, we come to the Walrus, whose strangely elongated upper canine teeth can be used for just the same purposes as the scaling-fork or climbing-spur. As, however, reference has already been made to these tusks, in connection with another department of this work, there is no necessity for occupying space with a second description.

DEFENCE OF FORT.

So much for attack; now for defence.

The simplest mode of defending a fort, or even a mountain pass, is by throwing or rolling rocks and heavy stones against the enemy.

Simple as it may appear, it is a very effective one, as can be well understood by those who have rolled a huge stone down a long and steep slope. The stone goes gently enough at first, but rapidly gains speed, until at last it makes great bounds from the earth, tearing and crashing through everything as if it had been shot from a cannon.

I have seen a stone which was too heavy to be lifted, and had to be prised over the edge with levers, spring completely through the topmost branches of a high tree, scattering the boughs in all directions, and then, alighting on another stone, split into many fragments, just like the pieces of a burst shell. That one stone would have swept off a whole party of soldiers had they encountered it while trying to ascend the slope.

THIS invention has also been anticipated in Nature.

Putting aside the obvious reflection that the most primitive warriors must have noticed the effects of stones falling over a precipice, we have, in Captain Hall's "Life with the Esquimaux," a curious account of the Polar Bear and its mode of capturing the Walrus. Gigantic as is this animal, and terrible as are its tusks, the Polar Bear will sometimes attack it, as is evident by the scars left on those Bears which have been fortunate enough to escape from their assailant.

Still, the combat is sure to be a severe one, and so the Polar Bear will, if he can, secure his prey by some other method.

"The natives tell many most interesting anecdotes of the Bear, showing that they are accustomed to watch his movements closely. He has a very ingenious method of killing the Walrus.

"In August, every fine day, the Walrus makes its way to the shore, draws its huge body upon the rocks, and basks in the sun. If this happen near the base of a cliff, the everwatchful Bear takes advantage of the circumstance to attack

BEAR KILLING WALRUS.   WARRIORS DEFENDING A PASS.

his formidable game in this way. The Bear mounts the cliff, and throws down upon the animal's head a large rock, calculating the distance and the curve with astonishing accuracy, and thus crushing the thick, bullet-proof skull.

"If the Walrus is not instantly killed, or simply stunned,

the Bear rushes down to it, seizes the rock, and hammers away at the head until the skull is broken. A fat feast follows. Unless the Bear is very hungry, it eats only the blubber of the walrus, seal, and whale."

IMITATION.

As is the case with the Norwegians, the Esquimaux have the greatest respect for the intellectual as well as the bodily powers of the Bear, and avowedly imitate it in its modes of hunting. One of these methods will now be mentioned.

It must first be premised that the Seal is a most wary animal, and when it lies down on the shore to sleep, it takes its repose by snatches, lifting up its head at very short intervals, looking all round in search of foes, and then composing itself to rest again. To approach so cautious an animal is evidently a difficult task, but the Bear is equal to it. The following is Captain Hall's account :—

"From the Polar Bear the Innuits (*i.e.* Esquimaux) learn much.

"The manner of approaching the Seal, which is on the ice by its hole, basking in the sunshine, is from him. The Bear

POLAR BEAR HUNTING SEAL.   ESQUIMAUX HUNTING SEAL.

lies down and crawls by hitches towards the Seal, 'talking' to it, as the Innuits say, until he is within striking distance, when he pounces upon it with a single jump. The natives say that if they could 'talk' as well as the Bear, they could catch many more Seals.

"The procedure of the Bear is as follows.

"He proceeds very cautiously towards the black speck, far off on the ice, which he knows to be a Seal. When still a long way from it, he throws himself down and hitches himself along towards his game. The Seal, meanwhile, is taking its naps of about ten seconds each, invariably raising its head and surveying the entire horizon before composing itself again to brief slumber.

" As soon as it raises its head, the Bear ' talks,' keeping perfectly still. The Seal, if it sees anything, sees but the head, which it takes for that of another Seal. It sleeps again. Again the Bear hitches himself along, and once more the Seal looks around, only to be ' talked' to and again deceived. Thus the pursuit goes on until the Seal is caught, or till it makes its escape, which it seldom does."

It is remarkable that while this " talk " is going on, the Seal appears to be charmed, raises and shakes its flippers about, rolls over on its side and back, as if delighted, and then lies down to sleep.

Now, the Esquimaux hunters imitate, as nearly as they can, the proceedings of the Bear, but are not so successful. Captain Hall mentions several instances where the native hunter failed even to come within gunshot without alarming the Seal, which instantly plunged into its hole and was lost.

THE same author mentions another instance where the Esquimaux hunter has copied the Bear.

When an Esquimaux hunter catches a young Seal, he takes

POLAR BEAR CATCHING SEAL, ETC.

care not to kill it at once, as he wishes to use it as a decoy. He ties a long line round one of the hind flippers, and then drops the little Seal into the hole through the ice by which it enters and leaves the water. The struggles of the young are nearly sure to attract the mother, and when she has discovered its condition the young Seal is cautiously drawn up on the ice. The mother follows, too intent on rescuing her young to think about herself, and, as soon as she is within reach, she is struck with the harpoon.

The Polar Bear, however, preceded the Esquimaux in this mode of hunting. The young Seal lives in a hemispherical dwelling scooped out of the snow, and communicating with the water by means of a hole through the ice. This dwelling will be described and figured when we come to the subject of Architecture.

Finding out, by scent or some other means, the habitation of the young Seal, the Polar Bear leaps upon the snow, bringing his feet together, and with his enormous weight breaking through the roof of the dwelling. He instantly captures the young Seal before it can make its escape. Then, driving the talons of one paw into its hind flipper, he lets it into the hole, and allows it to flounder about in the water. When the mother is attracted to her young, he draws his prey slowly up on the ice. The anxious mother follows, and is at once secured by the talons of the other foot, as is represented in the illustration.

The Fall-trap.

This is a stratagem which is often employed in War and Hunting, though its use is mostly confined to the latter. Schoolboys often avail themselves of this principle when they wish to play a practical joke, and to amuse themselves by setting a " Booby-trap." This trap is easily manufactured, and consists of a partially opened door, with a basin or jug of water balanced upon it. The natural result is, that any one who opens the door without proper precautions receives the jug and its contents upon his head, and is thoroughly drenched.

On the right hand of the illustration is seen a curious spear, the butt of which, instead of being lighter than the head, is very much heavier. The weight, however, is exactly where it is wanted, and indeed, in actual use, is trebled by a mass of tenacious clay, kneaded upon it. This figure is taken from a very perfect specimen in my own collection.

It is an African weapon, not used for war, but for hunting,

and, as far as I know, exclusively employed against the elephants. These animals have a way of forming roads or tracks for themselves through the woods, very much like those almost invisible paths which are made by the half-wild sheep of the great Wiltshire Downs, except that they traverse thick forests instead of broad downs.

The native hunters know all the elephant paths, and if a herd of elephants be seen approaching, the path which they will take is tolerably certain.

Armed with this knowledge, the native hunters climb the trees, and seat themselves on the branches which overhang the path, each hunter being supplied with one of these spears. As the elephants pass beneath him, the experienced hunter

MANGROVE SEEDS.

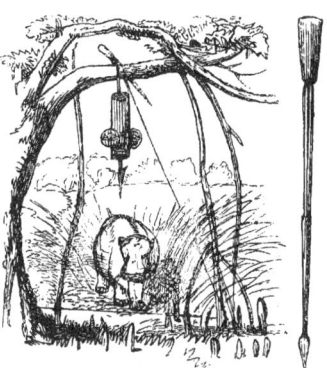

HIPPOPOTAMUS TRAP AND ELEPHANT
SPEAR.

selects a bull elephant with good tusks, and, taking a careful aim, drops the spear on its back.

On receiving the stroke, the elephant rushes off in mixed terror and rage. As the animal uses the legs of each side alternately, it sways its huge body from side to side at every step. With each movement, the spear also sways about, its weighted end giving it such a leverage, that the sharp edges of the head cut the poor animal to pieces.

ANOTHER kind of Fall-trap, which is common in many parts of Southern Africa, is not dependent upon the skill of the hunter, but, like the "booby-trap" above mentioned, is set in motion by the victim.

A figure of this trap is given in the illustration.

If the native hunter can find a spot where the Hippopotamus path passes under an overhanging branch, he makes a simple but most effective trap. He takes a heavy log of wood, and into one end of it he drives a spear-point. The log is then hung with its point downwards to the branch, the rope which is connected with its trigger or catch being stretched across the path at a few inches from the surface of the ground, and carried at right angles across the path.

The Hippopotamus takes no notice of the cord, which is usually made of one of the creepers or "bush-ropes" that are so common in hot countries. No sooner, however, does its foot strike the cord, than the trigger is released, and down falls the heavy log, driving its iron point deeply into the back of the victim. Even if the weapon were simple iron, such a wound must be mortal, but, as it is almost invariably poisoned, the wounded animal can scarcely travel forty or fifty yards before it lies down and dies.

One of these traps is shown in the illustration. In the foreground is shown the Fall-trap, pointed with iron, and weighted with large stones at the lower end, so as to bring it down with more force, and to prevent it from falling transversely.

The Spring-gun, once so formidable a protector of our coverts, was managed in a similar manner, except that the missile was discharged horizontally, and not vertically. The gun, loaded with shot, was fixed some eighteen inches from the ground, and a long and slight wire fastened to the trigger. The opposite end of the wire was made fast to a tree or other fixed object, and, as the gun was directed on the line of the wire, it is evident that any one who stumbled against it would discharge the gun, and receive the contents in his legs.

In France the gun was generally loaded with little pieces of bay salt, and I very much pity the unfortunate poacher who came across one of these guns. The pain would prevent him from escaping, and I think that the hardest-hearted of game preservers could not bring himself to prosecute a man who had already suffered so much.

Of a similar character are the Spring-bows which were once common in this country, and are still used in various parts of Asia. A bow and arrow are substituted for firearms, and the bow, after being drawn by the united efforts of several men, is held in its position by a stick, one end of which presses against the centre of the bow, and the other against the string.

A large arrow is then placed on the bow, and a cord is tied to the middle of the stick, led forwards in a line with the direction of the arrow, and fastened, as in the case of the spring-gun. As soon as the line is struck, the stick is jerked from its place, and the arrow is discharged, piercing the body of the trespasser. Tigers, bears, and leopards are the usual victims of this trap.

IT is remarkable that in the same country there is a pro-duction of Nature which may in all probability have given to the native hunter the idea of the Fall-trap. This is the Man-grove-tree, which is remarkable for the wonderful extent of ground which it will cover, and the nearly impenetrable thickets which it forms. In the present part of the work we have nothing to do with the aërial roots, several of which are shown in the illustration, and only restrict ourselves to the Seeds, and the curious manner in which they are planted by Nature.

In the illustration, on the left hand, the growth of the Mangrove is seen. The drawing is taken from a sketch by the late Mr. Baines, and generously placed at my disposal, as were all his drawings and journals.

The Mangrove is a wet-loving tree, never flourishing unless rooted in mud; and whether the moisture of the mud be attributable to fresh or salt water seems to make little dif-ference to the Mangrove, which, of the two, appears to prefer the latter. Now, the seeds of the Mangrove look very much like elongated skittles, except that one end comes to a sharp point. As they hang on the tree, the point is downwards. When they are ripe, they fall from the branch, and by their own weight are driven deeply into the mud, where they develop roots and leaves, and become the progenitors of the future Mangrove race.

I cannot but think that the native hunter, having seen the tremendous force with which the Mangrove seed buries itself in the mud, has applied the same principle to a weapon which shall bury itself in the body of an elephant.

WAR AND HUNTING.

CHAPTER VIII.

CONCEALMENT.—DISGUISE.—THE TRENCH.—POWER OF GRAVITY.—MISCELLANEA.

CONCEALMENT.

WE will first take Concealment by means of Covering.

If History repeats herself, so does Warfare. I have already shown the repetition of History in the Fortress—I shall now show it in the Field.

In former days, when arms of precision were not invented, concealment was not needed. No soldier ever was visited with a dream so wild as that of taking definite aim at the enemy, and reserving the fire until the aim was certain. I have in my collection several of the French and English muskets used about the time of Waterloo, and, though a fair rifle-shot, would not engage to hit a haystack with either of them at a distance of a hundred yards. With the Snider or Martini-Henry in the hands of a skilful adversary, he would be a bold man who would offer himself for a target at a thousand yards. Indeed, if the first shot happened to miss, the marksman would be tolerably sure to notice the failure, and to correct his aim with fatal certainty.

In those days, therefore, concealment was rather ridiculed than praised, the power of the new arm not being as yet appreciated. I well recollect, in the earliest days of the Volunteer movement, hearing a Volunteer captain declare, amid the cheers of his company, that " he had never sneaked behind a tree in all his life, and was not going to begin now."

In the present day, the power of the missile has been developed with such astounding rapidity, that to be exposed to the fire of rifles or cannon is almost certain death. Indeed, the only safety of the defence lay in the fact that the smoke soon rendered very accurate shooting impossible at long ranges, and that at short ranges, if a man got a bullet through his body, it mattered little to him whether the missile were a spherical musket-ball or a conical rifle-bullet.

Just, then, as forts have latterly sunk into the earth for the purpose of strength, so have our modern soldiers found that the true principle of modern warfare is never to lose sight of

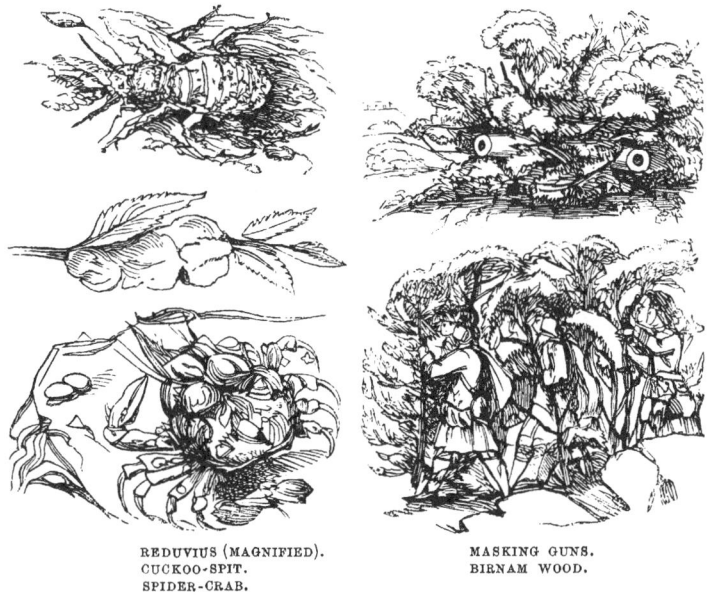

REDUVIUS (MAGNIFIED).
CUCKOO-SPIT.
SPIDER-CRAB.

MASKING GUNS.
BIRNAM WOOD.

the enemy, and never to allow the enemy to see yourself or the disposal of your troops.

L

Everything must be revealed to the commander-in-chief, everything must be concealed from the enemy.

In the late Franco-German war the principle of conceal-ment was largely used, and when cannon were brought into the field by the Germans for the purpose of attacking fortresses, they were always hidden under branches of trees, so that the enemy should not distinguish them from the ordinary features of the country, and that the sparkle of the sunbeams upon them might not be seen.

It would be almost superfluous to remind the reader of Malcolm's stratagem when besieging Dunsinane Castle :—

> " Let every soldier hew him down a bough,
> And bear 't before him ; thereby shall we shadow
> The numbers of our host, and make discovery
> Err in report of us."

Precisely similar modes of concealment are to be found in the animal world.

There is a certain insect belonging to the Heteroptera, and scientifically named *Reduvius personatus*. I am not aware whether it has any popular name. It is insectivorous, and ought to be welcomed in houses, as it is particularly fond of the too common bed-bug. So carnivorous are these insects that one of the Reduviidæ killed and sucked a companion of her own sex, her own mate, and, after only a few days' fast, her own young, and then sucked her own eggs.

During its larval and pupal stages of existence, the Reduvius covers its body and limbs with dust and any other refuse which it can find. In this manner it disguises its form so completely that it scarcely looks like an insect. Occasionally it seems to be dissatisfied with its coat of dust, throws it off, and sets to work at a new one.

One of these creatures, as it appears when covered with its dusty coating, is seen in the upper left-hand corner of the illustration. It is slightly magnified.

Below the Reduvius is the common Cuckoo-spit (*Aphrophora spumaria*), whose frothy masses are so plentiful in our hedge-rows and gardens.

If one of these masses be carefully opened, there will be found in it a little green creature with small, round, dot-like eyes. This is either the larval or pupal state of the Frog-hopper, as the insect is called in its perfect state, from its habit of taking long and sudden leaps when alarmed.

I well remember my delight when, as a child, I set to work at examining these froth-masses, and succeeded in tracing the insect through all its changes. The froth is derived from the sap of the tree, which is sucked through the proboscis, passed through the digestive organs, and then ejected in a succession of little bubbles. After awhile a little drop of clear liquid is seen to collect at the bottom of the froth, to increase, and then to fall, when another immediately begins to be formed. One species of Cuckoo-spit, which inhabits Madagascar, acts almost like a siphon on the tree, and pours out large quantities of clear water during the hottest part of the day.

Within this froth-mass the insect lies concealed, and, though utterly helpless, is safe from most of the enemies that would attack it if it were left exposed.

Beneath the Cuckoo-spit is the common Spider-crab, sometimes called the Thornback-crab, from the numerous spines with which its body is covered. Its scientific name is *Maia squinado*.

When the Spider-crab attains to a tolerable size, its rough surface forms attachment for various marine beings, chiefly those belonging to the zoophytes. In some cases these zoophytes grow to such a size that the Crab is completely covered by them, and its original shape effectually concealed. When one of these creatures is seen in a living state it presents the curious spectacle of a large bunch of zoophytes and corallines moving about from place to place without any perceptible limbs, the whole of the surface of the Crab being covered with extraneous growths.

DISGUISE.

NEXT comes concealment by means of Disguise.

On the right hand of the accompanying illustration is shown a singular mode of concealment adopted by the Barea, a warlike and predatorial tribe of Abyssinia. When Mr. Mansfield Parkyns was resident in Abyssinia he fell in with the Barea, through whose country he had to pass.

" Scarcely had we passed the brook of Mai-Chena when one of our men, a hunter, declared that he saw the slaves. Being at that time inexperienced in such matters, I could see nothing suspicious. He then pointed out to me a dead tree standing on an eminence at a distance of several hundred yards, and charred black by last year's fire." Here I must explain that in Abyssinia, as in several other parts of the world, the ground is annually cleared of its superabundant vegetation by setting fire to it, and allowing the flames to burn themselves out.

" However, all I saw was a charred stump of a tree and a few blackened logs or stones lying at its feet. The hunter declared that neither the tree nor the stones were there the last

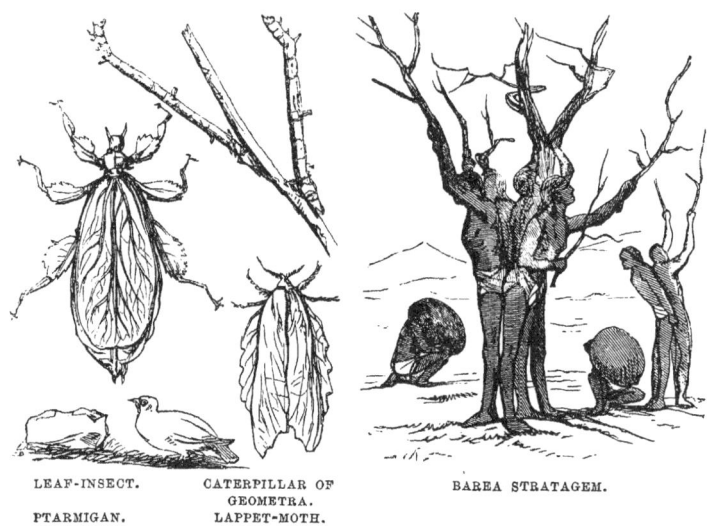

LEAF-INSECT.   CATERPILLAR OF
                         GEOMETRA.
PTARMIGAN.   LAPPET-MOTH.   BAREA STRATAGEM.

time we passed, and that they were simply naked Barea, who had placed themselves in that position to observe us, having no doubt seen us for some time, and prepared themselves.

" I could scarcely believe it possible they could be so motionless, and determined to explore a little. The rest of the party advised me to continue quietly in the road, as it was possible that, from our presenting a rather formidable appearance, we should pass unmolested; but so confident was I of his mistake, that, telling the rest to go on slowly as if nothing had been observed, I dropped into the long grass and stalked up towards them.

" A shot from my rifle at a long distance (I did not venture too close) acted on the trees and stones as powerfully as the fiddle of Orpheus, but with the contrary effect; for the tree disappeared, and the stones and logs, instead of running after me, ran in the opposite direction.

" I never was more astonished in my life, for so complete was the deception that even up to the time I fired I could have declared the objects before me were vegetable or mineral— anything, indeed, but animal. The fact was that the cunning rascals who represented stones were lying flat, with their little round shields placed before them as screens."

This stratagem is shown on the right hand of the illustration.

On the left are a few of the innumerable instances in Nature where Concealment is obtained by imitation.

The three examples which are here given are familiar to all entomologists.

The upper figure represents two of the Geometra or Looper Caterpillars, as they appear when at rest, and affixed to a twig. This appears to be a singular attitude of rest, but it is one in which they delight, and in which they remain for hours together, the claspers at the end of the body tightly grasping the branch, and the whole body held out so straight and motionless that it is hardly possible to believe that a veritable twig is not before the eye. The colour is that of the twig, and the different segments of the body look exactly like the little irregularities and projections of a young twig.

I have more than once seen a novice in entomology unable to distinguish these larvæ, even when the branch was pointed out, and there were several upon it.

Just below the Loopers, and on the left hand of the illus- tration, is shown the well-known Leaf-insect (*Phyllium*). These strange beings have the elytra and the flattened appendages of the legs so exactly like leaves that the most experienced eye can scarcely distinguish them from the leaves among which they are placed. Even when they have been on a small plant, such as a myrtle in a flower-pot, I have had the greatest difficulty in finding them, and have seen people examine the plant, and then go away declaring that no insects were on it.

On the right hand, and just below the looper caterpillar, is the common Lappet-moth of this country, shown in its position of rest.

When it assumes this attitude, it looks exactly like a withered leaf, the resemblance extending not only to the form, but the colour. All entomologists are familiar with many similar examples in insect life. The common Tortoise-shell Butterfly, for example, has a way of settling on patches of red soil, with which it harmonizes so well that it can hardly be seen. The various moths, also, are in the habit of resting on tree-bark, palings, and other objects, to which they instinctively know that they assimilate in hue. Many a beginner in entomology will pass a wooden fence or a wall, and not see an insect on either, while an adept will follow him and take twenty or thirty good specimens.

The last figure in the illustration represents a Ptarmigan (*Lagopus vulgaris*) in its winter dress. These birds have two differently coloured dresses, one for summer and the other for winter, and both adapted for concealment by imitation. In the former dress it is mottled with various shades of blackish brown, yellow, and white. As the bird is in the habit of settling among the grey lichen-covered stones on the sides of rocky hills, these colours harmonize so exactly with them that a Ptarmigan may almost be trodden upon before it is perceived.

In the winter, when the snow covers the whole country with one uniform sheet of white, except where the wind blows the snow aside, and exposes the underlying stones, the Ptarmigan assumes a different plumage, being almost entirely white, except a black streak over the eye, and the outer feathers of the tail, which are also black. Thus the bird becomes almost indistinguishable from a snow-covered stone, especially as it has a habit of squatting motionless and silent when it takes alarm.

The reader may, perhaps, remember that the common Stoat also has a summer and winter dress. The ordinary colour is rich reddish brown above, and white beneath, with a black tip to the tail. In the severe winters of Northern Europe the Stoat exchanges his ruddy coat for one of pure white, and is then known by the name of Ermine. It is remarkable that in the winter dress both of the Ptarmigan and Stoat the tail is black, while the rest of the coat is white.

THE TRENCH.

WE now come to a third mode of concealment in war, namely, that which is obtained by means of Trenches or Pits.

Even in hunting the pit or partial trench is largely used. In Southern Africa the hunter often employs such a trench, called technically a "Skärm." It is very simple in idea, and easily made, being based on the principle that lions, elephants, &c., look for their assailants on the level of the earth, and seldom, if ever, look above or below it. Accordingly the hunter, having marked some pool or lake whereunto the wild animals resort at night to quench their thirst, chooses a convenient spot, and there digs a trench some seven feet in length and four deep, and covers it in with stout tree-branches and logs of various size. The whole is roofed in with sods, and the only entrance is at one end.

Here the hunter sits and waits, and, as his ear is on a level with the surface of the ground, he can hear at a considerable distance sounds which would have escaped him had he been erect.

Waiting for a favourable opportunity, as the various beasts come to drink, the hunter chooses one, takes careful aim, and fires one of his heaviest guns. It is but seldom that the rest of the animals charge in the direction of the Skärm, but even

GALLERIA-MOTH (LARVA).   MILITARY TRENCH.

if they do, the hunter is quite safe under the shelter of his strong roof, which is able to resist even the heavy tread of an elephant.

In modern warfare, and especially during sieges, the trench is largely used, and is constructed on the most scientific principles, so as to shelter the assailants, while enabling them to proceed nearer and nearer to the fortress. A portion of one of these trenches is shown in the right hand of the illustration.

On the opposite side of the same illustration is shown the same principle as carried out in Nature.

There is a certain little insect, called the Wax-moth, or Galleria-moth (*Galleria alvearia*), which, although quite harmless in its perfect form, is in its larval state extremely injurious to beehives.

The mother moth contrives, aided by her tiny form and sombre colouring, to slip past the sentries at the mouth of the hive, and to lay her eggs among the combs. This done, she dies, but the evil of her visit lives after her.

Each of the eggs is hatched into a little caterpillar, having a soft grey body, but a hard, horny head of a black-brown colour. As soon as they are hatched they begin to feed, eating not only the waxen combs, but the honey and the bee-bread which were intended for the support of the legitimate inhabitants.

The reader may ask why the bees do not destroy this marauder on their premises. They would be only too glad to do so, but they cannot touch it. As it eats its way along, it constructs a strong silken tube, within which it lives, and which it gradually lengthens. This tube or gallery is exceedingly tough, and perfectly capable of resisting the bee's sting. Moreover, the caterpillar traverses its tube with such rapidity that the bee has no chance of knowing whereabouts the caterpillar may be when it makes its attack. When it feeds it only protrudes its armed head, the horny covering of which is an effectual protection against the sting.

When these creatures fairly get hold of a hive, the damage which they do is terrible, the whole of the combs being enveloped in the ever-increasing labyrinth of tubes. Even the bees themselves fall victims to the Galleria-moth, for the silken tunnels are driven through and through the combs, enveloping the broad cells as in the meshes of a net. Consequently, when the young bees are developed, they cannot escape from their cells, and perish miserably.

Nor do these tiresome insects confine themselves to hives; but they have an extraordinary facility for discovering bee-combs after they are removed from the hive. Some years ago I was making a collection of various insect habitations, and had brought together a carefully selected set of combs, showing the internal structure of the hive, and the different cells which are inhabited by the worker, the drone, and the queen bee.

One day, when about to arrange the collection in a glass case, I found that the whole of the combs had been destroyed by the Wax-moth. Scarcely a square inch of comb remained, and the contents of the box were little more than a congeries of Wax-moth galleries. Even the Wasp and Hornet nests which had been placed in the same box had been attacked, and, although they had not been so utterly destroyed as the waxen cells, they had been sufficiently injured to render them unfit for exhibition.

Many other insects work on the same principle. Certain Termites, for example, construct tunnels of clay, in order to conceal them on their travels, and have the art, even in the hottest and driest weather, of mixing their clay with some liquid which renders it, when dry, nearly as hard as stone. Indeed, there have been instances where the Termites have attacked the wooden beams of houses, and literally transformed them into beams of stone.

Then there are many Ants, notably several species of South America, which cover their approach by tunnels, and never venture into the open air.

GRAVITY AS A PROPULSIVE AGENT.

THE two figures on the accompanying illustration will almost speak for themselves.

We have already seen how the same force of gravitation which causes the avalanche to thunder down the precipice may be utilised as a means of projecting missiles in time of war. When, however, the stones or beams were once sent on their destructive mission, they were out of the control of those who

RAM.   HEAD OF BATTERING-RAM.

launched them. We now come to a modification of the force of Gravity, by which the missile, if we may so term it, is kept under control, its power increased or diminished at will, and its point of attack shifted according to the requirements of the moment.

Before the invention of artillery, the Battering-ram was by far the most formidable engine that could be brought against a fortified place. The principle of the Battering-ram was simple enough. A long and heavy beam, generally the trunk of a tree, was suspended by ropes at the centre of gravity, so that it could be swung backwards and forwards. Although a simple beam was an effective weapon, its value was much enhanced by loading the thickest end with a heavy mass of metal, usually iron, and, when there was time for adornment, roughly modelled into the form of a ram's head.

Generally the Battering-ram was mounted on an elevated platform, and the soldiers who worked it protected by a roof, which was called by the name of Testudo, or Tortoise. The force of this weapon was tremendous, and no wall, however strong, could resist it. Sometimes the beam was considerably more than a hundred feet in length, being composed of several pieces bolted and banded together with iron.

It may easily be imagined that such a weapon as this must have been a most terrible one, and, indeed, the whole success of the siege practically depended upon it. The assailants did their best to bring the Battering-ram into position under the walls, and the besieged did their best either to keep it away, or to neutralise its effects by catching it with nooses, dropping large stones upon it so as to break or dismount it, or, if they could not succeed in either of these attempts, they deadened the force of its blows as well as they could by interposing large sacks of wool between the wall and the head of the ram.

Considering the style of architecture which was then used in fortification, namely, a combination of height with thickness, the force of the Battering-ram would be even greater than that of artillery. The regular and rhythmical swing of the ram would soon communicate a vibratory motion to the wall, which would of itself tend to disintegrate the whole structure, while the blows of the iron head beneath broke away the stones, and rendered the downfall of the fort a mere matter of time.

The reader need hardly be reminded that the Battering-ram was so called because its mode of attack was practically the same as that of the animal from which it took its title.

MISCELLANEA.

By slow degrees, mankind, as they advance in civilisation, have robbed warfare of many horrors. Non-combatants, for example, are now left unharmed. Poisoned weapons have, by common consent, been abolished, and so have those instruments of warfare which, though they do not simply poison the blood by means of bodily wounds, do so by means of noxious vapours poured into the lungs.

It is sometimes rather unfortunate when civilisation and semi-barbarism meet in battle ; the former respecting the customs of honourable warfare, and the latter ignoring them. For example, in olden times, one of the most potent weapons in naval combat was the " stink-pot "—*i.e.* a vessel filled with sulphur and other ingredients, and emitting a smoke which was death when inhaled. Among the American Indians the well-known Chili-plant was much used for this purpose, the very first breath that was taken of the thin and almost invisible smoke causing the throat to contract as if clutched by a strong

BOMBADIER-BEETLE.

CHINESE STINK-POTS.

hand. If then any enemies had taken refuge in a cave, or were suspected of having done so, a fire was lighted at the entrance, a quantity of chilis thrown on it, and the rest left to time. No being could endure that smoke and live, and they must either stay in the cave and die, or come out and deliver themselves up to their foes. The former was the better part to take, as suffocation, however slow, is only an affair of a few minutes, while death by torture is prolonged through hours.

In the late Chinese war the stink-pot was extensively used, and our sailors took it in very bad part that the enemy should be allowed to employ such weapons, and they should be debarred from using them.

Whether this principle is still retained in the defence of fortresses I do not know. I recollect, however, some twenty years ago, going over a fortress in which suffocation was employed as a means of defence. A long gallery was so placed that the assailants were tolerably sure to force their way into it, thinking that it led to the interior of the fort.

It was, however, nothing but a trap, for it had no exit. As soon as a number of the assailants had poured into this trap, their exit was suddenly cut off by machinery provided for the purpose, and at the same time a quantity of sulphur and lighted charcoal was shot into the gallery from above, and the aperture instantly closed. It would be absolutely impossible that any one who had been enclosed in that terrible chamber should escape with life, for the first breath of that deadly vapour would render the strongest man insensible.

NATURE, as usual, has anticipated Art even in this particular.

In several parts of England, and especially along the shores of the Thames towards Gravesend, a little beetle is to be found under the flat stones of the river bank. Its scientific name is *Brachinus crepitans*. When this insect is alarmed, it has the power of ejecting a peculiar liquid, which, when it comes in contact with the atmosphere, bursts into a sort of pale blue-green flame, followed by a kind of smoke. Sometimes, when a tolerably large stone is lifted, the little explosions will go popping about in a most curious manner. Indeed, they carry reminiscences of school days, when it was a joy to distribute single grains of coarse gunpowder on the bars of the grate, and watch them melt, take fire, explode, and send forth little clouds of smoke.

Whether or not this capability was given as a means of defence I cannot say, but it assuredly answers that purpose.

There are several of the voracious Carabidæ, or Ground-beetles, which would be very glad to make a meal of the Brachinus. When, however, the Bombadier-beetle finds itself on the point of being overtaken, it elevates the abdomen with a peculiar gesture, and ejects the liquid. The effect on the pursuer is remarkable. It seems overwhelmed and stupefied by the sudden attack, moves about for awhile as if blinded, and, by the time that it has recovered its sense, the Bombadier-beetle is out of sight.

In some of the hotter parts of the world there are several species of Bombadier-beetles which attain considerable size, and their discharge is powerful enough to discolour the skin of the human hand.

I HAVE felt some little difficulty in classifying the curious invention which will now be described, but, as it is used for the purpose of making bullets, I have placed it in the category of War.

In the days of " Brown Bess," as the old musket used to be called, precision of aim was not required, for no commander

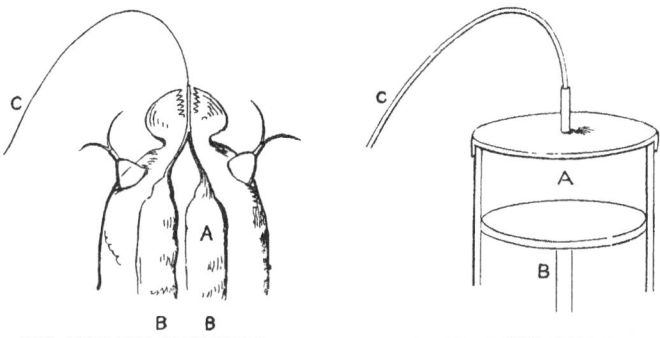

SILK APPARATUS OF SILKWORM.   BULLET-MAKING APPARATUS.

dreamt of opening fire until the enemy were at comparatively close quarters. In those days the bullets were spherical, and cast in moulds. After a time, when the Enfield rifle displaced the musket, and did double the execution at three times its range, bullets were still cast, though their shape was altered, and they took a sugar-loaf form instead of being spherical.

The rifle-testing machine at Woolwich, however, soon showed that at long ranges a cast bullet was nearly useless, one part being always lighter than another, and air-bubbles often taking the place of lead. After being cast, therefore, the bullets were placed in a " swedge," or " swage," *i.e.* a machine by which the lead was forcibly compressed until it was of a tolerably uniform density. Even this process, however, did not insure absolute exactness, and then a machine was invented by means of which the process of casting was superseded, and the bullets were pinched or squeezed, so to speak, out of cold lead.

On the right hand of the illustration is a plan of the ingenious apparatus by which the lead is supplied to the machine which actually forms the bullets. The sketch is not meant as a drawing of the actual machine, but is merely intended to show the principle.

The chief parts in this machine are a hollow cylinder, a piston, and a delivery tube. The cylinder is shown at A, and when used is filled with melted lead. The piston, B, is then forced upwards by hydraulic pressure, driving the lead through the delivery tube. As it issues into the air it hardens, and thus forms a solid rod of lead, C. This rod is then passed into the next machine, where it is cut into regular lengths, and these pieces are then placed in moulds, and forced into form by enormous pressure. Were it not for this ingenious machinery, the wonderful scores which are now made at long distances would be impossible.

Now let us compare Art with Nature, as seen on the left hand of the illustration, which is a chart or plan of the spinning apparatus of the Silkworm.

When I first saw the bullet-making machine at work, I at once perceived that it was nothing more than a repetition in metal of the beautiful mechanism which I had so often admired in this insect. In order to show the close analogies of the two objects, I have marked them with similar letters.

A represents the upper part of the reservoir or vessel which contains the silk in a liquid state. B B are the muscles which contract the reservoir and force the liquid matter out. It will be seen that both these vessels terminate in a delivery tube, identical in office with that of the bullet-making machine. As soon as the liquid silk passes into the air it is hardened, and is formed into a silken rod, C, just as is the lead in the machine. The only difference between the two, if it can be called a difference, is, that in the silkworm the rod is double, whereas in the machine it is single. The principle, however, is identical in both cases. The webs of spiders, and the threads by which so many caterpillars suspend themselves, and with which they make their nests, are all formed on the same design, namely, a reservoir containing a liquid which is squeezed through a tube, and hardens when it comes in contact with the air.

ARCHITECTURE.

CHAPTER I.

THE HUT.

THERE can be little doubt that mankind has borrowed from the lower animals the first idea of a dwelling, and it is equally true, as we shall presently see, that not only primitive ideas of Architecture are to be found in Nature, but that many, if not all, modern refinements have been anticipated.

To begin at the beginning. The first idea of a habitation is evidently a mere shelter or roof that will keep off rain from the inhabitant. When Mr. Bowdich was travelling in Western Africa, he was told that the Njina—another name for the Gorilla—made huts for itself from branches, the natives also saying that it defended these huts with extemporised spears. A more truthful account is given of the Mpongwe and Shekiani, namely, that the animal builds a hut, but lives on the roof, and not under it.

Although this information has since proved to be false, there was a foundation of truth in it, for there really is an ape in that part of Africa which makes huts, or rather roofs, for itself. This animal is the Nshiego Mbouvé (*Troglodytes calvus*).

This remarkable ape has a curious way of constructing a habitation. Choosing a horizontal branch at some distance from the ground for its resting-place, the animal erects above it a roof composed of fresh branches, each laid over the other in such a way that rain would shoot off them as it does from a thatched roof. M. du Chaillu gives the following account of this habitation :—

" As we were not in haste, I bade my men cut down the trees which contained the nests of these apes. I found them made precisely as I have before described, and as I have always

NEST OF NSHIEGO MBOUVÉ.   AFRICAN TREE-HUT.

found them, of long branches and leaves laid one over the other very carefully and thickly, so as to render the structure capable of shedding water.

" The branches were fastened to the tree in the middle of the structure by means of wild vines and creepers, which are so abundant in these parts. The projecting limb on which the ape perched was about four feet long.

" There remains no doubt that these nests are made by the animal to protect it from the nightly rains. When the leaves begin to dry to that degree that the structure no longer sheds water, the owner builds a new shelter, and this happens generally once in ten or fifteen days. At this rate the Nshiego mbouvé is an animal of no little industry."

The roof which this ape builds is from six to eight feet in

diameter, and is tolerably circular, so that it looks something like a large umbrella. When the animal is at rest it sits on the branch with one arm thrown round the stem of the tree, in order to support itself during sleep. In consequence of this attitude the hair is rubbed away on one side, thus earning for the ape the specific title of *calvus*, or bald.

It is rather remarkable that the Orang-outan of Borneo is likewise a house-builder, though not in the same manner as the African ape which has just been mentioned. This animal has a way of weaving together the branches of trees, so as to make a platform on which it can repose, its enormously power-ful arms being of great service in this task. The animal seems to make its platform in quite a mechanical manner, and it has been noticed that when an Orang-outan has been mortally wounded, it has expended its last energies in twisting the branches together so as to form a couch on which it can lie down and die.

Putting aside those cases where huts have been erected in trees by way of amusement, we may find instances where human beings have been forced to make their habitations in trees.

In some places, such as certain parts of South America, the natives are forced to make their houses in trees, partly on account of the climate, and partly for the purpose of avoiding the mosquitoes.

The delta of the Orinoco River is nearly half as large as England, and for a considerable part of the year is deep in water. Yet this tract is inhabited by the Warau tribe, who find in it their only mode of escape from the tiny but terrible mosquito. We in England know but little of the miseries inflicted by these insects, which are so plentiful in some parts of America that they are gathered in bags, pressed into thick cakes about as large as ordinary dinner-plates, and an inch in thickness, and then cooked and eaten.

Now it is found that although the mosquito infests the banks of rivers, it cannot venture far from land. The Waraus, therefore, make for themselves habitations which are far enough from land to baffle the mosquitoes, and near enough to be easily reached in canoes.

Fortunately for them, there is a tree called the Ita Palm, belonging to the genus Mauritia, which loves moisture, and grows abundantly in this delta. The Waraus, therefore, make their habitations in these trees, connecting several of them together with cross-beams, and laying planks upon them so as to form the flooring of their simple huts. Here they maintain themselves chiefly by fishing, but are sometimes obliged to visit the mainland, in spite of the mosquitoes. When, however, they return, they halt at some distance from the shore, and with green boughs carefully beat out every mosquito from the canoe before they dare to approach their dwellings.

The once-celebrated Lake Dwellers of Switzerland evidently lived after a similar fashion.

In this case insects drive human beings into trees, but there are instances where nobler animals have produced the same effect.

Some years ago there lived in Southern Africa a powerful chief called Moselekatze, who spent his whole life in warfare, converting all the male inhabitants into soldiers, dividing them into regiments, ruling them with the extreme of discipline, and by their aid devastating the neighbouring countries. He swept off all the cattle, which constitutes the wealth of the Kafir tribes, and either killed the male inhabitants or pressed them into his service.

The land was in consequence deprived of its natural defenders, and the wild beasts, especially the lions, increased rapidly, so that the position of the survivors was a really terrible one. They had no cattle to furnish the milk which is the chief food of the Kafir tribes; their weapons had been taken by Moselekatze; and they were forced to live almost entirely on locusts and wild plants. By degrees the lions became so numerous and daring, that the slight Kafir huts were an insufficient protection during the night, and the disarmed and half-starved inhabitants were perforce obliged to make their habitations in trees.

Dr. Moffat, the well-known missionary, saw one tree in which there were no less than twenty huts. They were conical, and made of sticks and grass, the base resting upon a platform or scaffold laid upon the fork of a horizontal branch.

The only mode of approach to these huts was by notches cut in the trunk of the tree.

How needful were these precautions was shown by the fact that the missionary himself spent a night in one of these aërial huts, and had the pleasure of hearing a number of lions snarl and growl all night over a rhinoceros hump which he had placed in an oven made of a deserted ant-hill. The oven, however, was too hot for the lions, and they had to retreat at daylight.

Passing from the tropics to the polar regions, we now take an instance where man has acknowledgedly copied an animal in the construction of his dwelling.

In Esquimaux-land, where no trees can grow, where for months together the sun never rises above the horizon, where the temperature is many degrees below zero, and where the land and ice are alike covered with a mantle of snow so thick that every landmark is abolished, it would seem that no human

SNOW-HOUSE OF SEAL IN ESQUIMAUX-LAND.   SNOW-HOUSE OF ESQUIMAUX.

beings could support life for one week. There is neither timber for house-building nor wood for fuel, so that shelter, warmth, and cookery seem to be equally impossible, and as these are among the prime necessities of human life, it is not easy to see how mankind could exist.

Yet these very regions are inhabited by sundry animals, and it is by copying them that Man can keep his place. We have already seen how the Esquimaux hunter copies the Polar Bear, and we have now to see how he copies the Seal in the material and form of his dwelling-house, and not only contrives to live, but to enjoy life all the more for the singular conditions in which he is placed. Captain Hall mentions, in his "Life with the Esquimaux," that one of the natives, named Kudlago, who was returning to his native country after visiting the United States, died while on board the ship. Towards the end of his life he was yearning for ice, and his last intelligible words were, " Do you see ice ? Do you see ice ? "

On the vast plains of ice that are formed in the winter-time the snow lies thickly, and yet upon such an inhospitable spot the mother seal has to make a home for her tender young. This she does in the following manner :—

She has already preserved a " breathing hole " in the ice, through which she can inhale air. How she finds so small a hole under the surface of the ice, where there are no landmarks to guide her, is a marvel to every swimmer. She has to chase fish and follow them in all their winding courses, and yet, when she is in want of air, is able to go straight to her breathing hole, and there take in a fresh supply of oxygen.

When she is about to become a mother, she enlarges this breathing hole so as to make it into a perpendicular tunnel. She then, with the sharp nails of her fore-paws, or flippers, scoops away the snow in a dome-like form, as shown in the illustration, taking the snow down with her through the ice, and allowing it to be carried away by the water. By degrees she makes a tolerably large excavation of a hemispherical shape, and when her young is born she deposits it on the ice-ledge around the tunnel. From ordinary foes the young Seal is safe, and nothing can discover the position of the house unless guided by the sense of smell.

How the Polar Bear and the Esquimaux hunter discover the dwelling and capture the inmates we have already described in the chapter treating of War and Hunting. Our present business is with the dwelling itself. Comparatively few of these snow-houses, or *igloos*, as they are called, are discovered, and they remain intact until the summer sun melts the roof and exposes the habitation. By this time, however, the young Seal has grown sufficiently to shift for itself, and no longer needs the shelter of a dwelling.

THE winter hut, or igloo, of the Esquimaux is made of exactly the same shape and of similar materials to the dwelling of the Seal, the chief difference being that it is built instead of excavated.

In order to save time, the igloo is generally erected by two men, one of whom supplies the material, and the other acts as bricklayer and architect in one. Each begins by tracing a suitably sized circle in the snow, which he clears away to some depth, so as to preserve a firm surface, either as a floor or as the material for the wall. In this work both men are equally valuable, for the skill required to cut the slabs of snow into such a shape that they can be formed into a hemispherical dome is quite as much as that which is needed for putting them together. I will call them the cutter and the builder. Sometimes a young hand is employed by way of labourer, and passes the snow slabs to the builder as fast as they are cut.

The builder receives the slabs, and arranges them in regular order, always taking care to "break the joints," just as do our bricklayers of the present day. Always remaining within the circle, he gradually builds himself in, and when he has quite finished the house, he cuts a hole through the side, emerges, and, by the help of his partner, puts on the finishing touches. He usually also adds a sort of tunnel to the door, through which any one must creep on his hands and knees if he wishes to enter the igloo. This part of Esquimaux architecture will presently be noticed more in full.

Perhaps the reader may wish to know what provision there is for ventilation. The answer is simple enough. There is none, the Esquimaux not requiring ventilation any more than they require washing. The two, indeed, generally go together; and it may be observed, even in our own country, that those who object to fresh air, and are always complaining of draughts, have a very practical aversion to the use of fresh water, and but little confidence in what Thackeray calls the "flimsy artificies of the bath."

The Esquimaux never washes, and knows not the use of linen. Consequently, it is no matter of surprise that a sailor of Captain Hall's crew could not make up his mind to enter an igloo. "Whew!" exclaimed the man, "by thunder, I'm not going in *there!* It's crowded, and smells horribly. How it looms up!"

Considering that there were inside that igloo a dozen

Esquimaux, all feasting on a raw, newly killed, and yet warm seal, the sailor had reason enough to decline a visit. Captain Hall, however, determined, in his character of explorer, to brave the strange odours, and moreover to join the inmates in their feast, knowing that as he would have to live among the Esquimaux for some two years, he would be forced to live as they did, and might as well begin at once. Consequently on this resolve, he drank the still steaming blood, and quaffed it from a cup which an Esquimaux woman had just licked clean.

Floors and Pillars.

One decided step in Architecture is the invention of the Pillar, and its capabilities of aiding to sustain another floor above it. We see this principle carried out in our great cathedrals, where the use of the Pillar is almost infinite. Take, for example, Canterbury Cathedral. A heedless visitor might easily pass through the nave, enter the choir, visit the various

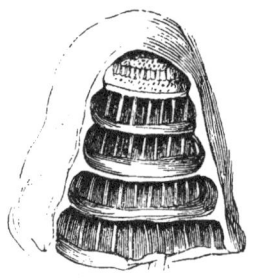
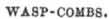
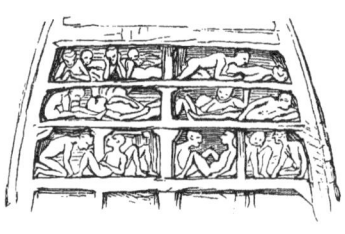

WASP-COMBS.   SLAVE SHIP.

side-chapels, and "Becket's Crown," without thinking that under his feet is a vast chamber, and that the floor on which he stands is, in fact, the roof of a great crypt.

The weight of the Cathedral, with its lofty towers, is so tremendous, that the building could not be erected simply upon the ground, but rests upon a complicated substratum of pillars and arches, whereby the weight is spread over a large surface. In fact, the Cathedral is really two buildings, the one erected upon the other.

In Nature there are many instances of pillars supporting different floors. One of the most beautiful examples is to be seen in the common Cuttle-bone, as it is called, this being the internal skeleton, if it may be so termed, of the common Sepia (*Sepia officinalis*), which is so often found on our coasts, especially after a gale. This year (1875) I found eight of these Cuttle-bones on the Margate sands, and all within a space of some twelve feet square.

This so-called bone is really composed of the purest chalk, for which reason it is in great request as a dentifrice, being easily scraped to almost impalpable powder when wanted, and not liable to be spilled, as is the case with any ordinary tooth-powder.

It is exceedingly light—so light, indeed, that it floats like a cork, even in fresh water. Now, as chalk is very much heavier than water, we may naturally ask ourselves how this lightness is obtained. If the upper surface be examined, it will be seen to be traversed by a vast number of wavy lines, something like the markings of "watered" silk. These show the lines of demarcation between the multitudinous rows of pillars of which the whole structure is formed.

If the "bone" be sharply snapped in the middle, and the particles of white dust blown away, a wonderful structure presents itself, which can be partially discerned by the naked eye, though a microscope is required to bring out its full beauties.

Even with an ordinary pocket lens we can make out some of its wonders. The object looks like a vast collection of basaltic columns, except that the pillars are white instead of black, and they are arranged in rows with the most perfect accuracy, just as if the place of each had been laid down with rule and compass. They are scarcely thicker than ordinary hairs, but they are beautifully perfect, and rise in tier after tier as if they were parts of a many-storied building. As a definite space exists between the pillars, the reader will understand why the whole structure should be so much lighter than water. In order, however, to see these wonderful pillars in perfection, a very thin section should be taken, and viewed with polarised light.

ANOTHER excellent example of Pillars and Flooring is to be found in the nests of various Wasps, including that of the Hornet.

In these nests the combs are arranged horizontally, and not vertically, like those of the bees, and in consequence they have to be supported in some way. This object is achieved by means of multitudinous pillars made of the same papier-mâché of which the combs are formed, and attached to the successive rows of combs. There is, however, one curious point of difference between the Wasp-comb and human architecture, namely, that the pillars do not support floors, or rest upon them, but sustain the weight of those which hang from them. The mouths of the cells are all downwards, and the combs are therefore suspended from the pillars, instead of being supported by them.

TUNNEL ENTRANCE TO THE DWELLING.

WE have already found occasion to treat of the snow-house, or igloo, of the Esquimaux, and have now to speak of a subsidiary, though necessary, part of Esquimaux architecture.

Perhaps the reader may have been unfortunate enough to travel by rail in the depth of winter, and to be associated with fellow-passengers who will insist on closing every window, even though the carriage be crowded. Suppose that on such a day, the weather being perfectly fine, the train stops at a station, and the guard outside opens the door to see if another passenger can be accommodated with a place.

No sooner is the door opened than a shower of snow at once fills the carriage. This is simply the moisture suspended in the air and generated by human lungs. The rush of cold air at once freezes this moisture and converts it into snow, thus showing those who will condescend to learn, that they have been breathing and re-breathing the air that has passed through a variety of human lungs, and is charged with their different moistures. I have seen the same phenomenon at a dinner party, where, after the withdrawal of the ladies, one of the windows was opened.

Now, in Esquimaux-land, it is absolutely necessary to conserve every atom of heat, for the cold is so intense that if a cask of water be near a coal fire, only the part next the fire will be thawed, the rest being ice. Cold, therefore, is a foe which has to be fought and kept away from the household. Then there are other foes—such as Polar Bears, for instance—

which would be only too glad to get into an igloo and make a meal of its inhabitants. The Esquimaux architect, therefore, avails himself of an ingenious device by which he can set both foes at defiance.

In summer-time he contents himself with a hut made of skins, and merely hangs a skin over the entrance by way of a door. But in the winter, when he is driven to his snow-house

NESTS OF FAIRY MARTIN.
TOWERS OF SAND-WASP.

HUTS OF ESQUIMAUX.

for shelter, he acts in a very different manner. Instead of merely cutting an aperture for a door in the side of the igloo, he constructs a long, low, arched tunnel, so small that no one can enter the igloo except by traversing this tunnel on his hands and knees. Sometimes a number of huts are connected with each other, one or two tunnels leading into the air, and the rest serving merely as passages from one hut to the other.

In Nature are several examples of tunnels constructed on the same principle.

There are, for instance, the curious nests of the Fairy Martin of Southern Australia (*Hirundo Ariel*), which bear a singular resemblance to oil-flasks, the body of the nest being rather globular, and the only entrance being through a tolerably long, tunnel-like neck.

Then there are the various Weaver-birds of Africa, with their long-necked nests. Some of these strange edifices look

almost like horse-pistols suspended by the butt, so round is the
nest, and so long and narrow is the tunnel-like entrance.

PASSING to the insect world, we find the same principle
carried out by the now familiar Mason-wasp (*Odynerus mura-
rius*), some of whose nests are represented in the illustration.

This insect makes a burrow, and at the bottom of it deposits
an egg, together with a number of little caterpillars on which
the grub, when hatched, will feed. The mother Wasp is not
allowed to pursue this task without taking precautions against
the admission of enemies to her burrow, especially the ichneumon-flies. As may be inferred from its popular name, the
Sand-wasp always selects a sandy spot for its burrow, and
generally chooses a piece of tolerably hard sandstone, which it
is able to bite into little pellets, aided by a kind of liquid which
it secretes.

The following account of the manner in which the Mason-
wasp forms and defends its home is taken from the invaluable
"Insect Architecture," by Rennie.

The author begins by describing the form and depth of the
burrow, and the soil in which it is made. He then proceeds to
show the wonderful manner in which the mother Wasp purveys
food for the use of her future young whom she will never see.
Guided by instinct, she places in the burrow exactly the
number of caterpillars which the young Mason-wasp will have
to consume before it attains its perfect condition. It is believed
that she partially paralyzes them with her sting before
placing them in the burrow. At all events, when they are
once packed away, they never move, so that the tiny Wasp grub
can feed upon them quite at its leisure.

Here is Rennie's account of the Sand-wasp and her burrow-
making :—

" When this wasp has detached a few grains of the moistened
sand, it kneads them together into a pellet about the size of
one of the seeds of a gooseberry.

" With the first pellet which it detaches, it lays the foundation of a round tower, as an outwork, immediately over the
mouth of its nest. Every pellet which it afterwards carries off
from the interior is added to the wall of this outer round tower,
which advances in height as the hole in the sand increases in

depth. Every two or three minutes, however, during these operations, it takes a short excursion, for the purpose probably of replenishing its store of fluid wherewith to moisten the sand. Yet so little time is lost, that Réaumur has seen a mason-wasp dig in an hour a hole the length of its body, and at the same time build as much of its round tower.

" For the greater part of its height this round tower is perpendicular, but towards the summit it bends into a curve, corresponding to the bend of the insect's body, which, in all cases of insect architecture, is the model followed. The pellets which form the walls of the tower are not very nicely joined, and numerous vacuities are left between them, giving it the appearance of filigree-work.

" That it should be thus slightly built is not surprising, for it is intended as a temporary structure for protecting the insect while it is excavating its hole, and as a pile of materials, well arranged and ready at hand, for the completion of the interior building,—in the same way that workmen make a regular pile of bricks near the spot where they are going to build. This seems, in fact, to be the main design of the tower, which is taken down as expeditiously as it has been reared.

" Réaumur thinks, that by piling in the sand which has previously been dug out, the wasp intends to guard its progeny for a time from being exposed to the too violent heat of the sun; and he has sometimes even seen that there were not sufficient materials in the tower, in which case the wasp had recourse to the rubbish she had thrown out after the tower was completed. By raising a tower of the materials which she excavates, the wasp produces the same shelter from external heat as a human being would who chose to inhabit a deep cellar of a high house.

" She further protects her progeny from the ichneumon-fly, as the engineer constructs an outwork to render more difficult the approach of an enemy to the citadel. Réaumur has seen this indefatigable enemy of the wasp peep into the mouth of the tower, and then retreat, apparently frightened at the depth of the cell which she was anxious to invade."

It is no wonder that the Sand-wasp should be so anxious to insure the safety of her nest, for her foes are multitudinous. Putting aside the ordinary Ichneumon-flies, we have the predatory Tachinæ, which are always hovering over such nests, and trying to deposit eggs therein.   For many years I have been in the habit of receiving letters from novices in entomology, wanting to know whether I am aware that the common House-fly is in the habit of acting as a parasite.   Of course, the writer has mistaken the Tachina for a house-fly, but I cannot regret the fact that some one has really begun to observe Nature, and not only to read books.

DOORS AND HINGES.

HAVING seen that both in Nature and Art the entrances to dwellings are guarded by tunnel-like approaches, we come naturally to another mode of guarding the entrance, namely, by a door moving on hinges. As to the multitudinous examples of doors and hinges in modern civilisation, we need hardly dis-cuss them, except to show the exact analogies which occur in Art and Nature.

Doors moving on hinges are very plentiful in Nature, even where we should least expect them. Take, for example, an egg, especially the egg of an insect, and we shall see that it is just about the last object in which we should expect to find a hinged door. Yet, if the reader will refer to the illustration on page 7, he will see that the tiny eggs of the common Gnat, numerous as they may be, are each furnished with a door which opens as soon as the inmate is hatched, and allows the little larva to escape into the water.

Another still more remarkable instance of a hinged door in an egg is to be found in one of the Rotifers, or Wheel-Animal-cules, so called because they possess an apparatus of movable cilia, which, when set in motion, looks exactly like a wheel running round and round. As the full-grown creature is barely one thirty-sixth of an inch in total length, the structure of its eggs must be infinitesimally beyond the range of human vision.

Yet, just as the telescope sets at partial defiance the vast spaces that intervene between our earth and her sister planets, so the microscope performs a similar task in the infinitesimally minute.   And, under the all-revealing lens of the microscope, the little egg of the Brachionus, though absolutely invisible to the unaided eye, yields up its secrets.

Fortunately, the shell is so transparent that the interior of the egg can be seen through it as if it were a mere film of glass. The astonishing division and re-division of the yolk take place before our eyes, being divided first into two, then into four, then into eight, then into sixteen, then into thirty-two, and so on, until the whole mass of the yolk is cloven into divisions too numerous to count.

By degrees, the form of the young Brachionus is developed within the egg, even to the very teeth, which work away as persistently as if large stores of food were being passed through them.

When the young is ready to take its place in the world, a new development occurs, which has been well related by Mr. Gosse :—

" All these phenomena have appeared in the egg we are now watching ; and at this moment you see the crystalline little prisoner, writhing and turning impatiently within its prison, striving to burst forth into liberty.

" Now, a crack, like a line of light, shoots round one end of the egg, and in an instant, the anterior third of the egg is forced off, and the wheels of the infant Brachionus are seen rotating as perfectly as if the little creature had had a year's practice.

" Away it glides, the very image of its mother, and swims to some distance before it casts anchor, beginning an independent life. At the moment of escape of the young, the pushed-off lid of the egg resumes its place, and the egg appears nearly whole again, but empty and perfectly hyaline (i.e. all but transparent), with no evidence of its fracture, except a slight interruption of its outline, and a very faint line running across it."

To pass from the egg to a more advanced stage in life. All practical entomologists have been greatly annoyed, in their earlier years of collecting, to lose larva after larva, from the attacks of Ichneumon-flies. It is certainly rather beyond the limits of ordinary patience to discover, watch over, and secure successfully a rare caterpillar, and then to find that it has been " stung " by an Ichneumon-fly.

The veteran entomologist, however, troubles himself very little about such minor misfortunes, and, as a rule, more than compensates for them by preserving the intrusive Ichneumon-fly, and giving in his diary full details of the insect on which it was parasitic, of the plant on which the caterpillar lived, the date of its appearance, and its numbers.

Now, there are many of these parasitic insects, notably those belonging to the genus Microgaster, which invariably make doors in their cocoons. I have now before me groups of cocoons made of the two commonest British species, namely, *Microgaster glomeratus* and *Microgaster alvearius*, and in both of them each tiny cocoon is furnished with a hemispherical, hinged door. I have also some exquisitely beautiful groups of Microgaster cocoons found in the West Indies. They are the purest white, shine with a satiny lustre, and are arranged round a hollow centre, much as if they had been gummed to the outside of a very large thimble. There are many hundreds of them, and every one has its little door still open as it was when the fully developed insect first made its escape.

ANOTHER curious example of a natural door may be seen by those who will look for it.

On plants infested with aphides, or "green blight," as the gardeners quaintly term them, may often be seen dead aphides much larger than the rest, globular, brown, and shining. These aphides have been "stung," as it is called, by a little Ichneumon-fly belonging to the genus Ophion, and having, like all its congeners, a flat and sickle-shaped abdomen. The egg which has been laid in the aphis soon hatches, and the young Ophion absorbs into itself all the juices of the aphis. It remains within the body of its involuntary host until it is fully developed, when it cuts a tiny, but beautifully perfect circular door in the skin, and emerges, leaving the door open and still attached by its little hinge.

Considering the small size of the aphis, and that the diameter of the door is only one-eighth of the length of the insect, the perfection of its form is really remarkable.

ONE of the achievements of modern Architecture is the Self-closing Door, especially where it must of necessity close by its own weight, and when the fitting is so exact, that even the most experienced eye can scarcely detect it. Such a door is to be found guarding the nest of the Trap-door Spiders, several species of which are found scattered over all the warm parts of the earth. A side view of one of these extraordinary nests is

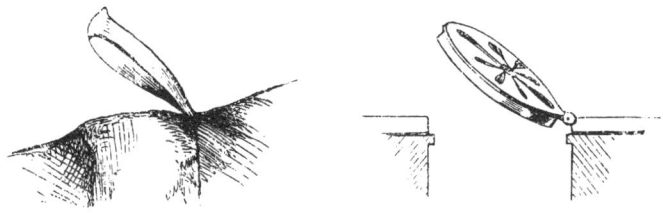

DOOR OF TRAP-DOOR SPIDER.   TRAP-DOOR OF COAL-CELLAR.

given in the accompanying illustration, and on the other side is the common trap-door of our cellars.

The Spiders which make these extraordinary dwellings generally begin by excavating a nearly perpendicular tunnel in the ground. They line it with a silken web, and construct a door which exactly fits the orifice, and which is bevelled so that it shall not sink too far, and thus betray itself. I have seen and handled one, where the burrow had been sunk among lichens and mosses, and the trap-door of the nest had been most ingeniously covered with the same growths. Although the surface of the slab of earth in which the nest was made is only a few inches square, it is almost impossible to detect the entrance, so admirably do the mosses on the door correspond with those outside it.

Almost invariably the nest is sunk in the ground, but I have a specimen sent to me from India, in which the Spider must absolutely have carried the clay to a fluted pillar, burrowed in it, and then made its beautiful habitation. The nest and its inhabitant were sent to me by an officer in the 108th Regiment, accompanied by the following letter :—

" The packet contains a large Spider and the upper portion of its peculiar nest, the history of which is as follows.

" On the thirtieth of last month (September, 1870), while searching for caterpillars on a bush growing close to one of the pillars of my verandah, which is a very low one, reaching to within a foot of the ground, I saw in part of the chunam masonry at the foot of the pillar what I at first sight took to be a couple of seeds sticking to a stone. On trying to pull one off, I found that it came up with ease, bringing with it what I thought was the stone.

"But I had scarcely got it up when it was smartly pulled back. This excited my curiosity, and I raised it again with a little force. I now saw, to my wonder and admiration, that what I had fancied was a stone was a small circular door with a pretty broad hinge, made all of silk; and then distinctly observed a large black spider dart down the hole to which the above door gave an entrance. But, not knowing the depth, I broke it.

"This piece I send to you, together with its original owner, who, at the beginning of my digging operations, ran up suddenly, shut the door in my face, and hung on to it like grim death when I tried to reopen it. He soon came away with the upper piece, still keeping the door resolutely closed."

ARCHITECTURE.

CHAPTER II.

WE now come to the Walls of the house, in which there is more variety than might be imagined.

Take, for example, our modern houses of the "villa" type. They are nothing but the merest shells, made of the flimsiest imaginable materials. Some years ago, while walking through a suburb where some very showy houses were being built, I amused myself by going over them and testing them. There was scarcely a room in which I could not thrust an ordinary walking-stick through the wall. When they were "finished" and "pointed," the houses looked beautiful, but their heat in summer, cold in winter, and moisture in wet weather, can easily be imagined, especially as the sand with which the mortar was mixed had been procured from the banks of a tidal river.

There is not the least necessity for such buildings. It is absurd to run up such edifices as that, and then charge £120 per annum for rent. The whole system is as rotten as the houses, and there is nothing but prejudice and trade-unionism to pre-vent our houses being cool in summer, warm in winter, and dry in all weathers.

It is well known that air is practically a non-conductor of
heat, and that therefore a layer of air between two very slight
walls is just as warm as if the wall had been made of solid
stone.   Now, there are several inventions whereby the present
brick could be made half its present weight, twice its present

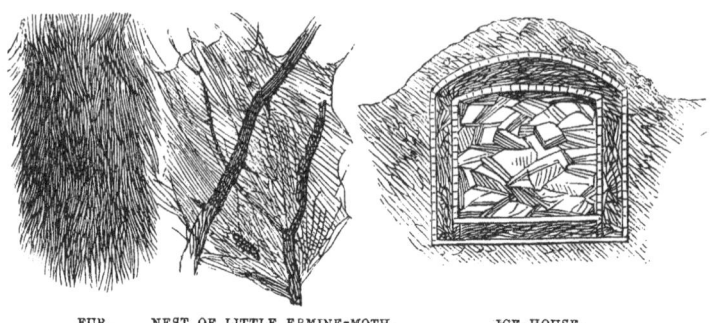

FUR.   NEST OF LITTLE ERMINE-MOTH.   ICE-HOUSE.

strength, hard and smooth as earthenware, so that it could not
absorb water like our common brick, and pierced with holes
through which air could pass.

Unfortunately, however, there is a stringent rule among brick-
makers and bricklayers that they are to play into each other's
hands, and that no bricklayer is to touch a brick which has not
been made in some definite district.   Should he do so, he is a
marked man, and will stand but little chance of getting even
a day's work.

The power of the double wall may be seen in many ways.
For example, in the old days of coaching, when one had to pass
hour after hour on the roof of the coach, it was known by
practical experience that double body linen, and two pairs of
stockings, worn one over the other, formed the best preparation
for the journey.   The reason was, that air became entangled
between the layers of fabric, and acted as a non-conductor of
heat.

Another mode of utilising the principle of the double wall is
seen in the refrigerators which add so much to the comfort of
the household in a hot summer.   The one principle of these
refrigerators is, to keep a layer of air between the ice and the
surrounding atmosphere.   The same principle may be used in a
reverse way, and heat be preserved instead of repelled.   Those

cooking-pots are now well known, where half-cooked meat can be inserted in the morning, and at luncheon-time be turned out quite hot and perfectly cooked. The fact is, that the vessels in question are covered with a very thick layer of felt. The felt, however, is only a device for entangling air, and a double wall would answer the purpose as well, if not better.

The now well-known fire-resisting safes are made on this principle, and after they have been for hours in a raging fire, and the outer case has become red-hot, the interior is quite safe, the papers uninjured, and even a watch continuing to go.

Then there is the ordinary Ice-house, a sketch of which is given in the illustration. A pit is first dug in the ground, and thickly lined with dry branches, straw, &c. The roof is constructed in the same manner, only the non-conducting power is increased by a thick coating of earth over the sticks and straw. The door, which is approached by a shelving cutting, is similarly protected, the covering only being removed when the door is opened.

I once made a very effective refrigerator out of two hampers, putting a small hamper inside a large one, and packing the space between them with straw.

In Nature we find many examples of this principle, which enables the inhabitants to bid defiance to frost.

A familiar example may be found in the cocoon of the common Silk-worm (*Bombyx mori*), and indeed in that of almost any silk-producing insect. When the caterpillar is about to make its cocoon, it begins by a number of rather strong threads attached to different points, and making a sort of scaffolding, so to speak, for the cocoon itself. Upon these is spun a slight outer cocoon of very loose and vague texture—the "floss silk" of commerce, and within that is the cocoon proper, in which the insect lies enclosed. It will be seen, therefore, that there are really three cocoons, one within the other, namely, the scaffold cocoon, the floss cocoon, and the silk cocoon itself, so that the inmate is protected from variations of temperature.

The cocoon of the emperor-moth, which has already been described, is made on the same principle.

There are several caterpillars which are social in their early stages, and which construct a common habitation. The Little

Ermine-moth (*Hyponomeuta padella*) affords a familiar example of this structure. The caterpillars are great roamers in search of food by day, and travel from branch to branch on their strong silken threads. At night, however, they return to a large white silken habitation which they have spun, and which they divide into many compartments, as may easily be seen by cutting the nest open with very sharp scissors. Within this habitation the caterpillars spin their separate cocoons, so that the system of double walls is thoroughly carried out.

There is another insect, very common on the Continent, but, happily for us, not introduced into England. It is called the Processionary Moth, from its curious habit of marching in exact lines, the head of the second caterpillar touching the tail of the first, and so on. These insects have likewise a common home, and spin their own separate cocoons within it.

THERE are two other sociable British Moths which make nests on a similar principle. These are the Gold-tailed Moth (*Porthesia chrysorrhœa*) and the Brown-tailed Moth (*Porthesia auriflua*). They are both beautifully white insects, but may

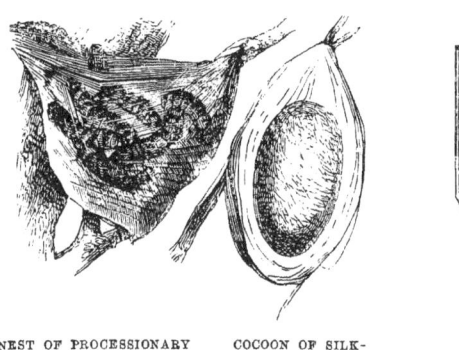 

NEST OF PROCESSIONARY   COCOON OF SILK-   FIREPROOF SAFE.
MOTH.   WORM.

easily be distinguished from each other, the Gold-tailed Moth having some brown-black spots on the upper wings, and a tuft of golden-yellow hairs at the end of the body; while the Brown-tailed Moth is without spots, and the tail-tuft is brown.

In habits they are very similar, and the description of the nest made by one will answer for that made by the other. I believe that broods of these two species have been known to construct a common nest. The nest is extremely variable in form, because it depends much on the number of twigs which it includes. Interiorly, it is divided into a considerable number of chambers, each containing one or several individuals.

As the caterpillars are hatched late in summer, they have to undergo the frosts of winter before they can attain their perfect state. Accordingly, before the winter-time comes on, they strengthen both the external walls and internal partitions of their nest, and then wait until the spring brings forth the leafage of the new year.

The nest is a beautiful structure, and I strongly recommend the reader to look for one in a hedgerow, take it home, and cut it up carefully. I would, however, advise him, if, like myself, he be subjected to a very sensitive skin, to be cautious in his handling of the nest. The hairs with which the pretty black, red, and white caterpillars are studded are irritant in the extreme.

I have several times suffered from them, and would much rather be severely stung by nettles than undergo the fierce irritation, mixed with dull heavy pain, which always accompanies the presence of these hairs. With me, as I suppose would be the case with persons of similar organization, these hairs cause large, hard tubercles to rise, just as if potatoes had been placed under the skin. The hairs of the Processionary Caterpillar have a similar effect, and in France the authorities have several times been obliged to close the public gardens for months, so severe was the pain which the caterpillars inflicted on persons who passed through the spots infested by them.

MUD WALLS.

THERE is a mode of wall-building which is much in vogue in some parts of England, and has much to commend itself. This is the Mud or Concrete Wall.

At first sight, the very name of a mud house gives an idea of poverty and misery, and is apt to be connected with hovels and pigsties. Mud walls, however, if properly built, are far warmer and drier than those of brick, and are even preferred to those of stone, when the latter can be easily and cheaply obtained. In Devonshire, for example, where even the cattle-sheds, or "linhays" (pronounced *linny*), and the pigsties are

made of the rich red stone of the county, it is a common thing
to see village houses built of mud. Sometimes the houses are
built of stone to the height of some ten or twelve feet, and the
upper parts made of mud.

If the builders are in any way fastidious, they make their
walls of a uniform surface by placing two rows of planks on

NEST OF TERMITE.   MUD WALL.

their edges at a distance from each other proportionate to the
thickness of the wall, pouring the mud between them, and,
when it has sufficiently hardened, shifting the planks. This,
however, is not necessary, and detracts much from the picturesque look of a genuine mud wall, especially when it is of
that rich red which characterizes the Devonshire soil. These
mud walls are locally known by the name of Cob.

WE have not to go very far in Nature to find good examples
of the strength which can be attained by mud walls.

In all parts of the world where Termites, popularly but
wrongly called White Ants, are to be found, the strength and
endurance of the mud wall can easily be tested. Of gigantic
dimensions when compared with the size of the architect, they
not only endure the rain-torrents which wash over them, but
can sustain the weight of the wild cattle, which are in the habit
of using them as watch-towers, and this although they are
hollow, and filled with chambers and galleries.

In Southern Africa these nests are much utilised. There is
an animal called by the Dutch settlers the Aard-vark, which

feeds almost wholly on Termites. At night it issues from its burrow, and, being armed with large and powerful claws, tears a great hole in the side, and devours the inmates.

These deserted nests are sometimes used as ovens, as we have already seen, a fire having been kindled within them for some time, the meat, well enveloped in leaves, being thrust into them, and the opening closed with clay. Sometimes they are used as graves, the corpse being placed in them, and the hollow filled up with earth, while the wall of the Termite nest, when pounded and mixed with water, is found to be the most tenacious clay that can be used for building or flooring huts.

PORCHES, EAVES, AND WINDOWS.

WE now come to some of the appendages of a house, namely, the Porch by which the rain is kept from a doorway, the Eaves by which it is kept from the walls, and the Windows which will admit light and air, but will prevent the entrance of intruders.

We first take the Porch, two examples of which are shown in the accompanying illustration, one being the work of human hands, and the other that of an insect.

The figure on the right hand represents an old-fashioned Porch, such as is often to be seen attached to old village churches, and which, being furnished with seats, serves also as a resting-place for those who are weary.

NESTS OF MYRAPETRA, WITH PORCHES.   PORCH.

The figure on the left hand of the illustration is a wonderful example of the Porch, as constructed by insects. It is the nest of a honey-making Brazilian wasp named *Myrapetra scutellaris.* The peculiarity of this nest consists in its exterior being covered with a vast number of projections made of the same material as the walls of the nest, but more solid and much harder. The colour of the nest is blackish brown.

The object of all these projections has not been ascertained, but there is no difficulty as regards some of them. Without a very careful examination, it is exceedingly difficult to see any opening by which the inhabitants of the nest can go in and out. It will be found, however, that there are many entrances, which are set in a row round the nest, each opening being situated under a projection, which thus performs the office of a porch as well as that of concealment.

Another hymenopterous insect carries out the principle of the Porch in its nest. This is the *Myrmica Kirbyi,* a tiny reddish Ant which inhabits India. It makes its nest of cow-dung, which it works up into a texture very like that of an ordinary wasp-nest. A series of large flakes of this substance overhang the entrances, so that the inhabitants can enter freely, while rain is kept out. For the purpose of greater security, one very large flake covers the roof in umbrella fashion. The whole nest is globular, and about eight inches in diameter.

NEXT we come to the projecting Eaves, like those of our houses, and serving to preserve the body of the edifice itself from wet. On the right hand of the illustration there is an example of the eaves as they are still to be seen in some of our country places, where the less picturesque slates have not yet superseded the old thatch. In some places these eaves extend considerably beyond the walls, and I know of several instances, especially in North Devon, where a supplementary set of eaves extends, like a penthouse, throughout the length of the building, and just above the windows of the ground-floor.

The reader will remark that the projections upon the Myrapetra's nest may very well fulfil the office of eaves as that of porches, and not only shelter the entrances, but serve to shoot the wet off the walls of the nest.

ON the left hand of the illustration are several instances of eaves as existing in Nature.

In the centre is the compound nest of the Sociable Weaver-bird of Southern Africa (*Philetærus socius*).

This is a dwelling constructed very much after the fashion adopted by many hymenopterous insects, namely, that each pair of birds make their own individual nest, but unite with their

DWELLING OF SOCIABLE WEAVER-BIRD, WITH THATCH.   THATCHED HOUSE.
THATCHED EGGS OF GOLD-TAILED MOTH.

companions in constructing a common roof or covering. More than three hundred nests have been found in a single habitation, and sometimes the birds miscalculate, or rather, do not calculate the resisting power of the branches, and, when the rainy season comes, the additional weight of water brings down the whole edifice with a great crash.

The thatch which covers this congeries of nests is made of the Booschmannees-grass, whose long leaves and tough wiry stems are admirably adapted for throwing off water, even though they be not bound together like our more regularly constructed thatch.

Perhaps the reader may be aware that in the Orang-outan, the Chimpansee, and other large apes, the hairs of the arms are very long, and point in different directions, so that if the creature should be caught in a rain-storm, and, after the manner of its kind, fold its arms on its breast, with the hands resting on the shoulders, the rain is shot clear of its body, the hairs performing the duty of eaves.

Both Japan and China have a rain-cloak, constructed on

exactly the same principle as the thatch of the Sociable Weaver-bird. They are nothing more than successive rows of long grass-blades fastened to a network of the proper shape. No amount of rain or snow can wet them through, and they have the advantage of being pervious to the exhalations of the body, though impervious to external moisture.

In this respect they are greatly superior to our waterproof coats, for, if the wearer has to undergo much bodily exertion, or is obliged to wear it for any length of time, he finds his clothing nearly if not quite as wet as if he had allowed the pure rain to fall on him from the clouds. I possess specimens of each kind of cloak.

When I procured them they were quite blackened with London smoke, and, on account of their resistance to water, washing them was a very long and troublesome business.

Above the nest are two patches of the Booschmannees-grass, as they appear when laid by the bird.

BELOW the nest is a group of the eggs of the Gold-tailed Moth, whose nest has already been described. Perhaps the reader wonders where the eggs are. Owing to the mode in which they are arranged, only a few can be seen, and are represented by the little white spots in the lower part of the figure. When the Gold-tailed Moth is ready for the great business of laying her eggs, she seeks a suitable place, and then piles them up in the form of a shallow cone. Her task, however, is not yet finished. Having arranged her eggs, she scrapes off the long downy hairs of the tail-tuft, and arranges them carefully on the eggs so as to cover them with a conical thatch, very much resembling that of an ordinary corn-rick.

The Brown-tailed Moth acts in a similar fashion.

Furs of various kinds act in the same manner, being impervious to wet during the life of the animal. Such, for example, is the fur of the Beaver, that of the Capybara, and that of the Seal, which are animals living in our time. These, however, are exceeded in their thatch-like powers by the three successive coatings of hair that were worn by the ancient Mammoth, the outermost being very long and very coarse, and hanging down in heavy tufts so as to shoot the water from them.

BEING on the subject of roofs, we will take a few more examples of the roof as anticipated in Nature.

That parallel fibres, whether animal or vegetable, can throw off rain when properly arranged, has already been shown. Much more is it evident that flat or partly flat plates will have

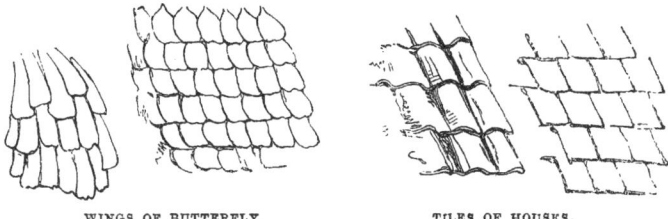

WINGS OF BUTTERFLY.　　　　　　TILES OF HOUSES.

the same effect, if they be arranged so that the joints are "broken," as masons and bricklayers say, *i.e.* so that the broad part of the upper row of plates overlaps the junction of two of the plates in the row immediately below it.

ON the right hand of the accompanying illustration are given two sketches of a modern roof, one slated and the other tiled. The figures on the left show that this formation has been anticipated by Nature, in the wonderful system of scales which cover the wings of butterflies and moths, and to which all their brilliancy of colour is owing. In spite of their minute size, most being too small to be distinguished by the unaided eye, they are arranged as regularly as the best workman could lay the slates or tiles on a roof, and on exactly the same principle.

The shapes of these scales vary in almost every species, but they are always arranged on the same plan, namely, being placed in successive rows, each overlapping the other.

In consequence, it is almost impossible to wet a butterfly's wing with water. The insect may be plunged beneath the surface, and the long hairs of the body will be soaked and cling together in a very miserable fashion. But the water rolls off the wings like rain off a slated roof, and even if a few drops remain on the surface, they can be shaken off, and the wing will be perfectly dry.

Mostly these scales are flat, but sometimes they are curved.

I have among my microscopic objects a piece of wing from a South American butterfly, the scales of which are oblong and bent, just like the curved tiles shown in the second right-hand figure of the illustration. These beautiful scales are deep azure or warm brown, according to the direction of the light.

Perhaps my readers may call to mind that some architects dislike the flat, square form in which slates are usually put on roofs, and try to make them less formal.

Sometimes they take their square slates, and fit them with one of the angles uppermost, so that each slate looks something like the ace of diamonds in a pack of cards. Sometimes they are still more ambitious, and certainly succeed in producing a

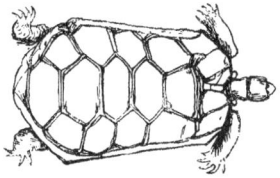

SHELL OF TORTOISE.

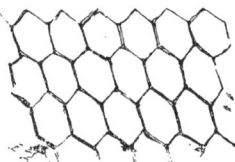

HEXAGONALLY TILED ROOF.

better effect, by cutting the slates in hexagons instead of squares, and fixing them as shown in the right-hand figure of the illustration. Putting aside the familiar hexagons of the honeycomb, and the apparent hexagons of an insect's compound eye, we have in the common Tortoise an example of hexagonal plates that exactly resembles the slate roofing.

In the next illustration we have a variety of the same principle exhibited in differently shaped tiles and scales. The figures on the right hand show the pointed, the square, and the oblong tiles. These also would answer very well as representations of different forms of scale armour, the one being intended to throw off rain, and the other to repel weapons.

On the other side of the illustration are examples taken from the animal kingdom. First comes the Bajjerkeit, or Short-tailed Manis, which has already been mentioned, and whose imbricated scales will resist the blows of any spear or sword. As to my own specimen, when it is struck, it resounds as if it were a solid plate of metal, and I should think that during the lifetime of the animal a reasonably strong axe would not easily make its way through that coat of mail.

Below the Manis are a pair of fish, whose scales, though not so strong as those of the mammal, yet are arranged in the same

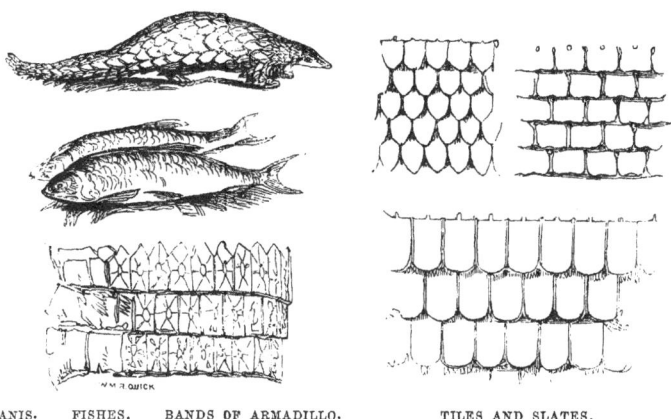

MANIS.   FISHES.   BANDS OF ARMADILLO.   TILES AND SLATES.

manner, and answer the same purpose. The last figure represents three scale-bands of the Armadillo, an animal which has already been mentioned. I may as well state here that in several anthropological museums there are various portions of defensive armour made from the scale-clad skin of the Crocodile, Manis, and similar animals.

ARCHITECTURE.

CHAPTER III.

THE WINDOW.—GIRDERS, TIES, AND BUTTRESSES.—THE
TUNNEL.—THE SUSPENSION-BRIDGE.

The Window, and its Modifications according to Climate.—Bars and Tracery.—
The Wheel-window and the Caddis.—Curious Structure of the Caddis-tube.
—Object of its Window.—The Girder as applied to Architecture.—The
Radius and Ulna.—The Tie as applied to Architecture, and its Value.—Com-
bination of the Tie and Girder.—Structure of the Crystal Palace.—Leaf of
the Victoria Regia.—A Gardener turned Architect.—The Buttress in Art
and Nature.—The Tunnel used as a Passage of Communication.—Natural
Tunnel of the Ship-worm.—The Thames Tunnel.—The Piddock, or Pholas.—
The Driver-ant.—The Suspension-bridge.—The Palm-wine Maker and his
Bridge.—Suspension-bridges of Borneo and South America.—The Creepers
and the Monkey Tribes.—The Spider and Little Ermine Caterpillar.

THE WINDOW.

HAVING traced, though but superficially, the chief parts of
a building, such as the walls, the door which is opened
through the walls, and the roof which shelters them, we
naturally come to the Windows by which light is admitted to
them, and enemies excluded.

There are, perhaps, few points in Architecture in which such
changes have been made as in the Window, which, instead of
being a difficulty in the way of the architect, is now valued as
a means of increasing the beauty of the building. Taking for
example even such advanced specimens of Architecture as those
furnished by Egypt, Greece, and Rome, we find that the
Window is either absent altogether, its place being supplied by
a hole in the roof, or that, when it is present, it was made
quite subordinate to the pillars and similar ornaments of the
building.

This fact is, perhaps, greatly owing to the influence of climate.
In the parts of the world which have been mentioned in con-

nection with this subject, light and heat appear to be rather
enemies than friends, and the object of the architect was to
enable the inhabitants of his houses to avoid rather than to
welcome both. Consequently, the Windows were comparatively
insignificant. They were not needed for the purposes of light
or air, those being generally furnished by the aperture in the
roof, and consequently were kept out of sight as much as
possible.

But when architects had to build for a sterner, a colder, and
a darker clime, where the sun never assumed that almost
devouring heat and light which in hot countries drive the
inhabitants to invent endless devices for obtaining coolness and
shade, a different style of Architecture sprang up. In this the
Window became nearly the most prominent part of the building:
the elements were excluded by glass instead of stone, and the
principal modifications of light were obtained by staining the
glass in various rich colours. Perhaps the Window has
attained its culminating point in the Crystal Palace, which is all
window except its foundations.

Partly in order to enable the glass to be inserted, and partly
to increase the beauty of the building, and to avoid the mean
appearance of Windows filled in with plain iron bars crossing

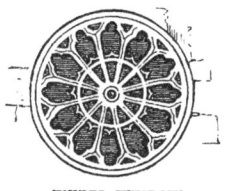

CADDIS GRATING. WHEEL-WINDOW.

each other at right angles, the interior of the Windows was
adorned with stone " tracery," varying much according to the
epoch of the building.

One of the most beautiful forms of the Window is that
which is called the Wheel. The window itself is circular, and
the tracery is disposed so as to bear an exact resemblance to an
ornamental wheel, the lines of the tracery running from the
circumference to the centre, just like the spokes of a wheel.
One of these Wheel-windows is shown on the right hand of the
illustration.

On the other side is an object, which at a hasty glance
might be taken for another Window of the same character. It
is, however, the work of an insect, and not of man, and is
magnified in order to show its structure better.

Any of my readers who may happen to be entomologists or
anglers, or both, are familiar with the Caddis-worm of our fresh
waters. Most of us know that the Caddis is the grub or larva
of the Stone-fly (*Phryganea*), an insect haunting the water-
side, and so moth-like in its general aspect that many persons
think that it is really a brown moth. The changes or meta-
morphoses of these insects are well worthy of notice.

In one respect the Caddis resembles the larva of the Wax-
moth, mentioned on page 151, inasmuch as it has a soft, defence-
less body, while the first three segments are comparatively
hard. Like the Wax-moth also, the Caddis lives in a tube
constructed by itself. Instead, however, of having a long and
fixed tube, up and down which it can pass at pleasure, the
Caddis makes a tube only a little longer than its body, and
light enough to be carried about, just as the hermit-crab carries
its supplementary shell. There are many species of Caddis-fly.

The Caddis inhabits fresh waters, and cares nothing whether
they be ponds or running streams. In order to defend its
white, plump, and helpless body from the fishes and other
enemies, it constructs a tube around its body, strengthening it
by a wonderful variety of material according to the locality.

Mostly the tubes are covered with little pieces of stick or
grass, or leaves, while some species use nothing but sand-grains,
constructing with them a tube very much resembling in shape
an elephant's tusk, and reminding the conchologist of the
dentalium shell. But they seem to use almost anything that
comes to hand. Taking only examples found by myself in a
single pond, these cases are formed of sand, stones, sticks, grass-
stems, leaves, shells of small water-snails, mostly the flat
planorbis, the opercula of the water-snail, empty mussel-shells,
a chrysalis of some moth which had evidently been blown into the
water from an overhanging tree, and acorn-cups. The larva,
however, does not seem to be able to fasten together any objects
with smooth surfaces, and though it has been known, when in
captivity, to make its cases out of gold-dust or broken glass, it
could not use either material when in the form of beads.

When it is full-fed, and about to enter the pupal state, it proceeds to prepare its habitation. As a larva, when it desired to feed, it protruded its head and the front of its body from the mouth of the tube, and then crawled about in search of nourishment, dragging the tube with it, and holding it firmly by means of the claspers with which the end of the body is furnished. But when it becomes a pupa it is no longer able to defend itself, and is instinctively compelled to secure its safety in some peculiar manner.

It cannot fasten up the entrance entirely, because it would not be able to breathe unless water could pass over its body. Accordingly, it constructs a grated window precisely like those of the old castles, so that water can pass freely, while no enemy can gain admittance. Unlike, however, the grated windows of the castle, which had no pretence to beauty, the Caddis always constructs its barriers in some definite pattern. Each species appears to have its own peculiar pattern, but all agree in making their window, if we may so call it, exactly like a wheel-window before the glass is inserted.

When the pupa is about to make its final change into the perfect form, it cuts away the tracery with a pair of sharp jaws, with which it is furnished for this sole purpose, emerges from the water, throws off the pupa-skin, and issues forth as a Stone-fly.

GIRDERS, TIES, AND BUTTRESSES.

NEXT in order come the means by which walls are supported internally by Girders and Ties, and externally by Buttresses.

OF late years the Girder, in its many varieties, has come into general use, especially in the construction of railway bridges and similar edifices.

RADIUS AND ULNA OF HUMAN ARM.   GIRDER (FROM A HOUSE IN BERMONDSEY).

On the right of the accompanying illustration is shown the Girder in its simplest form. The figure was taken from a Girder which is used in supporting the walls of a large

building in Bermondsey. Sometimes a transverse stay connects the centres of the two curved beams ; but it is seldom
needed.

The reader will see that if the interval between the curved
beams were to be filled up, we should obtain a form very like
that of the engine beam described in page 25 ; while, if we
could imagine two such girders intersecting each other at right
angles throughout their length, a section of the two would
exactly resemble the section of the engine beam as given in
the uppermost figure in page 25.

In the human body there are four admirable examples of
the natural Girder, namely, in the bones of the arms and legs.

On the left hand of the illustration are shown the two bones
of the fore-arm, technically named the " radius " and " ulna."
It will be seen that these bones are arranged on the principle
of the girder. In men who are especially powerful of grasp,
it has been noticed that the curve of the radius and ulna has
been exceptionally bold, while we have it developed to the
greatest extent in the fore-arm of the Gorilla, an animal whose
arms are simply gigantic.

The two bones of the legs, from the knee to the ankle, are
arranged in a similar manner, and are called the " tibia " and
" fibula." The last named signifies a brooch, and is given to
the bone because it is very slender, nearly straight, and when
in its place bears no small resemblance to the pin of the fibula,
or ancient Roman brooch.

Nature, however, has exceeded Art in her girder. Those of
man's manufacture can only exert their strength in one direction, and would be of little use if force were to be applied to
them in any other direction. Those of the human body, however, have the capability of partial revolution on each other at
their points of junction, thus enabling the Girder to apportion
its strength according to the direction of the resistance which
it has to overcome.

We now come to the Ties, *i.e.* those internal beams, whether
of metal, wood, stone, or brick, which prevent walls from falling
outwards. There is no danger of the walls falling inward, but
there is very great danger of their falling outward, especially

when the weight or "thrust" of the roof tends to force them apart.

In some buildings, such as an old country church which I attended for many years, the architect had openly acknowledged the tendency of the walls to fall outward, and had counteracted it by a series of great beams extending completely across the nave and aisle. As he had not even troubled himself to hide their office, so he did not trouble himself to conceal the fact that they were tree-trunks, but left them roughly squared with the axe, lest, if he had squared them throughout their length, he should have diminished their strength.

The effect of the partially squared beam is, of course, far more picturesque than that of a completely squared one. The architect, however, need not have been so careful about strength, for if the beams had been only half their diameter they would have been just as effective. The strain on them is by pulling, and not by pushing. Now, as any one can see by trying the experiment with a splinter of wood—say a lucifer-match—an enormous power is required to break it by tearing the ends asunder, while it can be easily broken by pushing them towards each other.

But for this power of resistance, we should never have had our Crystal Palace. That apparently intricate, but really simple (and the more beautiful for its simplicity), intersection of beams and lines diminishing in the distance to the thickness of spiders' webs, is nothing more than a combination of the Girder and Tie, the two together combining lightness and strength in a marvellous manner.

The story of the Crystal Palace is now so well known that it need not be repeated in detail. A vast building was required for the Exhibition of 1851, and not an architect was able to supply a plan which did not exhibit some defect which would make the building almost useless.

Suddenly a Mr. Paxton, who was a gardener, and not an architect, produced (on a sheet of blotting-paper) a rough plan of a building on a totally new principle, and not only fulfilling all the requisite conditions, but being capable of extension in any direction and to any amount. There have been very few bolder conceptions than that of making iron and glass take the place of brick, stone, and timber, and the result fully justified the expectations even of the inventor.

How a gardener suddenly developed into an architect remains to be seen; and, indeed, in this case the architecture was the result of the gardening, or rather, of practical botany applied to art. Some years before the invention of the Crystal Palace, that magnificent plant, the Victoria Regia, had been introduced into England. Its enormous leaves, with their wonderful power of flotation, caused a great stir at the time, and some of my readers may remember a sketch which was engraved in the *Illustrated London News*, and which represented a little girl standing on one of these leaves as it floated on the water.

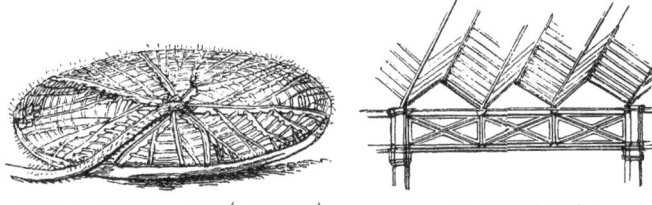

LEAF OF VICTORIA REGIA (REVERSED).   CRYSTAL PALACE.

Mr. Paxton saw how this power was obtained, and the result was that he copied in iron the lines of the vegetable cellular structure which gave such strength to the Victoria Regia leaf, and became more eminent as an architect than he had been as a gardener. The capabilities of the Crystal Palace had lain latent for centuries, but the generalising eye of genius was needed to detect it. A thousand men might have seen the Victoria Regia leaf, and not thought very much of it; but the right man came at the right time, the most wonderful building in the world sprang up like the creation of a fairy dream, and the obscure gardener became Sir Joseph Paxton.

I have no doubt that thousands of similar revelations are at present hidden in Nature, awaiting the eye of their revealer.

Now we come to the principle of the Buttress, *i.e.* giving support to the exterior, instead of the interior, and strengthening the walls by pushing them together, instead of pulling them together.

Putting aside the "flying" buttress, which is simply one buttress mounted on another to support the clerestory walls, the structure of the ordinary buttress is simple enough.

The most primitive form of the buttress is often found in country farms, where the farmer sees the walls of his barns and outhouses leaning suspiciously on one side, and, instead of going to the root of things, props them up by a stout pole or beam.

This, however, can be nothing but a temporary arrangement, especially as beams have a tendency to rot, and their ends to sink into the earth by the gradual pressure of the wall. The genuine buttress was therefore evolved, the basal part being very thick and heavy, and the upper part comparatively thin and slight. Simple as a buttress looks, much skill is needed in making it, and if it be not rightly built, it does infinitely more harm than good.

A case in point occurs within a short distance of my house. The walls of an ancient edifice having shown symptoms of yielding, and some ominous cracks made their appearance, a couple of very sturdy buttresses had been erected, in order to stop further damage. Unfortunately, the builder was ignorant

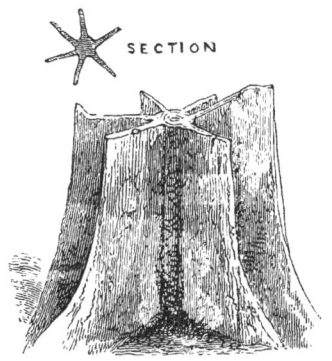

PADDLE-WOOD TREE.

BUTTRESSES.

of the principles of architecture, and though he made the buttresses very strong and massive, he omitted to make a solid foundation on which their bases should rest. Consequently he only hung the buttresses, so to speak, on the wall, and helped to tear it asunder by the additional weight.

NATURE, as well as Art, supplies her buttresses. In our own country we find the natural buttress more or less developed in our trees, as it is wanted.

Take, for example, any plantation, and examine the trees. It will be found that those in the centre, which are sheltered on all sides from the force of the wind, shoot up straight towards the light, have comparatively slight and slender stems, and occasionally display such energy in forcing themselves upwards, that when two branches find that there is not room for both, they form a sort of alliance, fuse themselves together, and force their united way towards the sky.

Take, however, the trees in the outside rows of the plantation, and see how they throw out their straight roots and branches towards the outside, and how, on the inside, their trunks are as smooth and their roots as little visible as those of the trees that grow in the centre of the plantation.

Almost any tree will develop itself in this fashion, showing that instinct can rule the vegetable as well as the animal world.

There is, however, a South American tree which far surpasses any of our trees in its power of throwing out spurs or buttresses, principally, I presume, because it may have to endure the fiercest storms from any quarter and at any time. So bold are these projections that several men would be hidden if standing between two of them, and so numerous are they that if a section of the tree were taken at the base of the ground, it would resemble a conventional star or asterisk, *, rather than an ordinary tree-trunk, O.

The scientific name of this curious tree is *Aspidomorpha excelsum*.

The natural buttresses are so thin and so wide that they look like large planks set on end, with one edge against the tree. Indeed, they are used as planks, nothing more being required than to cut them from the tree.

This is very easy, as, while the wood is green, it is so soft that a blow from a " machete," or native cutlass, is sufficient to separate it. With the same instrument the native makes these flat planks into paddles for his canoe, the soft wood yielding readily even to the imperfect edge of the rude tool. When the wood dries, it becomes very hard, light, and singularly elastic, all these properties qualifying it for its object. I have several of these paddles in my collection. They are much prized by the natives, and are always stained in various patterns with red and black dyes.

In consequence of the use which is made of this tree, it goes by the popular name of " paddle-wood."

THE TUNNEL USED AS A PASSAGE.

As to this division of the subject, I have not been quite sure where it should be placed, but think the present position a tolerably appropriate one.

We have already, in the igloo of the Esquimaux and the winter dwelling of the seal, found examples of the Tunnel

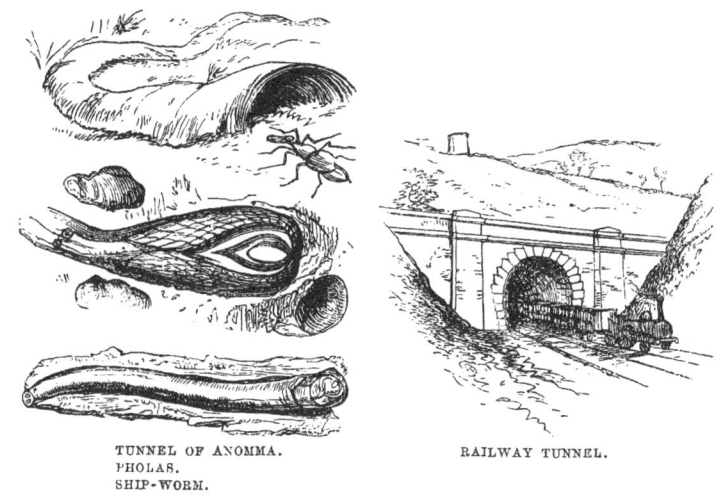

TUNNEL OF ANOMMA.
PHOLAS.
SHIP-WORM.

RAILWAY TUNNEL.

when used as an appendage to the houses and a means of security. We now come to the Tunnel as affording the means of locomotion.

Take, for example, our own railway system. Had it not been for the power of tunnelling, the railway would have lost nearly its value, for it would have been restricted to local districts, and could not have penetrated, as it now does, to all parts of the country, without reference to hill, dale, or level ground. Our present system of engineering has wonderfully developed the capability of tunnelling. In former times it was thought a most wonderful feat to drive a tunnel under the Thames, while in these days the tunnel through Mont Cenis has been completed, and we are hoping to make a submarine tunnel from England to France.

In Nature we can find many examples of Tunnels used for similar purposes. The silken tunnel of the Wax-moth larva has already been mentioned, and we now come to Tunnels where earth in some form, and not silk, is the material of which they are constructed.

The lowermost figure on the left-hand side of the illustration represents that well-known and most destructive burrower, the Ship-worm (*Teredo*), which, by the way, in spite of its popular name, is not a worm, but a mollusc. This creature has a peculiar interest for engineering, inasmuch as its mode of working gave Brunel the first idea of subaquatic tunnelling in loose, sandy soil, just as the Victoria Regia leaf gave to Paxton the idea which afterwards developed into the Crystal Palace.

The plan adopted by the Ship-worm is at the same time simple and effective. It feeds upon wood, and gradually eats its way through almost any timber that may be submerged. It does not, however, merely bore its way through the timber, but lines its burrow with a coating of hard, shelly material. Taking this hint, Brunel proceeded in the same fashion to drive his tunnel through the very ungrateful soils which form the bed of the Thames.

He built a "shield," as he called it, of iron, exactly fitting the tunnel, and divided into a number of compartments, each of which could be pushed forwards independently of the others. In each compartment was a single workman, and, as he excavated the earth in front of him, he pushed forward his portion of the shield, while the interior was cased with brick-work, just as a Teredo tunnel is cased with shell.

Above the Teredo is represented another marine tunnel-maker, as it appears in its burrow.

This is the mollusc popularly known as the Piddock, and scientifically as *Pholas dactylus*. It may be found abundantly in all our chalk cliffs, boring its tunnels deeply into the stone, and aiding the sea in its slow, but never-ending task of breaking down the cliffs on one side, while it gradually rears them up on another. As the material into which the Piddock burrows is so hard, there is no need for lining the tunnel, as is done by the Teredo. In this point, too, our engineers follow its example. When their tunnels pass through comparatively soft ground, they line it with masonry, proportioning the thickness of the lining to the looseness of the soil. But, when they come to solid rock, they are content with its strength, and do not trouble themselves about the lining.

The mode of action adopted by the Pholas has long been a disputed point, and even now appears to be not quite settled. I think, however, that William Robertson has proved by his experiments that the shell and the siphon are both brought into requisition. The shell perpetually rotates in one direction, and then back again, just like the action of a bradawl, and, by the file-like projections on its surface, rasps away the chalk, converting it into a fine powder. This powder, being of course mixed with water, passes into the interior of the animal, and is ejected through the siphon.

There are many species of Pholas which burrow into various substances, even in floating cakes of wax and resin. The same species, too, will burrow into different substances, and it is worthy of notice that those specimens which burrow into soft ground attain a much larger size, and their shells are in better preservation, than those which force their way through hard rock.

THE uppermost figure represents a very remarkable tunnel, having the peculiarity of being built instead of sunk. It is the work of an African Ant belonging to the genus Anomma, and popularly known as the Driver-ant, because it drives away every living creature which comes across its course of march.

There are many Ants which seem to rejoice in the full blaze of the tropical sun, running about with ease on rocks which would scorch and raise blisters on the hand if laid on it, and finding no difficulty in obtaining the moisture needful for the mud walls of their habitations. But the Driver-ants cannot endure the sun, and, unless compelled by necessity, will not march except at night, or at all events during cloudy days. Should, however, they be absolutely forced to march in the sunshine, they construct as they go on a slight gallery, which looks very much like the lining of a tunnel stripped of the surrounding earth. If their path should lead them to thick herbage, sticks, &c., which form a protection from the sun, the Driver-ants do not trouble themselves to make a tunnel, but take advantage of the shade, and only resume the tunnel when they reach the open ground.

Sometimes, when they are on a marauding expedition, they construct a tunnel in a very curious manner, their own bodies supplying the materials. The reader must know that there are several classes of these insects, varying in size from that of a huge earwig to that of the little red ant of our gardens. The largest class seem to care little about the sunshine, the protection being mostly needed by the workers. The following is Dr. Savage's account of their proceedings:—

"In cloudy days, when on their predatory excursions, or migrating, an arch for the protection of the workers is con-structed of the bodies of their largest class. Their widely extended jaws, long, slender limbs, and projecting antennæ, intertwining, form a sort of network that seems to answer well their object.

"Whenever an alarm is given, the arch is instantly broken, and the Ants, joining others of the same class on the outside of the line, who seem to be acting as commanders, guides, and scouts, run about in a furious manner in pursuit of the enemy. If the alarm should prove to be without foundation, the victory won, or danger passed, the arch is quickly renewed, and the main column marches forward as before, in all the order of an intellectual military discipline."

How they should be able to direct their course, and to chase an enemy, is not easy to understand; for, as far as is known, they are absolutely blind, not even an indication of an eye being seen.

THE SUSPENSION-BRIDGE.

THE mention of these Ants brings us to another point in architecture. We have already seen that they can not only build arched tunnels, but also can form their own bodies into arches, and we shall presently see how they can form them-selves into Suspension-bridges. We will, however, first take the Suspension-bridge, and its vegetable origin, before passing to the animal.

I have little if any doubt that the modern Suspension-bridge, with all its complicated mathematical proportions, was originally suggested by the creepers of tropical climates. There are few points in a tropical forest, no matter in what part of

CREEPERS.   SUSPENSION-BRIDGE.

the world, more striking than the wonderful development of the creeping plants. The trees are very much like those of our own forests, and are in no way remarkable, but the creeping plants form the chief feature of the woods.

They extend themselves to unknown lengths, crawling up to the very summit of a lofty tree, hanging down to the very ground, if not caught by a midway branch, running along the earth, making their way up another tree, and so on *ad infinitum*. They interlace with each other, forming almost impenetrable thickets, as has already been mentioned while treating of Nets, and there is scarcely a tree that is not connected with its neighbour by means of these wonderful creeping plants.

Of course the monkey tribes make great use of them in passing from one tree to another, thus being able to avoid the ground, which is never to a monkey's liking. Man, therefore, copies the example of the monkey, and makes use, either of the creepers themselves, or of ropes stretched from tree to tree in imitation of them.

In some parts of the world, where palm wine, or " toddy," is manufactured, the native has recourse to an ingenious device which saves a vast amount of exertion. As the calabash which receives the juice of the palm-tree is always fixed at a considerable height, and as each tree only yields a limited supply, the toddy-maker would be obliged to ascend and descend a great number of trees before he could collect his supply of palm-juice.

In order to save himself trouble, he has the ingenuity to connect the trees with each other by two ropes, the one about six feet above the other. He then has only to ascend once, and descend once, for he ascends one tree, and by means of the ropes passes from tree to tree without needing to descend.

The mode of traversing these ropes is simple enough, the lower rope serving as a bridge, along which the man walks, and the upper rope being held by the hands. Those who see these palm-wine makers for the first time are always greatly struck. At some little distance the ropes are quite invisible, and the man appears to be walking through the air without any support whatever.

In Borneo the Rattan is continually put in requisition as a bridge. It runs to almost any length, a hundred feet more or less being of little consequence; it is lithe and pliant, and so strong that it can hardly be broken. The " canes " formerly so much in vogue among schoolmasters, and now so generally repudiated, are all cut from the Rattan. Chiefly by means of this natural rope, the Dyak of Borneo flings his rude suspension-bridges across chasms or rivers, and really displays a wonderful amount of ingenuity in doing so.

The one fault of these bridges is their tendency to decay, or perhaps to be eaten by the multitudinous wood-eating insects which swarm in that country. However, the materials cost nothing at all, and time scarcely more, so that when a bridge breaks down, any man can fit up another at the expense of a few hours' work. As, moreover, the Dyaks have a curious way of building their houses on one side of a ravine, they find that a bridge of this kind saves them the trouble of descending and ascending the ravine whenever they wish to visit their house.

In many parts of America the Suspension-bridge is almost a necessity. The country is broken up by vast clefts, technically called "cañons." These canons are ravines in the rocky ground, with sides almost perpendicular. For the greater part of the year they are dry, but sometimes, and without the least warning, they become the beds of roaring torrents, rising to some thirty or forty feet in height, and carrying away everything before them.

Over these ravines are thrown suspension-bridges made almost entirely of creepers, and loosely floored with rough planks. Although they are very strong, they appear to be very fragile, and even under the tread of a human being swing and sway about in a manner that always shakes the nerves of one who is unaccustomed to them. Yet, even the mules of the country can cross them, the animals picking their way with the wonderful sure-footedness of their kind, and not in the least affected by the swaying of the bridge.

Passing from the vegetable to the animal world, we revert to the Driver-ants, which have already been mentioned. It has been seen that their soldier-ants can, with their own bodies, form a tunnel, under the shade of which the workers can pass, and we have now to see how they can, with the same materials, form a suspension-bridge.

It often happens that on their march they come to water, and, as they always advance with total disregard of difficulties, they must needs invent some very ingenious way of overcoming the difficulty. One of them climbs a branch which overhangs the water, clasps the undermost twig very tightly, and allows itself to hang from it. Another at once follows, and suspends itself from its comrade in like manner, the powerful and sickle-like jaws doing their duty as well as the legs. A chain of Ants is thus speedily formed. When the lowermost Ant touches the water, it merely spreads all its legs, and awaits the development of events. Another runs over it, holds to the first Ant by its hind-legs, and stands in the water, spreading its limbs as much as possible over the surface. Ant after Ant descends, until quite a long chain of the insects is formed, and is swept downwards with the stream. By slow degrees the chain is lengthened, until the Ants at its head are able to seize the bank on the opposite side of the water. When they have succeeded in doing so, the bridge is complete, and over that living bridge will pour a whole army of Driver-ants.

Even in those cases where this mode of travelling would be too perilous on account of the rapid torrent, the Ants contrive to suspend themselves in long strings until they effect a communication with the trees of the opposite bank.

It is, perhaps, needless to give more than a passing reference to the Suspension-bridges made by Spiders, by means of which they can traverse considerable distances. The similar bridge of the Little Ermine Caterpillar has already been mentioned, when treating of the subject of Double Walls.

ARCHITECTURE.

CHAPTER IV.

WE now come to some points in Architecture which cannot well be grouped together, and must therefore be treated as Miscellanea.

Our first example is one which was avowedly based upon an imitation of Nature, namely, the celebrated Eddystone Lighthouse, and we shall see that in two points—first its form, and next the mode in which the stones were fixed together—Nature had been closely followed by the architect.

Unlike ordinary lighthouses, this edifice had to be constructed so as to endure the full force of waves as well as wind. A few miles from the southern coasts of Devon and Cornwall there is a rock which in former times greatly endangered the ships which passed along the Channel. Several attempts were made to build a lighthouse on this dangerous spot. Winstanley's lighthouse, which was finished in 1700, was wholly swept away three years later, together with the architect himself, and some workmen who were engaged in repairs. So terrific is the force of the elements on this spot, that the lighthouse was entirely destroyed, and the only vestiges of it that were ever discovered were some iron bars and a piece of chain.

Another lighthouse was built a few years afterwards, but was burned down, it being of wood instead of stone. At last the work was put into the hands of Smeaton, who saw that he must build on a totally new plan. He took for his model the

TREE TRUNK.   EDDYSTONE LIGHTHOUSE.

trunk of a tree, and determined to build his lighthouse of the same form as the tree-trunk, and to fasten it into the rock just as a tree is fastened by its roots. Accordingly, he struck out a new principle in the construction of such edifices, and his model has been followed ever since. The reader will see, by a glance at the illustration, how close is the resemblance in external form. I may mention that the tree in question was sketched from one in a paddock opposite my house.

Having settled the form of the lighthouse, and made it like a tree-trunk, the next business was to fix it firmly in the rock, and, in fact, to give it roots of stone. For this purpose, he made the base of the edifice as wide as the rock would allow, so as to correspond with the wide base of a tree-trunk, and traced a circle of about ninety feet in circumference. Instead, however, of merely laying the stones as is usually done, or even letting them into holes cut in the rock, he hit upon a singularly ingenious device, whereby the building was practically a single stone.

Instead of cutting the stones square or oblong, as is usually done, he had them made so as to "dovetail" into each other, much after the fashion of a child's puzzle toy, or the junctions at the edge of a box. Thus, each stone fitted into those around it, while the lowest tier was dovetailed in similar fashion into the rock.

The stone employed was that which is called Moorstone, a very hard variety of granite. Each course of stones was carefully fitted together on shore, and their accuracy tested, and

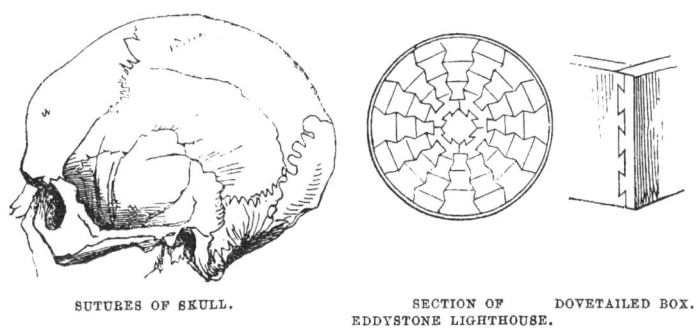

SUTURES OF SKULL.   SECTION OF   DOVETAILED BOX.
EDDYSTONE LIGHTHOUSE.

they were then taken to the Eddystone rock, and fixed in their places. Beside using these precautions, Smeaton fixed the stones in their place with the strongest cement, and furthermore fastened the stones together and united the several courses by strong oak treenails and iron clamps. As none of the stones weighed less than a ton, and some of them were double that weight, the strength of such an edifice may be imagined.

The accompanying illustration shows the arrangement of these dovetailed stones in one of the courses. It will be seen that the central stone must be laid first, and then the others arranged round it. The whole edifice is rather more than eighty-five feet in height, so that the elements have every chance of demolishing it, as they did that of Winstanley. More than a hundred years have now passed since it was built,

P

and, although the fury of the tempest has been such that the waves have washed completely over its summit, it stands as firmly as it did when it was finished in 1760.

WHETHER the original inventor of the "dovetail" took his idea from Nature I cannot say, but he certainly might have done so. On the left of the illustration is part of a human skull.

The skull is not, as many persons seem to think, made of a single bone, but it is composed of many bones, united by "sutures," which are, in fact, natural dovetails. Although in early life these sutures are comparatively loose, they hold the various parts together so firmly, that if the head be violently struck, the bones may break, but the sutures do not give way.

Perhaps some of my readers may ask how it is possible to take a skull to pieces without cutting it or fracturing the sutures. It is done in a way equally simple and ingenious. The skull is filled through the opening with dried peas, and then sunk under water. The peas expand with the moisture, and, as they exert an equable force in all directions, they slowly and quietly pull the sutures asunder, without injuring the bones.

THE DAM.

IN many human operations, where a certain depth of water is required in a running stream, the reasoning powers of man have enabled him to attain his object by building a dam, or obstacle across the stream, which forces the water to rise to its level before it can find a passage. Such, for example, are the Locks which render rivers navigable, and allow even the heavily laden barges to traverse miles of water which would otherwise have been closed to them.

Those mills, again, which are worked by water need that a sufficient amount of water should be ready in order that it may by its weight force the wheel round. Such a Dam is shown on the right hand of the illustration, the height to which it raises the water being shown by the level of the stream below the Dam, and that of the water as it tumbles over in a miniature cascade.

PUTTING aside the natural dams made by accumulations of the various débris that are washed down by a swollen stream, and which sometimes raise the water to a very great height, we have an example of a natural dam in the curious structure made by the Beaver, for the same purpose as that of the lock in the mill-stream, namely, to insure a depth of water sufficient for the needs of the beings that make them.

Every one has heard of the Beaver's dam, but there is so much misconception on the subject, that a few words will not be out of place.

Ingenious as is the animal in the construction of its dam, it is not nearly so accomplished an architect as was once supposed. We were told in the earlier books of Natural History that the Beaver felled trees, cut off their branches into convenient lengths,

DAM MADE BY BEAVER.   DAM MADE BY MAN.

and sharpened one end, like an ordinary stake. Then they were said to drive the sharp end of the stakes into the bed of the river, to set them side by side, to interweave smaller branches among them, and lastly, to fill up the interstices with mud, leaves, and similar materials. In fact, they were supposed to build a "wattle-and-daub" wall, like that which is in use at the present day in Southern Africa.

The Beaver does nothing of the kind. It needs a dam, and it makes one which is far stronger than the wattle-and-daub could be. It begins by felling a tree, and letting it lie across the stream, in some place where the banks are high and tolerably steep. A bend of the river is usually chosen for the new dam. Should not the tree be long enough for the Beaver's purpose, two trees are felled, one on either side, so that their branches meet in the middle.

These branches, and not any supposed stakes, are really the upright supports of the dam. The trees being thus laid, the Beaver cuts down branches from four to six feet in length, and lays them horizontally among the boughs of the fallen trees. Having thus made the foundations, so to speak, of its dam, the Beaver then proceeds to fill in the spaces with roots, grass-tufts, leaves, mud, and, indeed, almost anything on which it can lay its paws.

After this, the Beaver has to take but little trouble, for the stream itself becomes a silent, slow, but constant labourer, lodging floating débris against the dam, and making a sloping bank which much adds to its strength. By degrees, seeds that lodge on the dam spring into life, and their roots act like chains, binding the materials more closely together. Willow twigs too, if they lodge on the dam and be left undisturbed, are sure to " strike," as the gardeners say, and further to bind the structure together.

It is evident, from this short description, that the lower part of the dam is more solid than the upper. In fact, the floods are tolerably sure to wash away some eight or ten inches of the upper part every year, and the Beavers have to make it afresh. The height of these dams is not nearly so great as is generally supposed. Mr. Green, a practical trapper, states that the highest which he ever saw was only four feet six inches in height, and that the average is under three feet.

The house of the Beaver is made on the same principle as the dams. Every one knows that when sticks have been in the water for any length of time, they become saturated and sink. These sticks are chosen by the Beaver as the material for its house, and are laid horizontally in the water, the heaviest being reserved for the roof, so as to make it strong enough to ward off the attacks of predacious animals. As with the dam, mud, leaves, &c., are used to consolidate the edifice, but no mud can be seen from the outside, the animal always finishing off with a number of heavy logs laid on the roof.

SUBTERRANEAN DWELLINGS.

I do not intend in this place to take up the whole subject of Subterranean Dwellings, but only to point out cases where the use of the Subterranean Dwelling depends on the climate of the locality and the time of year, it being sometimes used and sometimes neglected, sometimes inhabited for the sake of warmth, and sometimes for that of coolness.

In various parts of India there are some most remarkable Subterranean Dwellings. They are more than mere dwellings, and are, in fact, magnificent palaces, sunk so deeply in the earth that very little more than their roofs appear aboveground. When, however, a visitor descends the stairs that lead to the interior of the palace, he finds it spacious, and with tiers of chambers one below the other, very much like the wasp-nest which has already been described. Nussur-ed-deen, the second King of Oude, had several of these palaces, but very seldom visited them, he having endeavoured to Europeanise himself as much as possible, and to cast off his native customs.

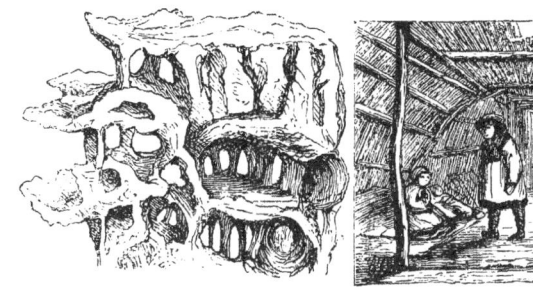

SUBTERRANEAN ANTS' NEST.   SUBTERRANEAN HOUSE OF KAMSCHATKA.

He used occasionally to visit them, but it was only out of etiquette, and he never really lived in them.

However much he might have rejected the ancient customs, it is evident that in this case, at least, he was punishing himself in rejecting these summer dwellings, which are always cool, and where, if one set of apartments is too warm, nothing is easier than to descend to the next.

THIS dwelling is made for the sake of coolness in summer. Another subterranean dwelling is made for warmth in winter, the non-conducting properties of the earth being in both cases brought into play. This is the winter dwelling of the inhabitants of Kamschatka.

During the summer-time the Kamschatdales live in comparatively slight huts mounted on poles, and having the floor some ten feet from the ground.

During the winter, however, they live in habitations of a very different character.

In order to make these houses, they begin by digging a large hole in the ground, about nine or ten feet in depth. This they line with poles and sticks, making, in fact, a wall as of a house. A stout conical roof is then raised over the hole, and upon the roof earth is thickly strewn and beaten down, just as has been mentioned when treating of the ice-house. The only access to this strange house is by a circular aperture in the centre of the conical roof, serving at once the purpose of a door, a chimney, and a window. A notched pole answers as a ladder, a low wooden dais placed against the wall serves as a bed or a chair, for there is no other, and a few stones placed together act as a fireplace.

In looking at both these subterranean dwellings, I could not but be reminded of a very common insect which has a double dwelling, one moiety being aboveground, and the other moiety below it. This is the common Wood-ant (*Formica rufa*), whose large, leafy hills are so plentiful in some of our woods. On account of its size, this species is sometimes called the Horse-ant.

At first sight the nest looks something like a small haycock, made entirely of chopped grass. When examined more nearly, it will be found to consist mostly of grass-stems, little bits of stick, and leaves. Those of the fir are in great request, for when they are dry they are very light, and their form enables the Ant to interweave them with each other, so as to form the necessary tunnels and galleries which line the interior of the nest. The materials seem most unpromising, but they are used with wonderful skill, such as no human fingers could equal.

After a little while a number of entrances into the nest are visible. They are almost invariably sheltered by projecting leaves, which act as porches, so that when the nest is viewed from above, they are almost entirely hidden. Each of these openings runs into one of the main galleries of the nest, and from thence issues a perfect labyrinth of passages.

This, however, is only half the nest, for the galleries and tunnels extend far beneath the surface of the earth, and have sundry enlarged portions or chambers wherein the immature pupæ may lie during their period of helplessness.

Owing to the very loose structure of the upper nest, and the tendency of the earth to fall into the galleries of the lower nest, it is very difficult to obtain a trustworthy view of the interior. Perhaps I may here be allowed to extract a passage from my "Insects at Home," the description of the nest and its interior having been written almost on the spot :—

"I have, however, succeeded in obtaining an excellent view into the interior of a Wood-ants' nest, though it was but a short one. Accompanied by my friend Mr. H. J. B. Hancock, I was visiting some remarkably fine Wood-ants' nests near Bagshot. We took with us a large piece of plate glass, placed it edgewise on the top of an Ant-hill, and, standing one at each side, cut the nest completely in two, leaving the glass almost wholly buried in it.

"After the expiration of a few weeks, during which time the ants could repair damages, we returned to the spot, and, with a spade, removed one side of the nest as far as the glass, which then served as a window through which we could look into the nest. It was really a wonderful sight.

"The Ant-hill was honeycombed into passages and cells, in all of which the inhabitants were hurriedly running about, being alarmed at the unwonted admission of light into their dwellings. In some of the chambers the pupæ were treasured, and these chambers were continually entered by Ants, which picked up the helpless pupæ, and carried them to other parts of the nest where the unwelcome light had not shown itself.

"Unfortunately this view lasted only a short time. Owing to the partial decomposition of the vegetable substances of which the Ants' nest is made, the interior is always hot and always moist. Now, the day on which we visited the nest happened to be a cold one, and, in consequence, the moisture of the nest was rapidly condensed on the inner surface of the glass, and in a few minutes completely hid the nest from view, leaving me only time to make a rapid sketch. Unfortunately some one discovered the plate of glass and stole it.

"Next time that I examine a Wood-ants' nest, I shall take

care to insert the glass exactly east and west, and shall open
ts southern side towards noon on a hot sunshiny day, so that
the rays of the sun may warm the glass and prevent evaporation."

Many other creatures make subterranean dwellings, but the
Wood-ant is remarkable for possessing a double dwelling, the
two portions communicating with each other, and capable of
being used according to the degree of heat required.

THE PYRAMID.

WE have already seen how the Eddystone lighthouse was
the precursor of many similar buildings, all, like their predecessor, having their form copied, with more or less strictness,
from the outlines of a tree-stem.

NATURAL MOUNTAIN.   ARTIFICIAL MOUNTAIN, OR PYRAMID.

Another form of building which was intended for endurance,
and, indeed, is the most enduring of all shapes, is the Pyramid.

We are all familiar with the simple, yet grand outlines of
the Pyramids of Egypt, whose vast antiquity takes us back to
the times of Isaac and Joseph, and which seem capable of
resisting the effects of Time, the universal destroyer, for thousands of years yet to come.

We may ask ourselves what was the natural object from
which the Pyramid was copied. The name itself, which is
formed from a Greek word signifying fire, shows that a flame
was thought to have furnished the idea of this form of building.
I cannot, however, but think that the flame had little, if anything, to do with it, and that the real model may be found in
the hills which have been formed by Nature.

Examples of the Pyramids and the Hills are given in the
accompanying illustration.

SUBAQUATIC MORTAR.—PAINT AND VARNISH.

HAVING now disposed of the chief points in Architecture, we take some of the subsidiary details.

Of late years, when the traffic between different continents has so largely extended itself, and when shipping has increased both in the numbers and dimensions of the vessels, it is absolutely necessary that we should have harbours and docks enlarged and multiplied sufficiently to meet the calls upon them.

Now, it is comparatively easy to construct a building on shore, for all the mortars and cements which are used for the

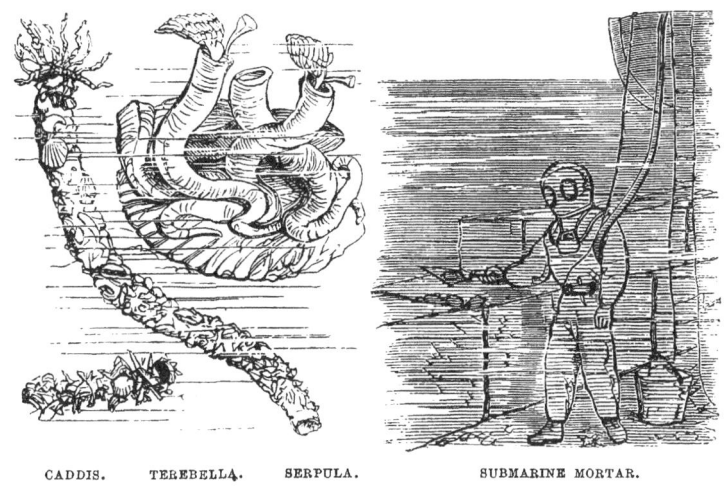

CADDIS.   TEREBELLA.   SERPULA.   SUBMARINE MORTAR.

purpose of fastening the stones together are applied when wet, and incorporate themselves with the stones as they dry. But to make a mortar which could be applied while the stones were under water, and would "set" while beneath the surface, was a task not easily to be overcome. Yet it has been done so effectively that at the present day we can build beneath the surface of the water as securely, though not as rapidly, as if the stones had been laid on dry ground.

Several such mortars are now known, and, as is so often the case with human inventions, have been anticipated in Nature.

We have already seen how the Caddis-worm of the fresh

waters can cement together, while under water, the various materials of which its tubular house is formed. The different Sticklebacks perform similar feats, no matter whether they inhabit fresh or salt water.

All those who take an interest in the productions of the sea-shore will have noticed upon our coasts the flexible tube of the Terebella, with its curiously fringed ends. This tube, as any one may see at a glance, is composed of grains of sand and similar materials, fastened strongly together by a kind of cement exuded from the worm, and possessing the property of hardening under water. As on some of our coasts fragments of shell are used for the tube, the worm goes by the popular name of Shell-binder.

If one of these worms be taken out of its tube, placed in a vessel with sea-water and a quantity of sand, broken shells, and little pebbles, the mode of building will soon be seen. At the extremity of the head are a number of extremely mobile tentacles, and these are stretched about in all directions, seizing upon the particles of sand and shell, seeming to balance them as if to decide whether they are suitable for the tube, and then fixing them one by one with the cement which has already been mentioned.

Generally speaking, the Terebella works only in the evening, but, if it be hastily deprived of its tube, it cannot help itself, and is perforce obliged to work while it can. It is worthy of remark that the Terebella, although, as a rule, it lives in a tube all its life, is capable of swimming with the usual serpentine motion of marine worms, and, when taken out of its tube, rushes about violently, and soon exhausts itself by its efforts.

Along most of our rocky seashores may be seen vast quantities of a sort of hardened sand, penetrated with small tubes. On a closer examination this sand-mass is resolved into a congeries of tubes, matted and twisted together, and each being the habitation of a marine worm called the Sabella. This name is derived from a Latin word signifying sand, and is given to the worm in allusion to the material of which it makes its habitation.

Like the Terebella, the Sabella uses its tentacles for the purpose of building the tubes, which are much stiffer than those of the Terebella. They are strong enough, indeed, to give the feet a firm hold while traversing the rocks, and this is a matter of no small moment when the tide is coming in, and the shore has to be regained without loss of time.

Then we have other marine worms, known as Triquetra and Serpula, which make tubes in a somewhat similar manner, but of very fine materials and very strong cement, so that the tube is nearly as hard as stone.

Space would fail me if I were to enumerate these creatures at greater length, but enough has been said to show that man's invention of subaquatic cement has been anticipated in Nature by the inhabitants both of salt and fresh water.

WE now come to the subject of Paint and Varnish. Putting aside their use as a means to increase the beauty of the object to which they are applied, we will view them in the light of preservatives, and acknowledge the truth of the old Dutch proverb, that "Paint costs nothing." Certainly, when the wood to which it is applied is thoroughly dry from within, it not only costs nothing, but repays itself over and over again as a preservative of the wood, and a defence against moisture from without.

The instances in which Paint is applied to wood are too numerous to be mentioned. Perhaps some of my readers may remember the case of the naval captain who, on taking command of his ship, was supplied, according to custom, with exactly half the amount of paint required for her. The invariable etiquette had been that the captain supplied the remaining half at his own cost. But the officer in question was not at all disposed to be "put upon," and was a thorn in the sides of the "Naval Lords."

Finding, by actual measurement, that the paint supplied to him was only half the amount which was really needed for the ship, he sent his respectful compliments to the Admiralty, asking whether they wished the port or the starboard side of the ship to be painted, for that there was only enough paint for one half of the ship, and he awaited instructions as to which side of the vessel it was to be applied. He was impervious to "minutes," "directions," &c., and, as far as I remember, this very impracticable man got his way, and was supplied with the requisite amount of paint.

Long before man ever invented paint or varnish the Hive
Bee had made use of it.

Every one who has kept bees knows how they always fasten
the edge of the hive to the board, and stop up any crevices
that may be left open. The material which they use for this
purpose is not wax, but a substance called "propolis." This

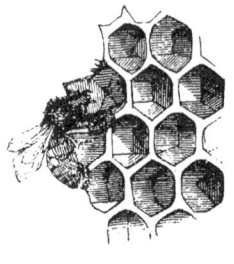

BEE VARNISHING CELLS.   PAINTER VARNISHING WOOD.

term is composed of two Greek words, signifying a suburb, or
the outskirts of a town, and is given to this stationary substance in consequence of the use which is made of it.

Not only do the bees use it for fastening the hives, but
also for strengthening their combs. Wax is a very precious
material, and the beautiful hexagonal structure of the bee-comb
is intended for the purpose of combining the greatest amount of
storing space with the least expenditure of material. The
plates of wax of which the cells are composed are so thin that
their edges would break down even under the feet of the bees
as they passed over it, and accordingly the bees strengthen
the edges of the cells with propolis, as any one may see by
examining a piece of bee-comb. The propolis is of a darker
colour than the wax, and has a peculiar varnish-like appearance.

The propolis, as distinguished from wax, is mentioned by
Virgil in his Georgics :—

> " Collectumque hæc ipsa ad munera gluten
> Et visco et Phrygiæ servant pice lentius Idæ."—*Georg.* iv. 40.

It is evident that the propolis cannot be obtained from the
same source as the wax. The latter is secreted by the bees

under little plates or flaps upon the abdomen, while the propolis is purely a vegetable exudation. It is obtained from many trees, the principal being the horse chestnut. All who have handled the buds of this tree are aware that they are covered with a viscous and very adhesive matter, which serves as a varnish or protection to the bud before the leaves are strong enough to break out. This is the material which the bees gather for their propolis, and at certain times of the year the chestnuts may be seen swarming with bees, all busily engaged in scraping off the varnish.

TOOLS.

CHAPTER I.

THE DIGGING-STICK.—SPADE.—SHEARS AND SCISSORS.—CHISEL
AND ADZE.—THE PLANE AND SPOKESHAVE.

The Use of Tools a Distinction between Man and Beast.—All Men, however
savage, use Tools, but none of the lower Animals can do so until taught by
Man.—Tools needed to break up the Ground.—The Digging-stick of savage
Life: its Use and its Efficacy in practised Hands.—Digging-sticks in Nature.
—The Heart-urchin, and its Mode of digging in the Sand.—The Spade:
its Shapes and Uses.—Natural Spades.—Fore-foot of the Mole and Mole-
cricket.—The Aard-vark, the Ant-eater, and the Mattock.—Shears and
Scissors a Sign of Civilisation, never being employed by Savages.—Mecha-
nical Principle of Scissors, the Inclined Plane, the Lever, and the Cutting
Edge.—Chinese Shears and the Pruning Scissors.—Use of the Inclined
Plane.—The Diagonal Knife of the Guillotine.—The Shears in Iron-works.
—The "Drawing Cut" of Swordsmen.—Jaws of the Turtle and Tortoise.—
The Snapping Turtle and the Chicken Tortoise.—The Locust, the Cockchafer Grub, the Great Green Grasshopper, and the Wart-biter.—The Leaf-
cutter Bees and their Nests.—The Chisel and Adze.—Structure of Rodent
Tooth and Chisel.—Use of the hard Plate of Enamel or Steel.—Combination
of hard and soft Materials.—Teeth of Hippopotamus and Hyrax.—Principle
of the Adze.—Self-sharpening and Self-renewing Tools.—The Plane and
Spokeshave.—Principle on which they are made.—The Spokeshave and
its Uses.—The "Guard" Razor.—The Hoop-shaver Bee and its Nest.—Its
natural Plane, and the Use which is made of it.

AMONG the many points of distinction between man and the
lower animals, we may consider the use of tools as one of
the principal lines of demarcation. Man stands absolutely
alone in this respect. There is no race of savages, however
degraded they may be, that does not employ tools of some kind,
and there is no beast, however intelligent, that ever used a tool
except when instructed by man.

As to the stories that are told of the larger apes using
sticks and stones by way of weapons, they are absolutely without foundation, no animal employing any tool or weapon save
those given to them by Nature. It is true that a monkey may

sometimes be seen to take a stone for the purpose of cracking nuts which are too strong for its teeth, and to perform that task with great deftness; but such animals have always been taught by man, and had they remained in their own country, not one of them would have used a stone, were the nuts ever so hard.

THE SPADE.

WE will begin our notice of tools by taking that which must have been the first tool invented by man. One of the principal duties assigned to man is the culture of the earth, and this he cannot do without tools, increasing their number and improving their structure in proportion to his own development in agriculture.

Before seed can be sown, it is necessary that the earth should be broken up, and, owing to the structure of the human frame,

HEART-URCHIN. DIGGING-STICK.

this task cannot be fulfilled by man without a tool which will enable him to rival many of the lower animals, *i.e.* make use of those digging appliances which have been furnished by Nature.

It is evident that the first earth-breaking tool must have been a pointed stick, and we find that in Southern Africa, in parts of Asia, and in Australia the Digging-stick is still in use for the purpose of breaking up the ground. The Australians are wonderful adepts in the use of the Digging-stick, which is one of the simplest of instruments, being merely a stick some two feet in length, pointed at one end, and the point hardened in the fire.

The mode of using it is by holding it perpendicularly, pecking it into the ground, and throwing out the loosened soil with the hands. In this way they can excavate with such rapidity, that a strong navvy, armed with the best spade, would not be able to keep pace with a black man armed only with his "katta," or digging-stick.

In Africa the Digging-stick is used in exactly the same manner, and is generally made more weighty and effective by having a perforated stone fastened on the handle.

HERE, again, man has been anticipated by Nature, and the savage of Australia or Africa digs in exactly the same manner as the common Heart-urchin of our shores, sometimes called the Hairy Urchin, in consequence of the number and fineness of the spines, which look just like hairs to the naked eye. The scientific name of this creature is *Amphidotus cordatus*.

Mr. Gosse, in his "Evenings at the Microscope," gives so admirable an account of the mode of digging employed by the Hairy Urchin that I cannot do better than employ his own words. After describing the variety of structure of the different spines with which the shell is so thickly set, he proceeds as follows :—

"But what is the need of so much care being bestowed upon the separate motion of these thousands of hair-like spines, that each should have a special structure, with special muscles for its individual movement? The hairs of our head we cannot move individually : why should the Heart-urchin move his?

"Truly, these hairs are the feet with which he moves. The animal inhabits the sand at the bottom of the sea in our shallow bays, and burrows in it. By going carefully, with the lens at your eye, over the shell, you perceive that the spines, though all formed on a common model, differ considerably in the detail of their form. I have shown you what may be considered the average shape, but in some, especially the finer ones that clothe the sides, the club is slender and pointed ; in others, as in those behind the mouth, which are the largest and coarsest of all, the club is dilated into a long, flat spoon ; while in the long, much-bowed spines, which densely crowd upon the back, the form is almost uniformly taper throughout, and pointed.

"The animal sinks into the sand mouth downwards. The hard spoons behind the mouth come first into requisition, scooping away the sand, each acting individually, and throwing it outwards. Observe how beautifully they are arranged for this purpose, diverging from the median line, with the curve backwards and outwards.

"Similar is the arrangement of the slender side spines; their curve is still more backwards, the tips arching uniformly outwards. They take, indeed, exactly the curve which the fore-paws of a mole possess,—only in a retrograde direction, since the Urchin sinks backwards,—which has been shown to be so effective for the excavation of the soil, and the throwing of it outwards.

"Finally, the long spines on the back are suited to reach the sand on each side, when the creature has descended to its depth, and by their motion work it in again, covering and concealing the industrious and effective miner."

The reader will notice that this mode of digging is exactly like that which is followed by the users of the Digging-stick, the earth being first broken up, and the loosened portions thrown aside. The whole of the description of the spines is exceedingly interesting, but, as it does not bear directly on the present subject, I cannot admit it into these pages.

Now comes another development in digging tools.

We have already seen how effective an instrument a mere piece of stick can be in the hands of a skilful workman, and the manner in which it can tear up a given depth of soil. But, for agricultural purposes, something more is needed, and the ground must not only be broken up, but a certain regularity must be observed, in order to allow space to be accurately measured, and the crop apportioned to the area.

Out of the Digging-stick, then, the Spade was developed, its chief advantage being that it dispensed with the use of the bare hands, and not only tore up the ground, but threw out the loosened soil.

The reader will remember that in the preceding description of the Heart-urchin it was mentioned that many of the spines are shaped at their ends something like spoons, and that their comparatively wide blades are used in scraping the sand and shovelling it aside. In fact, these flattened spines are natural spades, used on the same principle as the modern spade of civilisation.

On the right hand of the illustration are shown two forms of

spade, the one being the ordinary garden tool, and the other a
rather curious implement which is in great use among the
metal mines of Cornwall.   The use of the ordinary spade is too
familiar to need explanation, and we come to the Miner's spade.
This implement is used rather as a shovel than as a spade, the
peculiar bend near the blade preventing the foot from being
used as a means of forcing the instrument into the ground.
In fact, it is not meant for the same office as that which pertains
to the ordinary spade, neither can it be handled in the same
way.

In Devonshire there is a kind of spade in general use very
much resembling the mining spade, but having a very long

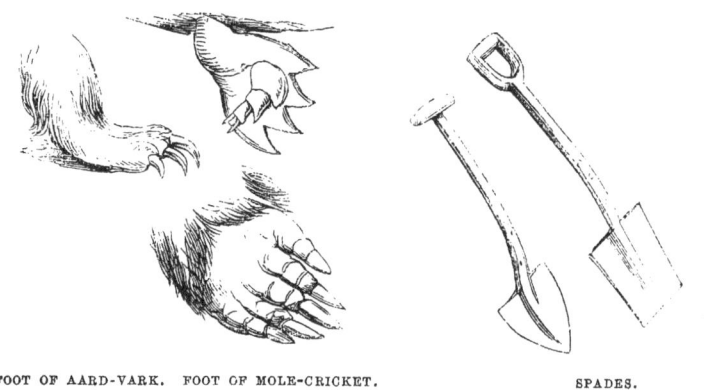

FOOT OF AARD-VARK.   FOOT OF MOLE-CRICKET.   SPADES.
                     FOOT OF MOLE.

handle without any crutch at the end.   The natural consequence of this shape is, that the spade cannot be used in the
ordinary way, neither can it penetrate the earth to any
depth.   It can "peel" the ground, so to speak, and can cut
away successive layers of soil.   But as for digging "two
spits deep," or even one spit, the spade would be absolutely
incapable of such a task, no matter how strong might be the
hands that wield it.   As for the foot, it may be put out of the
question.

WE will now turn to a few examples of spades in the world
of Nature.

The lowest figure represents the fore-paw of the Mole,

with its powerful armature of strong and sharp claws, and its broad blade of a palm. The reader will easily see that in this animal the digging powers are wonderfully developed. The peculiar form of the fore-foot closely resembles that of the miner's spade, while the curvature of the palm serves, almost without exertion, to throw out the earth which has been scooped away by the sharp claws.

To watch a Mole burrow is really a curious sight, the only drawback being that the animal sinks itself so rapidly beneath the earth that a long inspection is impossible. I have kept several moles for the purpose of watching their habits, and have always been interested in their mode of burrowing. I can only define it by using the word "scrabbling." The animal scurries and hurries about, seeking for a tolerably soft piece of ground. When it has found one, it travels no further, but scratches away with its fore-paws with wonderful power and rapidity, seeming to sink, as it were, into the earth, rather than to excavate a tunnel.

THERE is an insect well known to entomologists, called the Mole-cricket, because its structure and many of its habits are strangely similar to those of the animal from which it derives its name. At the upper part of the illustration is seen a portion of the fore-foot of the Mole-cricket, and a better implement of excavation can hardly be imagined.

The reader will probably have noticed that in both these creatures the spade, if we may so call it, is not a mere flat plate, but is cleft into several points. It thus answers the purpose of a fork as well as a spade, the several points serving to break up the soil, and the flat palm to throw the earth aside.

This principle is carried out even more fully in the fore-paw of the African Ant-bear, or Aard-vark (*Orycteropus Capensis*), a figure of which is given in the illustration. This animal is a great excavator, living in burrows of such dimensions that the wild boar is in the habit of making its home in them after they are deserted.

Something more, however, than a digging apparatus is needed for the Ant-bear. This animal feeds almost wholly on the Termites, which it obtains by tearing down the walls of

their dwellings.   Now, as these wonderful buildings are nearly
as hard as brick, and, indeed, are composed of the same materials,
it is necessary that the claws of the Ant-bear should be modified
so as to be able to break through the walls.   Accordingly,
they are much more curved than those of the Mole and the
Mole-cricket, and so serve for tearing as well as digging, being
struck into the wall, and thus pulling it down, just as a
labourer breaks down a bank with his mattock.

Indeed, had we wished to extend these analogies still further,
we might easily have given the claws of the Aard-vark as a
prototype of our English mattock.   The same weapons as
possessed by the Ant-bear of tropical America are used in
exactly the same manner, but are even stronger, and extend to
such a length that when the animal walks, it cannot stretch its
claws out in front, but is obliged to double them under its feet.

SHEARS AND SCISSORS.

THESE instruments are sure signs of civilisation, no savage
nations having the least idea of them.   Even the Kafir and Esquimaux tribes, which are such admirable workers in skin, never
use scissors in shaping their garments, but invariably employ
knives for that purpose.   The Chinese, however, seem to have
known scissors from time immemorial, and to have shaped them
almost exactly like our own instruments.   I possess one pair
of tailor's shears from China in which there is only one ring,
namely, that for the thumb.   The place of the other ring is
taken by an elongated, slightly curved and moderately pointed
rod of steel, which is used for tracing the pattern on the material
preparatory to cutting it.

Simple as the scissors may seem, they combine several very
important principles, namely, the inclined plane, the lever, and
the cutting edge.   Were they to be merely two edges moving
directly upon each other, their effect would be comparatively
slight; but, owing to the manner in which the blades are fixed
at one end, they are drawn as it were over the object between
them, and so divide it with comparative ease.   In some instruments, such as the pruning scissors, there is only one cutting
blade, the other being used merely as a support for the branch
which is being cut.

A well-known example of a single cutting blade is found in the guillotine. In the earliest times of this invention an ordinary axe-head was suspended above the neck of the criminal. It was found, however, that its operation was very uncertain, simply because the blow was a direct one, and not oblique. The blade was then set obliquely, as in the present machine, and its effect was absolutely certain.

Perhaps some of my readers may be swordsmen, and therefore know the power of the "drawing cut," by which a great effect may be produced with very little apparent exertion. Even in the simple operation of cutting bread we always use the knife diagonally, though perhaps we may be ignorant of the principle of the inclined plane.

Next comes the principle of the lever, as exemplified by the handles of the scissors. By lengthening these handles, the power of the blades is enormously increased, as may be seen in the various shears in any great iron-works, which cut through thick iron as if it were butter. Our own garden shears for trimming borders show very well the power of the long arms and short blade.

In the animal world we find many examples of natural shears, one of the best of which is afforded by the jaws of the Tortoise or Turtle. Owing to the manner in which they feed, whether they be vegetarians or carnivorous, their jaws are made for cutting, and not for lacerating or mastication. They have no teeth, but each jaw is furnished with a horny edge, as sharp as a knife-blade, and very strongly made. With these jaws the animal can shred to pieces the objects which it attacks, just as if it had been furnished with a pair of veritable shears. Any one who has possessed an ordinary Tortoise must have noticed the havoc which it will occasionally make in a garden. I had one of these reptiles for some years, and was obliged to keep it under restraint, in consequence of the power of its jaws.

Being a Tortoise of discrimination, it took a great fancy to the strawberry beds, and invariably picked out the ripest and best-flavoured fruit. Reversing the usual proverb of making two bites at a cherry, the Tortoise always took two bites at a strawberry, and sometimes three or four, according to its size.

At last, I was obliged to restrain it by boring a hole in the edge of its shell, passing one end of a string through it, and fastening the other to a peg driven into the ground. At first, I tied the string to a brick, but the Tortoise was so strong that it dragged the brick about the garden, leaving reminiscences of its progress in the channels which it had cut through all kinds of vegetation with its scissor-like jaws.

The reader, in comparing the illustration of the Turtle-jaws with that of the Shears, will see at once how exact is the analogy between the two.   The sharp-edged jaws correspond with the blades of the shears, the joint at the skull corresponds with the pivot of the shears, and the muscles which move the

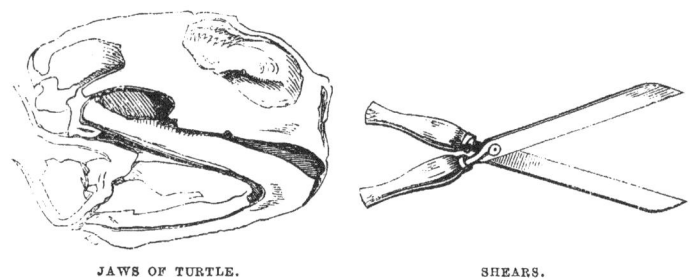

JAWS OF TURTLE.   SHEARS.

jaws, but which could not be shown in the present illustration, are the prototypes of the handles.

In some of these creatures, especially those which are car-nivorous, the power of the jaw is tremendous.   One of them, a Snapping Turtle, has been known to bite off several fingers of a man's hand as easily as if they had been carrots.   Some years ago I kept some Chicken Tortoises alive, and was much struck with the enormous proportionate power of their jaws.

They were quite little creatures, only a few inches in length, but their appetites were astonishing, and their mode of satisfying their hunger remarkable.   They were always ravenous after meat, and had a curious way of seizing their food in their mouths, placing one paw on either side of their jaws, and then pushing the meat forcibly away, so as to cut out a slice as large as their jaws.

They were very good-tempered little things, but, small though they were, I should have been very sorry to have one of them take a bite at my finger by mistake.

Knowing their general characteristics, I took care not to have any living creature in the same vessel. But I have heard, from those who have had practical experience, that Chicken Tortoises ought to be banished from any place wherein fish are kept, especially if they be gold fish, the Tortoise having a way of coming quietly beneath them, biting out a mouthful of their bodies, and then disappearing with its booty.

BESIDE the Tortoise, there are many creatures which possess natural shears, such as the Locust, whose ravages are only too notorious. Then, taking our own country, we have plenty of examples of insect shears. Such is to be found in the jaws of the Cockchafer larva, or " White Grub " as it is popularly called. It lives underground, and feeds chiefly on the roots of herbage, shredding them to pieces with its shear-like jaws. And, as it spends on the average three years in the one task of perpetual eating, the damage which it does can be easily imagined.

There is a very pretty English insect which admirably exemplifies the power of the natural scissors. This is the Great Green Grasshopper (*Acrida viridissima*), which is equally voracious in all its stages of existence. It is always ready to use these jaws, and I do not recommend the reader to allow his finger to get between them, or their points will probably meet.

One of these insects, indeed (*Decticus griseus*), has derived the name of Wart-biter from its supposed use in curing warts. All that was needful was to catch a Wart-biter, and hold one of the warts to its jaws. It was sure to seize the wart, and bite it smartly, and there was a firm belief that any one thus bitten would be freed from the unsightly excrescence. The bite of the shear-like jaws caused much pain at the time, and this very pain had in all probability something to do with the cure.

AN admirable example of the insect jaws used as scissors is to be found in the well-known Leaf-cutter Bees, insects belonging to the genus Megachile.

They make their nests in burrows, sometimes in wood, and sometimes in the ground, and form them in a very singular manner. After fixing upon a suitable burrow, the Bee goes off to a tree, generally a rose, and, using her jaws just as a tailor uses his shears, cuts off a nearly semicircular piece of leaf, flies away with it to her home, and, by dint of bending, pushing, and pulling it, she forces it to the bottom of the cell. Successive pieces of leaf follow, until she has made a thimble-shaped cell, and she then places at its end an egg and a supply of honey and pollen.

Cell after cell succeeds, each being introduced into its predecessor just as thimbles are packed. Judging from a specimen in my collection, there are about eight layers of leaves to form the walls of the cell, and the average length of each piece of leaf rather exceeds half an inch. The entire length of the cell-group is two inches and a half. The leaf-slices are always cut from the edge, and, in my specimen of the nest, the serrated outer edges of the leaves are all in one direction.

Should any of my readers find one of these nests, it will be as well for them to dip a needle point into diamond cement, and introduce it under the outermost coating of leaves. Otherwise, when the leaves are dry, and the insects break their way into the open air, the cells will probably fall to pieces.

These Bees are much more abundant than is usually thought. In summer-time it is hardly possible to find a rose-bush on which are not a number of leaves from which pieces of variable size and shape, but always with a curved outline, have been cut as with scissors. While cutting them, the Bee seems to trace out her pattern, as it were, by using her feet like one leg of a pair of compasses, and her head as the other leg. As soon as she has nearly finished the operation, she poises herself on the wing, to prevent her weight from tearing away the leaf irregularly, and then, while still on the wing, makes the last few bites, and severs the leaf entirely.

THE CHISEL AND THE ADZE.

ALREADY we have seen how exact is the analogy between the scissors and the turtle-jaw. As we are upon the subject of cutting instruments, we will continue it, trying to discover some further analogies.

On the right hand of the illustrations we see three cutting tools made by human hands—*i.e.* the Chisel, the Stone Adze of

Polynesia, and the Steel Adze of this country. We begin with the Chisel.

All those who have even a slight knowledge of anatomy know how curiously exact is the resemblance of the Chisel of civilised life to the front tooth of any Rodent animal. The head of the Beaver is here given as an example, but the tooth of a mouse, rat, or rabbit, which can easily be obtained, is quite as good an example. These teeth are made after a very beautiful fashion. Their outer surface is covered with a plate of very hard enamel, while the rest of the tooth is of bony matter, and comparatively soft. Consequently, when the tooth is used, the enamel plate forms a sharp edge, while the rest of it is worn away, thus keeping the chisel-like end in its proper form.

The power of these teeth may be appreciated by any one who has been bitten even by so small a rodent as a mouse, the

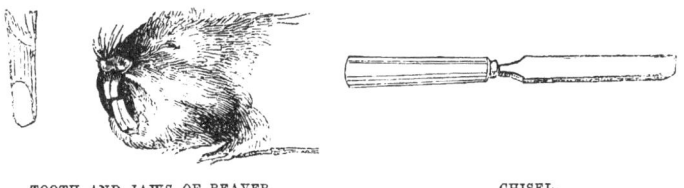

TOOTH AND JAWS OF BEAVER.   CHISEL.

sharp edges meeting in the flesh, and causing a very painful wound. When the teeth are large, as in the Beaver, and the jaws powerful, their force is something wonderful, tree-trunks of considerable size being cut down quite easily.

Perhaps some of my readers may not be aware that the Chisel is constructed on exactly the same principle as the tooth of the Rodent animal. It is not entirely made of steel, as is generally thought. In the first place, a valuable material would be needlessly wasted, and, in the next place, the tool would not keep its edge except with infinite labour in grinding.

The principal part of the Chisel-blade is therefore made of soft iron, a very thin plate of steel running along the back. This plate answers the same purpose as the enamel in the tooth, while the soft iron takes the place of the soft bone. Axe-blades, which are, in fact, formed like two chisels placed back to back, are made on a similar principle, except that the steel

plate occupies the centre of the blade, and the soft iron is on
either side. Thus the thin plate of steel is easily brought to an
edge, while the soft iron can be ground away without any difficulty.

I do not mean to state that the inventor of this combination
of thin steel and soft iron had taken his idea from the Rodent
tooth, but only to show that the invention, beautiful, simple,
and ingenious as it is, has its prototype in Nature. I may
here mention that the Plane-iron, which is, in fact, a modified
Chisel, is made in exactly the same fashion.

NEXT we come to the Adze.

In some respects there is much resemblance between the
blade of the Adze and the teeth of the Rodent, especially in
their curve, which is almost identical in both. This form is

ADZE-TEETH OF HIPPOPOTAMUS.

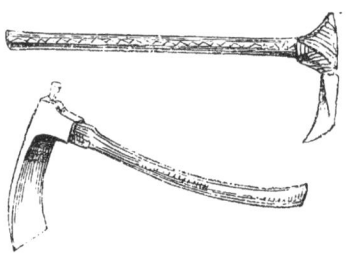

STONE ADZE OF POLYNESIA.
STEEL ADZE.

seen in the structure of other teeth than those of Rodents.
There is, for example, the tooth of the Hippopotamus, which is
not only curved, like that of the Rodent, but bevelled off in a
similar way at the tip. With these formidable teeth, one of
which is now before me, the Hippopotamus makes terrible
havoc among the herbage, mowing it down, so to speak, and
stowing it away wholesale in its enormous stomach. A Hippopotamus indeed, when angered, has been known to sever a
man's body completely in two with a single bite, so trenchant
are the teeth, and so powerful the jaws.

Then there is a little animal called the Hyrax, or Rock-
rabbit, which is the coney of Scripture. This creature is really
one of the pachydermatous group, although its small size, hairy
coat, its activity among the rocks, and its apparently rodent

teeth, have induced many persons to place it among that group. These teeth, however, like those of the Hippopotamus, are bevelled off at their tips, and, as they perform a similar office, they take a similar curve.

It is worthy of notice that in the Stone Adze the bevelled edge much more resembles the rodent tooth than does the Steel Adze, the reason being evidently that stone is more fragile than steel, and requires greater thickness. Still, the principle is the same in both, only the metal is more attenuated than the stone.

The Rodent or Hippopotamus tooth has still a great advantage over any chisel or adze made by man, whether of stone or metal. As our tools are blunted, we are forced to spend much time in sharpening them, and by degrees grind the tool away until it becomes useless. Now, the teeth are so arranged that their perpetual use, instead of blunting, only sharpens them, and in proportion as they are worn away in front they are supplied with fresh matter from behind, and perpetually pushed forwards, so that they are self-renewing as well as self-sharpening.

The Plane and Spokeshave.

I have already made mention of the Plane in connection with the Chisel, and shown that, like that tool, it is formed on the same principle as the Rodent tooth.

The use of this important instrument in carpentering cannot be overrated, as is shown by the numberless varieties which are used by carpenters, and the different uses to which they are put, sometimes merely smoothing a level surface, and sometimes forming a "moulding" where ornament is required.

In principle, a Plane is a cutting edge or chisel, pushed along the object to be worked, and, the edge being guarded, taking off a very thin shaving from the surface.

On the right hand of the accompanying illustration is shown the Plane in action, with the thin shavings falling from it in curled masses. Perhaps some of my readers may have visited some of the great iron-works, and been struck with the use of the Plane as applied to metal instead of wood, long iron shavings being taken off as easily as if they were deal, and curling in just the same manner.

THERE is an instrument very familiar to carpenters, called the Spokeshave, on account of its use in trimming the spokes of wheels. Different as it may be in appearance, it is identical in principle with the plane, having an edge guarded by a piece of wood, so that the blade cannot cut too deeply into the object on which it is employed. The chief distinction, indeed, is, that the workman, instead of pushing the blade from him, draws it to him.

When shaving was more in fashion than it is in these more

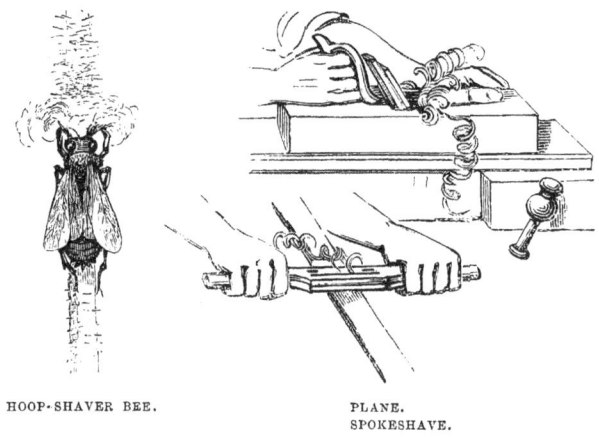

HOOP-SHAVER BEE.

PLANE.
SPOKESHAVE.

sensible days, there were many inventions to lessen the trouble, not to say the perils, of shaving. To use the razor in a hurry was anything but an agreeable occupation, especially if the weather were frosty, and the fingers so chilled that they hardly knew whether or not they had the razor between them.

In order to render this very unpleasant task less disagreeable, some ingenious individual invented the Guard Razor. The principal part of the invention consisted in a plate of metal sufficiently thin not to add materially to the weight of the razor, and sufficiently strong to resist a moderate amount of pressure. This was fixed along the blade of the razor in such a way that it just allowed the edge to show itself, and, in fact, converted the razor into a plane or spokeshave. The exact amount of edge which might be shown was regulated by screws, and the guard itself could be removed at pleasure, so as to allow of the razor being sharpened.

Now let us see if we can find any examples of the Plane or Spokeshave in Nature.

I TRACE at least one example of the Plane in the insect world. More than a hundred years ago, that very observant naturalist, Gilbert White, noticed a bee performing a curious task. She was running up the stem of the garden campion, holding her jaws extended, and stripping off the down with all the dexterity of a hoop-shaver. She collected a bundle nearly as large as herself, and then flew away with it. What she did with her burden he knew not, but the history of the insect has been told fully, though briefly, by Mr. F. Smith, in his " Catalogue of British Hymenoptera :"—

" Although the species belonging to this genus are numerous, and are found both in the Old and New World, there is only one found in this country, *Anthidium manicatum ;* this is truly a summer bee, not making its appearance before the latter part of June or beginning of July.

" This insect, so far as my own observation has enabled me to ascertain, does not construct its own burrow, but makes use of any hole which is adapted to its purpose. I once detected a bee entering the hole above the wheel of the sash-line in a summer-house ; but its nests are most commonly formed in the holes bored in old willow stumps by *Cossus ligniperda* (the Goat-moth) : formerly they were easily obtained in Battersea Fields, where the willows abounded.

" It is probable that when the parent insect has selected one of these ready-formed tunnels, she enlarges the end used as the depository of the nest, and this is easily effected, as the stumps in question, at the depth of a couple of inches, consist of soft decayed wood.

" The chamber being formed, the bee collects a quantity of down from woolly-stemmed plants, with which she forms an outer coating. She then constructs a number of cells for the reception of the pollen, or food of the larva ; they consist of a woolly material, mixed with some glutinous matter which resists the moisture of the food they contain, and in which the larva, being full fed, spins a brown silken cocoon. These bees pass the winter in a larva state, and do not appear until mid-summer.

"In one respect, the sexes of this genus differ from most other bees, the males being much larger than the females."

The reader will see from this account how exact is the analogy between the carpenter's plane and the jaws of the bee. In consequence of the simile employed by Mr. White, the insect has been popularly known by the title of the Hoop-shaver Bee. It is a tolerably common insect, and abounds in the South of England.

TOOLS.

CHAPTER II.

THE SAW AND ITS VARIETIES.

Cutting Tools and their working.—Structure of the Edge.—The Kris.—Edge of a Razor.—The Sword and the Apple.—Australian Saw.—Fretwork Saw.—Various Saw-flies.—The Pioneer's Saw.—Cutting Tools of Trichiosoma.—Side Teeth of the Saws.—The Cordon Saw, or Band Saw.—Tooth-ribbon of Whelks, Slugs, and other Molluscs.—The Dog-whelk, or Purpura.—The Circular Saw.—Sawyer-beetles and their Mode of Work.

STILL keeping to the Cutting Tools and their varieties, we come to the Saw, *i.e.* the cutting tool set with teeth upon its edge. Now, in plain fact, there is no cutting instrument that does not more or less partake of the character of the Saw; for, in the first place, it is absolutely impossible for man to grind an edge so fine that, when magnified, it will not appear to be deeply notched, and, in the next place, its cutting powers are greatly due to the notches and teeth, and the direction of their points.

We will take both these subjects in turn.

First, as to the notches, or serrated edge. I have now before me two instruments, each the best of their kind, and in both of which the serrations are essential to efficacy. The first is a Malayan dagger, or " kris," and the second is a surgeon's lancet, made by Ferguson, of London.

In the kris the edge is intentionally serrated, having been eaten away by means of acids until the required effect was produced. The Malayans know by experience that such an edge is most deadly in a weapon, and that it will cut certain vital parts which a smoother edge might pass without doing any damage.

Now we will take the lancet, and put it under the microscope, when it assumes the most curious resemblance to the kris. Its mirror-like surface looks as if it had been very roughly treated with a coarse file, while its thin and delicate edge, which is perfectly smooth to the eye, and which will pass through a piece of stretched wash-leather without any apparent opposition, becomes as rough and jagged as that of the Malayan weapon.

Take even, for example, the common butcher's knife, which is perpetually being sharpened on the "steel" that hangs at his belt. The reader may observe that the butcher does not rub the blade of his knife backwards and forwards on the steel, as unskilful persons do. Rapid as is the movement gained by constant practice, any one may see that the blade is always moved in one direction, so as to force the microscopical teeth to point one way, and so to act as a saw when the knife is drawn across the meat.

The power of these teeth or notches may be inferred from a well-known fact. If a razor, no matter how sharp, be pressed upon the human skin without any "draw," it will indent the skin, but not cut it, while the slightest drawing movement will cause a deep wound. It is the knowledge of this fact that enables an expert swordsman to sever an apple placed on the palm of the bare hand, without even scratching the skin. I have witnessed this feat, and at once saw that it was due to the absence of any "draw" to the cut. The apple was laid on the palm of the hand, which was opened as widely as possible, so as to flatten it. The sword was then brought down on the apple with a sort of chopping movement, so that, although it indented the skin, it did not even inflict a scratch.

By the use of the "drawing" movement, the same sword severed a gauze veil laid across it, the two halves floating in opposite directions. By the same cut, I have seen some astonishing feats performed with an Indian sword now in my collection, the objects of attack falling asunder as if by magic, without any apparent force being used.

HAVING now glanced at the principle of the Saw, we will proceed to some of its details.

The simplest form of Saw in existence is that which is in use among the Australian natives, and consists of obsidian flakes set along one side of a stick. It looks a rude and inefficient affair enough, but it can cut better than might have been thought, as I can testify from experiments on such a saw in my collection.

Many as are the varieties of the Saw, the principle is the same in all, and the chief distinction lies in the shape and arrangement of the teeth, according to the work which they have to do. Watch-spring Saws, for example, which have to cut metal, have their teeth so slight as to be hardly perceptible, and arranged nearly in a line with each other. The Fretwork Saws, which have to cut delicate patterns in wood,

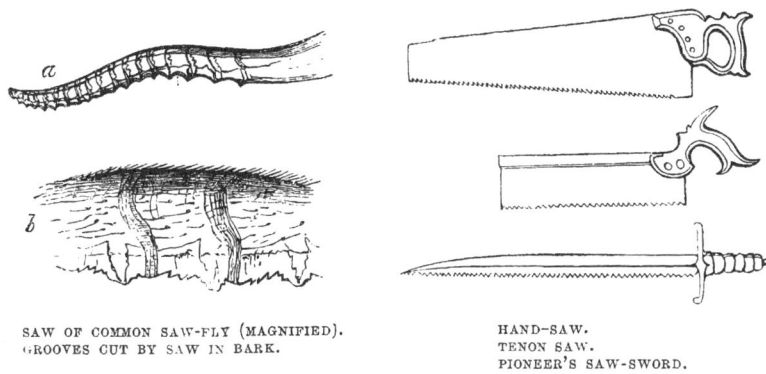

SAW OF COMMON SAW-FLY (MAGNIFIED).
GROOVES CUT BY SAW IN BARK.

HAND-SAW.
TENON SAW.
PIONEER'S SAW-SWORD.

with the slightest possible waste of material, are of the same character. Then we have the long curved teeth of the Circular Saws, which tear their way savagely through great tree-trunks, and fill the air with clouds of sawdust. There are also the Tenon Saw, with its thin blade and broad back; the pioneer's saw for cutting green wood, with its double array of teeth, so as to make a wide "kerf" in which it shall not be clogged; together with many others that we cannot enumerate here.

WE will now examine some Saws as found in Nature.

I need scarcely say that some of the best examples of natural saws are furnished by those insects which are known to entomologists as Tenthredinidæ, and to the general world as Saw-flies. These insects are supplied by Nature with a pair of

R

most remarkable saws, which aid them in depositing their eggs. Indeed, without these instruments, the whole race of Saw-flies would long ago have become extinct.

They haunt almost every kind of tree and many plants, and one valuable plant, the Turnip, is so devastated by them, that whole crops are sometimes swept away. As, therefore, the knowledge of the life-history of any insect will tell us whether to protect or destroy it, and the best method of adopting either course, we will cast a hasty glance at some of our commonest Saw-flies, the instruments which they employ, the mode in which they use them, and the analogies between them and the saws made by the hand of man.

In the first place, it must be observed that the use of these saws is to cut grooves in young bark, these grooves being the depositories of their eggs. It follows, therefore, that as a tolerably wide groove is needed, the saw-blade is a tolerably thick one, and the teeth set on the same principle as that which is employed in the saw-sword of the pioneer. When the microscope is applied to the cutting instrument of the Saw-fly, it reveals the fact that there are two horny saws, which work alternately in their grooves, and that they are strengthened by a thick plate of horn on their backs.

The system of toothing is very complicated. Not only are the sides as well as the edges of the saws toothed, but each tooth is furnished with smaller teeth, after the fashion of the shark's wonderfully effective cutting apparatus. These subsidiary teeth vary greatly in shape and size according to the species, and in some cases each tooth is quite a complicated structure. In *Trichiosoma lucorum*, for example, a bee-like insect, very common upon hawthorn, the teeth are extremely beautiful. It is difficult to describe them without diagrams, but I will try to give the reader an idea of them.

Each tooth is somewhat of a lancet shape, but is not terminated by a single point. At the tip comes the secondary tooth, which is conical and stands on a footstalk. The cone, however, is not simple, but is made of some seven or eight cutting plates, each smaller than its predecessor, and the last being a sharp conical point. The reader may imagine how effective such a saw would be in cutting green wood, the toothed sides and the subsidiary teeth alike preventing the blades from clogging, while the alternate movement of the saws enables them to do double work in the same time.

Mr. Westwood, who examined these insects very closely, throws out, in his "Modern Classification of Insects," the idea which forms the subject of this book. Writing of the cutting weapon of the Saw-flies, he remarks that "from its admirable construction it cannot be doubted that a careful examination of its various modifications might furnish ideas for improved mechanical instruments."

Mr. Gosse, in his "Evenings at the Microscope," points out that, beautiful and elaborate as these instruments are, they are but the sheaths of a still finer and more delicate pair of saws. These secondary saws have only a few teeth on the edge, and these near the point, whereas the sides are furnished with a number of sharp blades, set on their edges, slightly over-lapping each other, and directed backwards. There is a similar structure on the ovipositor of the Sirex, as we shall see when we come to treat of Boring Instruments.

Although the saws are made expressly so that they shall not stick in the wood, there are many instances known where female Saw-flies have been found dead on the branches, their saws still in the last groove which they have cut. I am inclined to think that these must be females which have deposited all their eggs, and which have died, as do nearly all insects under similar circumstances. This opinion is strengthened by some observations made by Mr. J. K. Lord on the Cicada, the female of which is furnished with a similar ovipositor :—

" I was curious to watch the female depositing her eggs.

" She first clasps the branch on both sides with her legs, and with the ends of the file very carefully slits up the bark. Then, placing the instrument longitudinally, she files away until she has obtained sufficient length and breadth. The *small* teeth of the files are now used crosswise of this fissure, until a trench is made in the soft pith.

" When large enough, slowly down the groove in the centre of the instrument glides a small pearly egg, pointed at both ends, and so transparent that the little grub within is clearly discernible. Gently she lays it within its bed, and then drops a thin gummy material on it, to secure it from moisture. This finished, she proceeds to deposit another, and so on, until a sufficient number are produced to fill the fissure ; then over all she drags the everted bark. It is easy to perceive where the Cicada has been concealing her brood, by the elevation on the branch.

" In this manner she deposits about seven hundred eggs, going from branch to branch, her marvellous instinct teaching her to select the most suitable wood for the purpose. The time occupied in constructing each nest was from fifteen to twenty minutes. Her earthly mission finished, she drops, fainting and exhausted, from the branch, and dies.

" The male, who is always trilling his refrain, goes on, indifferent, or unconscious, that the task of his faithful spouse is finished, singing even, until his time comes—then he too drops beside her. Thus the songs one by one cease,—not only the Cicada's, but all the forest choir, and give place to blasts that sigh in mournful music through the leafless trees."

The Sirex and several of the larger Ichneumon-flies are often found dead in like manner, and I have no doubt from the same cause. An elaborate description of the beautiful double saws of the Cicada is given by Mr. Westwood in the work already quoted, together with illustrations.

THE RIBBON SAW, CORDON OR BAND SAW.

PERHAPS some of my readers may be acquainted with a saw which has of late years come into extensive use—namely, the Ribbon Saw, Cordon Saw, or Band Saw. This is an endless steel band toothed on one edge, and passing over two wheels. It has the advantage of being of almost any breadth, some being several inches wide, while others are mere narrow ribbons, barely the sixth of an inch wide. The fretwork of pianos and other articles of furniture is cut almost exclusively by the Cordon Saw. A thick piece of wood is cut of the requisite shape, and the upper and under surfaces planed quite true to each other. The pattern is traced on the upper surface, and a very narrow Cordon Saw is then applied to it, cutting completely through the thick block, and adapting itself to all the intricacies of the pattern. The block is then cut into thin slices, so that a number of pieces of fretwork can be made with comparative ease. To those who have been accustomed to cutting fretwork with the slow hand-saw, the Cordon Saw is simply fascinating, the slender steel ribbon cutting through the wood with wonderful rapidity and very little sound.

BEAUTIFUL as this invention is, it was long ago anticipated in Nature ; and the Cordon Saws, which we shall now see, are armed with teeth many more in number, and far more complicated in detail, than those of any saw made by the hand of man. I allude to the Tooth-ribbon possessed by many of our common molluscs, such as the Limpet, the Whelk, the Periwinkle, the Slug, &c. The last mentioned of these creatures possesses a natural Cordon Saw with nearly twenty-seven thou-

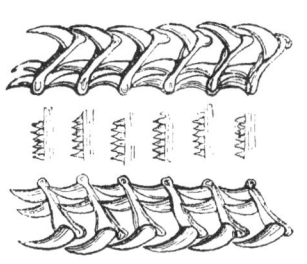

PORTION OF TOOTH-RIBBON OF WHELK
(HIGHLY MAGNIFIED).

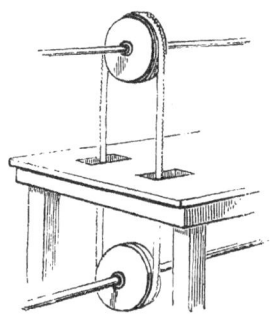

RIBBON OR CORDON SAW.

sand teeth, and scarcely a tooth that is not elaborately cut into secondary teeth.

As all these creatures have their teeth differently formed and set, according to the species, it will be impossible to describe them separately. I will therefore restrict myself to the Tooth-ribbon of the common Whelk, a specimen of which is now before me. When viewed through the microscope, it is found to consist of a flat membranous ribbon, on which are set three rows of teeth, those of the outer row being hooked, and those of the inner one plain.

The outer teeth are formed somewhat like the Hebrew letter כ, both of the points being very sharp, and the central part being furnished with two secondary teeth. All these teeth overlap each other, so that some care in manipulation is required before their form can be made out.

Along the centre of the tooth-ribbon run successive rows of small, lancet-shaped teeth, six in a row, so that altogether there are eight teeth in each row.

The power of this weapon is astonishing. Some of my readers may be aware that Whelks are carnivorous beings, and that they swarm upon any dead animal which may be found in the sea. Indeed, when we hear of the mutilations which take place on dead corpses after a shipwreck, and which are generally attributed to fishes, we may make up our minds that the real delinquents are the Whelks, together with various crustacea, and that the principal instrument in effecting such mutilation is the tooth-ribbon which has just been described.

The Whelks feed largely upon other molluscs, in spite of their shells. A periwinkle has a peculiarly hard shell, and yet Mr. Rymer Jones saw a Dog-whelk (*Purpura lapillus*) eat a periwinkle in a single afternoon, first boring a hole through its shell with the tooth-ribbon, and then, by means of the same weapon, licking it, so to speak, out of its shell.

The Periwinkle itself has a similar tooth-ribbon, and so have the Limpet and the pretty Top-shell. These creatures are vegetarians, but they are furnished with similarly armed tongues, and use them in the same way. Nothing is easier than to see these tooth-ribbons in use. When sea-water is kept in glass vessels, a green flocculence is sure to collect upon the glass and to render it opaque.

If, however, a few Periwinkles and Top-shells are placed in the tank, they immediately set to work at this confervoid growth, and by means of the tooth-ribbon sweep off the green substance, leaving the glass nearly clean. This movement can be seen with the naked eye, but with the assistance of a pocket lens the action of the tooth-ribbon is beautifully shown as it issues from its socket, makes its sweeping curve, with the tiny teeth glittering like specks of glass, and then is withdrawn ready for another sweep.

Should sea-water and living Periwinkles not be easily obtained, the same phenomenon may be observed in fresh water, and with the common Pond-snail, which may be caught by thousands in any stream and in most ponds.

THE CIRCULAR SAW.

IN one sense the Cordon Saw is a Circular Saw, but we now restrict the name to the tool which has a circular blade, more or less deeply toothed on the edge. The largest and coarsest of these saws are of enormous diameter, have teeth several inches in length, and can cut a large tree-trunk asunder in a wonderfully short time.

There is a huge saw of this kind in Chatham Dockyard. It is kept in a sort of cellar covered with flap doors, where it really has the air of some dread monster lying in wait for prey. A tree-trunk is brought for it to feed upon. The doors slowly open, the saw emerges, revolves so fast that the eye cannot detect the teeth, seizes on the tree-trunk, tears its way through

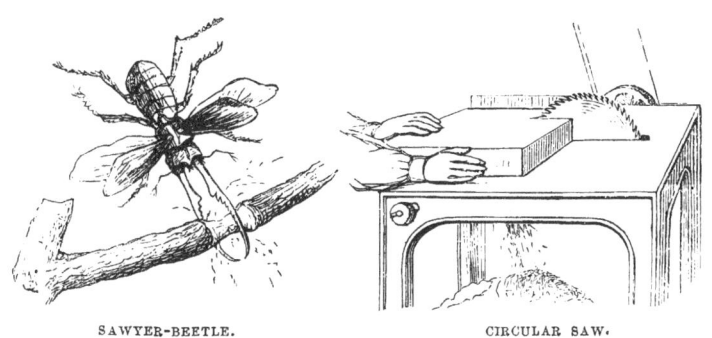

SAWYER-BEETLE.   CIRCULAR SAW.

with a scream and roar, and then sinks back into its cellar. I have often watched this saw in action, and have never been able to get over a kind of feeling that it was alive.

Now, if we suppose the saw to be pierced in the centre, and to have teeth on the inside instead of the outside, it would be equally efficacious ; and, indeed, we have several tools used for cutting iron bars or pipes, that are constructed on a similar principle, though the cutting tooth revolves slowly instead of rapidly, and is urged by a lever handle.

THERE is in Nature a Circular Saw of just such a character, the teeth having their points directed inwards, and not out-wards.

In tropical America there are several large beetles which, like our Stag-beetle, feed upon the sap of trees, and obtain it by wounding the young branches with their jaws.

One or two of them are pointed out as having the power of cutting a branch completely off by seizing it in their deeply toothed jaws, and flying round and round the branch so as to convert themselves into a circular saw. The late Mr. Waterton showed me a branch which had fallen on his head, and which was said to have been cut off by the Sawyer-beetle, as the insect is called. He did not actually see the insect at work, but he had no doubt that the natives were right who told him that it was the work of beetles' jaws. Certainly the cut looked exactly as if it had been made in the way described. The branch was somewhat thicker than an ordinary walking-stick.

TOOLS.

CHAPTER III.

BORING TOOLS.—STRIKING TOOLS.—GRASPING TOOLS.

The Bradawl and the Gimlet defined.—Natural Bradawls.—The Ichneumon-flies.
—A Pimpla engaged in Boring Operations.—Principle of the Wedge.—
Resisting Power of Earth.—Pitching Tents in Sand.—Hidden Forces of
Nature.—The Aloe-leaf and its Growth.—A cruel Punishment.—Natural
Gimlets.—Ovipositor of the Sirex, and its Analogy to a Carpenter's Gimlet.—
The Auger and the Gad-fly.—Striking Tools.—The Hammer.—Origin and
Development of the Tool.—The Axe.—The Woodpecker and the Nuthatch.—
The Ivory-billed Woodpecker.—Grasping Tools.—Pincers and their Modi-
fications.—Sugar-tongs and Coal-tongs.—Natural Pincers.—Bivalve Mol-
luscs.—The Clam's Grip.—The Earwig.—Crab and Lobster Claws.

BORING TOOLS.

NEXT in importance to the edged tools which cut, come the
pointed tools by which holes can be bored. We have
an abundance of such tools, but they can all be reduced to
two types, namely, those which, like the Bradawl, are forced
between the fibres, and those which, like the Gimlet, cut away
the material as they pass through it.

They may, again, be shown to be different modifications of
a single principle—*i.e.* that of the Wedge or Inclined Plane,
which, as has already been shown, is identical with that of the
screw. The Bradawl is, in fact, a sharp wedge, which is forced
through the fibres, sometimes being merely forced between
them, and sometimes cutting them, and thus forcing aside the
severed fibres.

A natural example of the Bradawl is to be found in various
Ichneumon-flies, especially those with very long ovipositors,
which are intended for boring into wood.

All the Ichneumons are parasitic, laying their eggs in the
larvæ of other insects, mostly those of moths and butterflies.

Generally these larvæ exist in the open air, and the Ichneumon-fly has little difficulty in piercing them. But there are some which live either in wood or underground, and, in order to reach their hidden bodies, the Ichneumon is furnished with an extremely long and sharply pointed ovipositor.

This wonderful instrument is not so thick as an ordinary horsehair, although it is composed of three portions, and seems to be utterly inadequate to the task which it has to perform. Ascertaining by its instinct the exact locality of the caterpillar which it desires to pierce, the Ichneumon-fly clings firmly to the tree, bends the body so as to bring the point of the ovipositor against the wood, and, by moving the abdomen backwards and forwards, gradually works the instrument into the wood, sometimes piercing it to a considerable depth.

Mr. Westwood once saw an Ichneumon-fly thus boring its way into a dry post, the wood of which must have been very hard. When she had bored far enough, she partially withdrew the ovipositor, and then re-plunged it into the hole that she had made, as if she were depositing eggs. While engaged in this operation, she stood very high on her long legs, resting only on the extremities of the feet. She belonged to the genus Pimpla.

THE principle of the Wedge or Inclined Plane is admirably shown by objects which we pass unheeded every day, and yet afford wonderful examples of the power of the wedge.

Scarcely any vegetable growth is so plentiful as grass, which has been used in that sense by the highest of all authorities, " which to-day is, and to-morrow is cast into the oven." Grass forces its way everywhere—not only in cultivated grounds, but in the wildest of lands, where there is scarcely any nurture for it. Even among the habitations of mankind the grass will have its way, and clothes deserted housetops with verdure, and forces itself between the stones that pave neglected streets.

Place side by side some of these stones, together with a very young and tender Grass-blade, and it will seem to be impossible that so fragile an object should be able to exert any influence on the solid stone. Let any one try to push a sharp skewer between the stones, and he will find that he has to exert power sufficient to crush a thousand grass-blades. Yet these slight and delicate objects will force themselves between the stones, and sometimes to such an extent as to cover the whole roadway with verdure.

The force which is employed is simply marvellous, and can only be appreciated by those who know the resisting power of earth, however dry and loose it may be. Even sand has so strong a resistance that tents can be pitched in the desert without difficulty. Of course the ordinary tent-peg would be useless, but the desert dwellers can pitch their tents with perfect security. They fasten the tent-rope to a branch or piece of bush, scrape a hole in the sand, put the bush into the hole,

GRASS-BLADES.   WEDGE.

cover it up again, and it will withstand almost any strain, though it be only covered with a few inches of sand.

When miners blast rocks with gunpowder, they take advantage of the resisting power of sand. They bore a suitable hole, place a charge of gunpowder at the bottom, and then merely pour loose sand into the hole until it is filled. When the powder explodes, the rock or coal is shattered to pieces, but the sand is not blown out of the hole. This operation is called " tamping."

Every one, again, knows how firm are gate-posts, and how they resist the weight, jarring, and leverage of a heavy gate, all because they are sunk a little way into the earth.

Considering, therefore, that such fragile things as young grass-blades can force their way through the superincumbent weight, we can but be amazed at the aggregate of active force which is in full operation in every pasture field and garden lawn.

As far as I know, not being much of a botanist, every seed that springs up does so on the wedge principle, though the form of the wedge may be varied.

A terrible example of the force which is exercised by this principle among the vegetables is shown in some parts of the world where the Aloe flourishes in a wild state. In our colder clime the Aloe, though it does live in the open air, is a slow-growing plant. But, in its own land, it shoots up with a surprising vigour, and its sharply pointed and saw-edged leaves are said to grow to the extent of six inches in a single night.

Taking advantage of this rapid, and, at the same time, powerful growth, the natives, when they want to punish a man with more than ordinary severity, tie him hand and foot, and bind him to the earth just over a sprouting aloe plant, and leave him there. In twenty-four hours the man is nearly certain to be dead, the aloe-leaf having forced itself completely through his body. Or, if he be not actually dead, he lives in frightful tortures, which are continually increased by the flinty point and notches forcing themselves slowly, but surely, through the body.

For an example of the Gimlet we may take the ovipositor of the Sirex, an insect which I believe has no popular name. It is coloured much after the same manner as the hornet, and is often mistaken for that insect by those who are not versed in entomology. And, as its long and straight ovipositor is generally taken for a hornet's sting, the insect assumes a double terror to the ignorant.

Now, the real fact is, that in its larval stage of existence the Sirex feeds upon the wood of the fir-tree—a diet which, to our ideas, is about as unsatisfactory as can well be imagined. In order that the young Sirex may be within reach of food, the egg must be introduced deeply into the body of the tree, and, for the egg to be so received, a channel must be cut for it.

This is done by means of the marvellously formed ovipositor. Many admirable descriptions have been given of the head of this instrument and its boring powers, but I am not aware that any one has noticed the secondary cutting blades that are set along the shaft of the principal borer, and which answer exactly the same purpose as the spiral cutting edge of the gimlet or auger.

Not being desirous of repeating my own observations in different words, I transfer to these pages a short account of the ovipositor of the Sirex, as examined by me when writing my work on British Insects, entitled "Insects at Home," and published by Messrs. Longmans and Co. :—

"I very strongly recommend any of my readers who may obtain a female Sirex to disengage the actual borer from its two-bladed sheath, and examine it with the aid of a microscope. A half-inch object-glass will give quite a sufficient power.

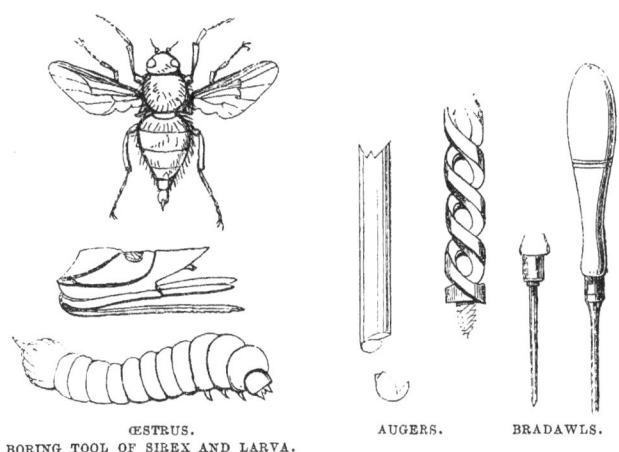

ŒSTRUS.   AUGERS.   BRADAWLS.
BORING TOOL OF SIREX AND LARVA.

"It is straight, stiff, and elastic, as if made of steel, and, if bent, will spring back to its proper form with the elasticity of a Toledo rapier.

"But the borer possesses an auxiliary cutting apparatus which places it far above the rymer in point of efficacy. Even with an ordinary magnifying lens, it is easy to see that the end of the borer is developed into a sharp head, very much resembling that of a boarding-pike, and that the outline of the shaft is broken into a series of notches.

"The half-inch glass, however, discloses a marvellous example of mechanical excellence. The head of the borer is then seen to be armed with long, sharp teeth, slightly curved inwards, and acting just as does the carpenter's ordinary centre-bit.

"So much for the head of the borer: we will now turn to the shaft.

"It appears that, in order to make a clean-cut hole for the reception of the egg, the shaft of the borer has to finish the task which the head begins. Accordingly, it is armed on each of its sides with a series of hard, sharp-edged ridges, running diagonally across it, and acting exactly as do the sharp ridges of a coffee-mill."

In point of fact, the ovipositor of the Sirex is the natural type of the improved gimlet of the present day. Instead, however, of having a single, spiral, sharp-edged groove running along the whole length of the shaft, it has a series of small, sharp blades, set exactly in the same line as is taken by the spiral groove, and acting in exactly the same manner—*i.e.* by cutting out successive portions of wood, and, by the diagonal position of the blades, throwing out the débris as fast as it is cut.

I cannot but think that, if any modern tool manufacturer could take as his model the saw-like ovipositor of the Tenthredinidæ, and the auger-like ovipositor of the present insect, he would produce a series of most valuable implements, possessing powers far beyond those of ordinary tools.

These short blades are arranged just like the "studs" on modern shells, and very much resemble them in shape, though not in material.

THE Auger finds also a natural representative in the ovipositor of an insect.

That of the common Gad-fly (*Œstrus bovis*) is most beautifully constructed. It is tubular in form, and is of a telescopic nature, consisting of four tubes of different sizes, the smaller fitting into the larger just as is done with the joints of a common telescope, or those of a Japanese fishing-rod.

The end of the ovipositor is developed into little projections, some of which are armed with hard, sharp points, which act exactly like the cutting edge of the auger. This elaborate appliance is necessary on account of the thick, tough skin of the ox, which the Gad-fly has to penetrate before it can deposit its eggs. Perhaps the reader may be aware of the fact that the modern system of cutting channels in stone with the diamond point, as was so well exemplified in the Mont Cenis Tunnel, is but an imitation, and an imperfect one, of the method adopted by the Gad-fly. We shall soon recur to this instrument.

Striking Tools.

IF we search the records of antiquity as left by races of men that have for countless ages vanished from the face of the earth, we shall find that in some shape or other the Hammer was a tool in constant use, and that in principle, though not in material, there was no difference between the Hammer of the Stone Age and that of a blacksmith of the present day.

The development of the instrument can easily be traced, especially as it is a tool which does not admit of much elaboration.

The original hammer was evidently a simple stone, and answered equally as a tool and a weapon. As, however, man progressed towards civilisation, he found that the stone itself was insufficient for his needs, and that he required much more force. The most obvious mode of doing so was to take a larger stone, but this expedient soon became valueless, inasmuch as a large stone was a cumbrous instrument to handle, and could not be directed with any certainty or delicacy.

The principle of the lever was then applied to the stone, which was affixed to a handle, and thus became elevated into the rank of a comparatively civilised tool. Sometimes the stone had a hole bored through it, into which the handle of the hammer was inserted, as is the case with most of our present hammers and pickaxes. Sometimes the end of the handle was enlarged, and the stone thrust through it, as is now done with the axes of Southern Africa. Sometimes a long, flexible rod was used by way of handle, the centre of it taking two turns round the stone, and the ends being lashed together. Handles thus made may be seen in any blacksmith's forge of the present day.

The tool thus made was soon developed into various forms for different uses. By lengthening and pointing the head, it became a pick for loosening the earth. By widening and flattening the head, it became a hatchet ; and, by performing the same alteration in the pickaxe blade, it became an adze. I possess a singularly ingenious tool from Borneo, in which the head is movable, so as to be used as a hatchet or adze at pleasure.

In Demmin's " Weapons of War" many such hammers and

axes are figured. One of them is very remarkable. It is an ancient war-hammer made of black stone, and is shaped exactly like a pickaxe, except that one end of the head is carved into a semblance of some animal's head. The handle is passed through an oval hole in the centre, just like our pick-axes of the present day. This remarkable example of the art of the Stone Age was found in Russia. The head was nearly a foot in length.

NATURE possesses many examples of this principle, of which I have chosen two, namely, the Woodpecker and the Nut-hatch.

The wonderful power of beak possessed by both these birds is familiar to every one, but it is not so generally known that

NUTHATCH.   WOODPECKER.   HAMMER.

they do not merely peck after the usual fashion among birds, *i.e.* delivering the stroke with the force derived from the neck alone. These birds have an additional leverage. Grasping the tree firmly with their feet, they not only peck, but swing their whole bodies with each stroke, bringing their weight to bear upon the object. They thus convert themselves into living hammers, the feet acting the part of the human hand, the body of the bird being analogous to the handle of the hammer, and the head playing the same part in both cases.

In England these birds are not known as well as they ought to be, partly because they are both very shy creatures, and partly because the gradual extinction of forests has deprived them, and especially the Woodpecker, of their undisturbed homes. Yet those who are early risers may see both birds in places where their presence is quite unsuspected, except, perhaps, by those who can recognise the signs which they have left behind them.

There is a common saying to the effect that "a carpenter is known by his chips," and the proverb is equally true of the Nuthatch and the Woodpecker. Nutshells scientifically split asunder, and jammed into the rough bark of a tree-trunk, betray at once the Nuthatch to the eye of a naturalist; while an accumulation of shattered bark, splinters of wood, and similar débris announces, in equally bold type, that a Woodpecker has been at work.

The power of the Woodpecker's beak may be gathered from Wilson's well-known account of an Ivory-billed Woodpecker, which he had wounded and was trying to rear. While staying at an hotel, he locked the bird in his room, and, on returning within an hour, found an astonishing state of things.

"He had mounted along the side of the window, nearly as high as the ceiling, a little below which he had begun to break through. The bed was covered with large pieces of plaster, the lath was exposed for at least fifteen inches square, and a hole large enough to admit the fist opened to the weather boards, so that in less than another hour he would certainly have succeeded in making his way through.

"I now tied a string round his leg, and, fastening it to the table, again left him. I wished to preserve his life, and had gone off in search of suitable food for him. As I re-ascended the stairs, I heard him again at work, and on entering had the mortification to perceive that he had almost ruined the mahogany table to which he was fastened, and on which he had wreaked his whole vengeance."

The beak of the Woodpecker was employed upon its new master quite as forcibly as upon walls and furniture, but Wilson was of too generous a nature to resent his injuries, and lamented sincerely when the bird died.

The reader will probably observe that the Hammer which has been given as an illustration of this principle is the ordinary geologist's hammer, and that it has been selected because its head is so formed that one end can be employed for the usual tasks of a hammer, while the other end, with its slight curve and sharp point, is, in fact, a sort of pickaxe, and used for the same purposes. Indeed, this instrument is an almost exact reproduction of the stone hammer which has already been mentioned, the blunt end being represented by the carved head, and the sharp end by the pickaxe point.

GRASPING TOOLS.

ALREADY we have spoken of the Shears and Scissors, together with their mode of action and dependence upon leverage. We now come to a set of tools which, although equally dependent on leverage, develop that power by grasping instead of cutting. Without these tools, the arts and sciences could have scarcely made themselves felt, as there are but few manufactures in which the artificer does not require a grasping power far superior to that of the human hand.

Perhaps the enormous power of the Pincers is never shown to better advantage than in the great iron-works, where enormous masses of white-hot metal have to be brought under the blows of the steam hammer. I do not know of anything which affords a more imposing realisation of the Divine command that man is to subdue the earth as well as to replenish it. There is the vast hammer, striking blows which are felt throughout a large area as if a succession of earthquakes had been let loose. In the furnace there is an enormous mass of iron, heated to such a degree that an unpractised eye could no more dare to look at it than to stare a midsummer sun out of face.

Where are the armies who are to cope with such forces? A few stalwart and grimy men come forward, each man with a curious but unmistakable air of one who wages a war of giants. The furnace door is opened, and out rushes a blinding light which strikes on the eyeballs like a shock of electricity. The men seize the handles of an enormous pair of Pincers, suspended in the middle by a chain, and though no unpractised eye can distinguish the glowing iron from the enveloping fire, they run the Pincers into the furnace, seize the iron, swing it to the anvil, and turn it this way and that way as easily as if it were a feather, while the blows of the gigantic hammer descend upon it, enveloping them in a torrent of sparks which spurt as if they were mere splashes of water, and seem to do them no more harm.

Taking the minor exposition of the Pincers principle and their use, we may mention the ordinary Pincers which are mostly used for drawing nails. Then there are the smaller Pincers called Pliers, all of which are constructed on the same principle, and the chief of which are the Round-nosed Pliers, the Long-nosed Pliers, and the Gas Pliers. Sometimes a mixture of the Hammer and the Pincers is ingeniously contrived, as in the tool which is represented on the right hand of the illustration.

Then we have the still smaller and feebler Pincers of civilised life, such as the Sugar-tongs and the ordinary Coal-tongs of

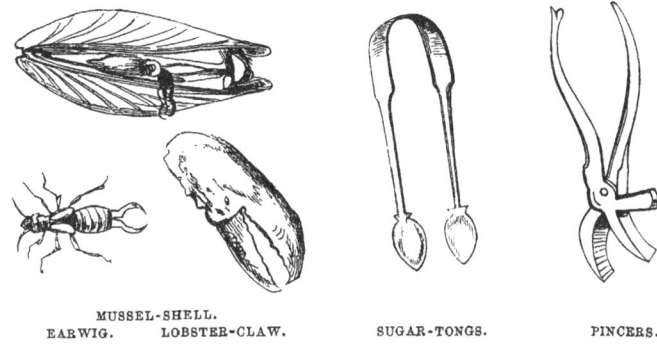

MUSSEL-SHELL.
EARWIG.   LOBSTER-CLAW.   SUGAR-TONGS.   PINCERS.

our firesides. Anatomists could have had no practical existence without the Pincers, of which their beautifully constructed and much-elaborated forceps are but variations.

Take, again, the dentist, with his series of shining instruments, which he so carefully keeps out of sight until he has got his patient safely in that awful chair, and which glide, as by a conjurer's trick, empty into an open mouth, and return in a few seconds with a tooth between their polished jaws.

ALL these instruments have their parallels in Nature, and in many instances the natural pincers might supply useful hints to modern tool-makers.

In the left-hand upper corner of the illustration is shown the common fresh-water Mussel, which is so plentiful in almost all our rivers and many of our ponds. Its scientific name is *Unio margaritiferus*. The latter title, which signifies " pearl-

bearing," is given to it because it furnishes the British pearls
which were at one time so highly valued.

Like other bivalve molluscs, this Unio has the two halves
of the shell fitting quite tightly upon each other, and, when
they are drawn together by the contraction of the internal
muscles, they can give a very severe pinch. In many uncivilised parts of the world the natives take advantage of
this property, and use them as tweezers, chiefly for the purpose of pulling out hairs which they are pleased to think are
not needed.

I need not state that with all bivalves the power is increased
in proportion to the size of the shell. Even an Oyster can
pinch most severely, while the Giant Clam, the shell of which
weighs some four hundred pounds, could nearly take off a man's
leg if it seized him.

Mr. J. Keast Lord, in his "Naturalist in British Columbia," relates an amusing story that was told to him by an old
settler respecting the power of the Clam's grip:—

" You see, sir, as I was a-cruising down these flats about
sun-up, the tide jist at the nip, as it is now, I see a whole
pile of shoveller-ducks snabbling in the mud, and busy as dog-
fish in herring time. So I creeps down, and slap I let 'em
have it. Six on 'em turned over, and off went the pack,
gallows scared, and quacking like mad.

"Down I runs to pick up the dead uns, when I see an
old mallard a-playing up all kinds o' antics, jumping, backing,
flapping, but fast by the head, as if he had his nose in a steel
trap; and when I comes up to him, blest if a large Clam
hadn't hold of him, hard and fast, by the beak.

" The old mallard might ha' tried his hardest, but may I never
bait a martin-trap again if that Clam wouldn't ha' held him agin
any odds till a tide run in, and then he'd ha' been a gone
shoveller sure as shooting. So I cracked up the Clam with
the butt of my old gun, and bagged the mallard."

Of course the reader will remember that this was only an
ordinary Clam, and not one of the giant race.

BELOW the shell are two very perfect instances of natural
Pincers, each acting in a different manner, but on the same
principle.

The Earwig is too familiar to need much description, but I may as well state that its pincers are not primarily intended as weapons, although they can be so used on occasion. (I was about to say, at a pinch, but refrain.) They resemble our ordinary pincers in that both blades move equally, and they are so completely under the control of their owner, that the insect uses them with a delicacy of touch that a lady's fingers could hardly surpass. They are really tools, and not weapons, and are employed for the purpose of folding the wide and delicate wings under the tiny elytra.

There is another insect called the Scorpion-fly (*Panorpa*), the male of which is furnished with a pair of pincers at the end of a long and flexible tail, articulated just like the tail of a scorpion, and moved in exactly the same manner. It is but a little insect, but its gestures are so menacing as it flourishes its tail about, that non-entomologists may well be pardoned for being afraid of it. Moreover, small as are the pincers, they really can give a smart nip, and make themselves felt on the human skin.

IF we want examples of exceedingly powerful pincers, we need only go to the Lobsters and Crabs, especially to the latter, whose claws are often of enormous thickness in proportion to the size of the animal. All those who have visited the seaside know how severe is the pinch of the common Green Crab, comparatively small though it be, and the same may be said of the river crayfish, which is, in fact, a lobster in miniature.

As to the lobster itself, fishermen are so well acquainted with the power of its claws, that they tie them together with string as soon as the animal is caught. Formerly they used to "peg" them, *i.e.* drive a wooden peg into the joint so as to prevent it from moving. This custom, however, is now prohibited by law on account of its cruelty.

The power of the Crab's claws is so great that a bite from a large Crab will inflict a severe injury, and render a hand helpless. It has more than once happened that men who have been feeling for Crabs in the recesses of the rocks at low water have been seized, and seriously imperilled, not being able to release themselves from the gripe.

Indeed, it is said that there have been instances where the

Crab has held so tightly, that the man has been drowned by the returning tide, no one having come to his assistance. I am, however, inclined to doubt this statement, thinking that the Crab would not be likely to remain in its hiding-place very ong after the water came up. Still, that such an idea should be currently believed in many parts of England shows the estimation in which the gripe of the Crab's claw is held.

TOOLS.

CHAPTER IV.

POLISHING TOOLS.—MEASURING TOOLS.

Files and Sand-papers.—The Sheffield File and its Structure.—The Equisetum, Mare's Tail, or Dutch Rush.—Beauty of its Surface when seen through the Microscope.—Sand-paper.—Skin of Dog-fish, Skate, and Shark.—Skate-skin used for Sword-handles.—Distinction between the File and Sand-paper.—Measuring Tools.—The Plumb-rule and the Level.—Their Use in Tunnelling.—The Measure and its Uses.—The Two-foot Rule and the Tape Measure.—Ovipositor of Gall-fly.—Tongues of the Woodpecker, Wryneck, and Creeper.—The Spirit-level and its Uses.—Theodolite and Callipers in Nature and Art.—The Contouring-glass.—Pincers of Earwig again.—Jaws of Insects.—The great Sialis of Columbia.

FILES AND SAND-PAPERS.

HAVING now examined the analogies between the cutting, boring, striking and grasping tools of Nature and Art, we come to those finishing tools which smooth and polish the surface.

The first is the File, an instrument which needs but little description. It consists of a surface of hardened steel, broken up into rough-edged teeth of infinite variety, according to the work which the file has to do. It is rather remarkable, by the way, that at present the English files are infinitely superior to those produced in any other part of the world; that their teeth are all made by hand; and that a genuine Sheffield file will first cut its way through a piece of iron in half the time that would be occupied by a file of any other nation, and then would easily cut its antagonist in two.

As long as the File is intended to work upon metal, there is little difficulty in its manufacture, except that no machinery has yet been invented which can give the peculiar edging of

the ridges, and to which is owing the unmistakable "bite" of
a real English file.

But there are occasions when the hand of the most cunning
file-maker is baffled, and when it is necessary to cut files so
delicate that the unaided human eye cannot trace their teeth.
Art, therefore, has recourse to Nature, and the cabinet-maker,
who cannot obtain any file made by human hands which will
answer his purpose in the higher branches of his trade, makes
great use of the "Dutch Rush," as he calls it.   It is not a rush
at all, but simply a species of Mare's Tail, or Equisetum, a
plant which fills in profusion almost every marshy spot in
England.

The peculiar fitness of the Equisetum for this purpose
cannot be appreciated even by those who use it until it has

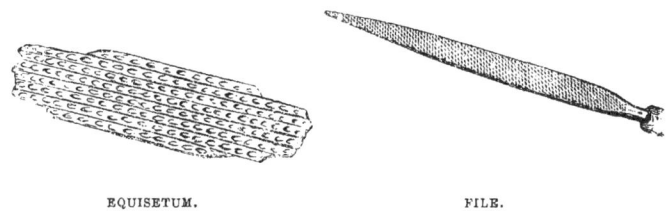

EQUISETUM.   FILE.

been viewed under the microscope.   I have now before me a
small piece of Equisetum, placed under a half-inch power, and
viewed by direct illumination, it being treated as an opaque
object.

The microscope reveals at a glance the source of the power
which the ingenuity of man has taken advantage of.   The
surface of the Equisetum is seen to be composed of myriads of
tiny parallel ridges, each ridge bristling with rows of flinty
spicules, looking very much like the broken glass upon the top
of a wall.   Minute as they are, these spicules can do their
work, and they enable the joiner to finish off work in a manner
that could not be accomplished by any tool made by human
hands.

I find, by recent inquiries, that modern joiners scarcely, if
ever, use the Equisetum, preferring emery-paper as cheaper
and more expeditious, and knowing that the popular eye is not
able to appreciate the difference of the surface obtained by

the Equisetum from that which is given by the finest emery-paper ever made. Wood-carvers, however, if they be of the conscientious kind, and love their work for its own sake, adhere to the Dutch Rush, and are all the happier for it.

PASS we now to the coarser kinds of polishers, the chief of which is popularly known as Sand-paper, and is made by coating some tissue with glue, and scattering upon it sand of different qualities, according to the work to be done. Sometimes, when the work is rough, the sand is large, rough, and coarse, and sometimes, when the work is fine, the sand is so carefully sifted before it is scattered on the glued paper, that there is little distinction between the sand-paper and emery-

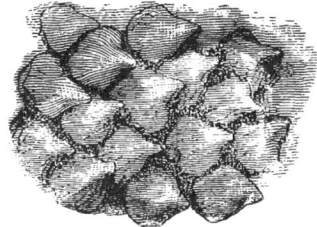

DOG-FISH SKIN.   SAND-PAPER.

paper. Linen, by the way, is generally used instead of paper, as being more enduring, less liable to crack, and capable of being folded so as to obtain access to crevices which paper could not touch.

AGAIN in Nature we find a parallel, and the coarse Sand-paper of modern Art has long been anticipated in the scale-clad skins of many fishes.

The accompanying illustration is taken from the skin of a Picked Dog-fish found by myself lying dead on the rocks in Bideford Bay. I cut off a piece for transmission to the draftsman, and found that not only did it feel exactly like cutting through a piece of very common sand-paper, but that it blunted the edge of a new knife in exactly the same manner as would have been done by the roughest of sand-paper.

This kind of skin is common to all the shark tribe (including the Dog-fishes, which are but sharks in miniature), and to the

Skate, Saw-fish, &c.   I have now before me a small, but perfect
example of the Saw-fish, the surface of which is covered
with flinty scales like those of the Dog-fish, but very much
smaller, requiring the aid of a magnifying lens to distinguish
them.   Even to guess at the number of them is impossible,
for they cover the whole of the body, and extend to the very
end of the beak, in some places glittering in a strong light as if
pounded glass had been sprinkled all over the fish.   One of the
most interesting points in their structure is the manner in
which they reach the rounded jaws, and there become converted into teeth powerful enough to crush the animals on
which the fish live.   The structure of these jaws will be
explained in a future chapter.

Some of the skates and sharks have these scales of great
size, so as to show their formation almost without the aid of a
magnifying-glass.   This is the case with a species of skate, the
skin of which is used by the Japanese for wrapping round the
handles of their best swords, and which is greatly valued by that
nation, the sword being an almost sacred article in the eyes of
a Japanese.

There is a well-known museum in which these swords are
labelled as having handles of "granulated ivory."   Now, in
the first place, there is no such thing as granulated ivory ; and,
in the next, a mere glance ought to tell the observer that
the so-called ivory is a skin of some sort, worked upon the
handle while wet, and kept in its place by copper studs.   Even
the junction of the edges is perceptible, and yet the authorities
of the museum in question, although they have been repeatedly corrected, still persist in calling the skate-skin by the
absurd title of granulated ivory.

However, if ivory could be granulated, it would certainly
look very much like the skate-skin.   When examined closely,
the scales, whether of Dog-fish, Skate, Shark, or Saw-fish, are
seen to resemble hexagonal cones, not coming quite to a
point, but truncated, so as to have an hexagonal flattened tip.
They are almost of a flinty hardness, especially at their tips,
and on inspection of them the observer is not surprised at the
use of Dog-fish skin in place of sand-paper.

Perhaps the reader may ask why the Equisetum should be
taken as the prototype of the file, and the skin of the Dog-fish

as that of sand-paper. The reason is this. The flinty points of the Equisetum are set upon parallel ridges something like those of a file, while the scales of the Dog-fish are without any apparent order, being crowded against each other like the cutting particles upon the sand-paper. That there should not be an order, and that a definite one, is out of the question. But it has not yet been detected by human eyes, and therefore may be practically treated as non-existent.

Tools of Measurement.

In many of the arts, more especially those which belong to engineering and carpentering as a part of architecture, it is absolutely necessary to make sure of a perpendicular line, *i.e.* a line which, if continued, would reach from any point of the earth's surface to its exact centre below and its zenith above. Were it not for the power of producing this line, none of the great engineering works of modern or ancient days could have been undertaken.

Take, for example, the wonderful tunnels which have been driven through the earth, of which the Mont Cenis Tunnel is one of the greatest triumphs of modern engineering. Beginning, as the workmen did, at opposite ends of a tunnel many miles in length, and labouring only by the lines laid down by the engineers, the men worked steadily on until they met in the centre.

A few blows, and the then narrow dividing wall was shattered, the men shook hands through the aperture, and then, after enlarging it, leaped wildly from one side to the other, having successfully solved the great problem. With such marvellous precision had the lines been laid, that only a few inches had to be smoothed down on either side, and the sides or walls of the tunnel showed no traces of the junction.

So rapid has been the progress of engineering that a tunnel of a mile in length would, within the memory of man, have been thought as daring a project as was the Mont Cenis Tunnel, which has just been given as an example. Indeed, I know of a railway tunnel, not quite a mile in length, where the engineers had committed some error, so that the two halves, instead of meeting exactly, overlapped each other so much that the mistake was only discovered by the workmen, who heard the strokes of their companions' picks on their sides, and not in front. Consequently, a great waste of time took place, and the centre of the tunnel had to be made with a double curve, like the letter S, and trains are obliged to slacken speed until they have passed it.

Those who have lived long enough to remember the current literature of the past generation will call to mind the ridicule that was cast upon the idea of a tunnel that should pass under the Thames. That it would be useful if it could be completed, no one ventured to doubt, but that such an idea could be conceived by any one out of a lunatic asylum was rather too much for the journalists of the day. However, the tunnel was made, and so proved the theorists wrong on the one side. And, when made, it was of very little use, which proved them wrong on the other side. Now the proposal to carry a submarine tunnel from England to France excites not half the opposition that was elicited by the comparative child's-play of a tunnel under the Thames.

The only mode of laying down the lines on which the men worked is by suspending very heavy balls to very fine wires, and then, by means of delicate optical instruments, ascertaining whether the wires are in line with each other.

Familiar instances of the use of this principle may be seen in the plumb-rule and level of the builder or carpenter. The latter, with a base of ten feet in length, is often used by the gardener when he wishes to lay the absolutely level lawns that are required for our modern game of croquet, where the hoops are scarcely wider than the balls, and the lawn has in consequence to be nearly as level as a billiard table.

I may here remark that the name plumb-rule is derived from the Latin word *plumbum*, or lead, in allusion to the leaden weight at the end of the string. The word "plumber" is due to the same source, and signifies a worker in lead.

THESE invaluable aids to the development of civilisation are due to one principle, namely, that which we call Gravitation, but which ought more properly to be termed Attraction, and which attracts all parts of the earth towards its centre. We are all familiar with the anecdote of Newton and the falling apple, which may be true or not, but which at all events bears on the present subject. No matter on what portion of the spherical earth a tree may be, every fruit becoming disengaged from it is attracted to the earth, the line which it takes, unless disturbed by external forces (such as wind, &c.), being that which passes from the zenith to the centre of the earth.

This imaginary line is a perfect perpendicular, and the visible line which is formed by the delicate wire of the tunnel-boring engineering instrument, or the comparatively coarse string of the plumb-rule and level, are approximations sufficiently close for practical purposes. So it is in a mathematical

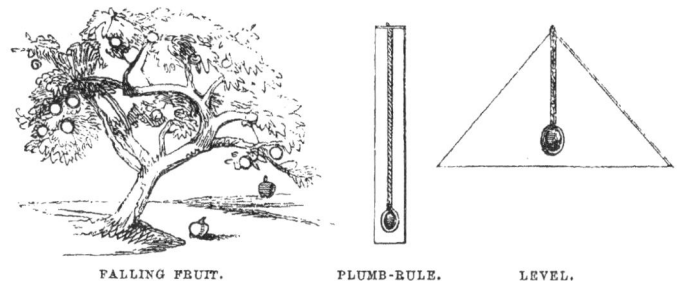

FALLING FRUIT. PLUMB-RULE. LEVEL.

proposition. As mathematical lines have no breadth, they are simply indicated or represented by the lines of the figure, the bodily eye being incapable of seeing what is perfectly visible to the mental eye, namely, length without width. So the wire and string perform in practical work exactly the same office which is fulfilled by the lines of a mathematical proposition drawn on paper.

We have already, when treating of the Fall-trap, seen how this principle is brought into operation by those who are utterly incapable of discerning the physical principle, though they can apply it materially with wonderful effect.

It is, perhaps, needless to mention the value of the Measure to any handicraftsman.

I well remember that when, some twenty-four years ago, I was taking lessons from a carpenter in the art of making ladders, gates, fences, hurdles, and other rough-and-ready work, my quaint old tutor related an anecdote of and against himself.

He very ingeniously set me to work at boring the auger-holes in the gate-posts which were to be united by the mortise chisel and mallet, and to sweeten the rather severe, because unaccustomed, labour, told me that, when he was a boy, he was doing just the same thing.

Being rather tired of twisting the auger handle (and no wonder either), he withdrew the instrument, and put his finger into the hole by way of ascertaining its depth. Immediately he found himself on his back, having received a tremendous box on the ear from his father, whose parental wrath was excited by the idea of his son condescending to use his finger by way of measure, when he had a two-foot rule in its own special pocket.

There are, however, many cases where even a two-foot rule would be insufficient for the work, and where a measure of thirty or forty feet is needed.

Now, there is no doubt that by means of a two-foot, or even a six-inch, rule any number of feet might be measured accurately ; but, considering the number of junctions that have to be made, it is not likely that any pretence to accuracy could be insured.

Then, a rod of forty, or even of twenty, feet in length would be awkward and unmanageable, and the only plan left is to take a string or cord of the requisite length.

Even here, however, is a difficulty. The string would not allow of short measurements, such as inches, being written upon it. Let, however, a broad tape of inelastic material be substituted for the string, and all is easy enough.

The next plan is to provide for the portability of the tape in question, to insure its reduction into the smallest possible compass, and to be sure that it is not twisted so as to damage its accuracy. These objects are all attained by the ordinary Tape Measure of the present day, which, whether it be a yard measure in a lady's workbox, or a surveyor's measuring tape, is a ribbon of comparatively inelastic material, coiled up when not wanted, and capable of being drawn out to its fullest extreme when needed.

Putting aside the breadth of the line, and consequently disregarding the liability to twist, we have in the Fishing-reel of the modern angler an exact case in point. So we have in the lady's yard measure, and in the gardener's or builder's tape, all these being modifications of the same idea.

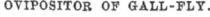

OVIPOSITOR OF GALL-FLY.

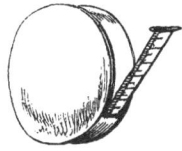

SPRING MEASURE.

SUPPOSE now that we pass to Nature, so as to ascertain whether any such provisions were in existence before it was imitated, however unconsciously, by man. This certainly was the case with one of the commonest and most insignificant of our insects, the little Gall-fly, belonging to the genus Cynips. It could not lay its eggs without the aid of a very long ovipositor, and, owing to structural details, it cannot carry that ovipositor in a straight line, as is done by many insects, some of which have already been mentioned. Accordingly, it is coiled up exactly like our measuring tapes, and can be unrolled when needed. The long, protrusible tongues of the Wryneck, Creeper, and Woodpecker are examples of a similar structure, the tendinous portions being coiled round the head when not needed.

THE SPIRIT-LEVEL.

HAVING now seen how the forces of Nature enable us to produce a perfectly perpendicular line, we will see how the same force, though applied in a different manner, enables us to produce a perfectly horizontal line, the intersection of the two lines producing a right angle.

The measuring tool in question is called the Spirit-level, and is represented on the right hand of the accompanying illustration. Its construction is very simple, consisting of a

FLOATING BUBBLE.

SPIRIT-LEVEL.

tube, nearly filled with spirit, and having just one bubble of air in it. Now, owing to the force of gravitation, the air-bubble must always be uppermost. Consequently, if the tube be a perfect·cylinder, whenever it is held so that the bubble is in the centre, the tube must be horizontal, a hair's breadth of deviation altering the line. I may here mention that, as far as the principle of the instrument goes, water would serve the purpose as effectively as spirit. But as in cold weather the water might freeze, and so burst the tube, as well as being useless until it was thawed, spirit is always substituted.

This instrument is used for various purposes. Sometimes it is employed for levelling billiard tables, or for ascertaining the exact level of walls and other parts of buildings. Surveyors could scarcely do their work without the Spirit-level, which forms an important part of their chief instrument, the theodolite. Indeed, the new science of land drainage, by which the tough, unproductive clay soil is converted into fertile earth, is entirely dependent on the use of the Spirit-level, which detects the slightest rise or fall in the ground.

A most ingenious modification of the Spirit-level is used by military engineers, and is known by the name of the "Contouring-glass," a term which requires some explanation.

It is of the utmost importance that a military engineer should be able, whether on foot or on horseback, to ascertain the approximate heights of the various points which he visits, the efficiency or failure of a battery very much depending on the comparative elevation of the spot on which the battery is placed, and that of the place against which its fire is directed. In an unknown country, of which no detailed maps exist, an invading force must of necessity depend on the extemporised surveys of their engineer officers, and one of the most valuable of their devices is the system of Contouring, invented, as far as I know, by the late Colonel Hutchinson, R.E.

The idea is simple enough. A hill is seen, and the engineer makes a sketch of it before he ascends. At the foot he halts, and marks the spot where his foot presses the earth. He then looks in front at a spot exactly on the level of his eye, marks it, and walks to it. He then draws a line across his sketch, at the exact spot on which he is standing, and that is the first "contouring line." Others follow, until he has reached the top of the hill.

Now, if he can trust himself to look exactly horizontally, he has ascertained the elevation of every part of the hill. He knows the height of his eye from the sole of his foot, and calculates accordingly. Suppose, for example, that it be five feet, and that ten contouring lines are marked, he knows that the entire height is fifty feet, and that each line means an elevation of five feet.

This is a very excellent theory, but one which is not reduced to practice so easily as it looks. There is nothing more deceptive than a contour, especially upon an irregular hill, the invariable mistakes being either greatly to overrate or underrate the height of the contour. When I took my first lesson in this art I caused much amusement to the professor under whom I was studying, by making Shooter's Hill consist of about seventeen contours. However, as many military students made very much the same mistake, I was not so humiliated as I supposed.

Of course, if a surveying officer be mounted, he takes the contour line as measured from his eye to the ground through the centre of the saddle.

After some practice the eye becomes so much accustomed to the contouring lines that they are taken almost mechanically ; but, until this result be gained, an absolute proof is needed, which is furnished by the Contouring-glass—which, by the way, is not a glass at all, after the common acceptation of the word.

It is a simple brass tube about three inches long, not thicker than a man's little finger, and open throughout. A small spirit-level is fixed on its lower surface, and on the very centre of the upper surface is a tiny steel mirror, which projects downwards like a knife-blade. In order to get a "contour," the observer looks through the tube, slightly depressing its end. He then gradually raises it, still looking through it. As the tube becomes exactly horizontal the bubble in the spirit-level is reflected in the little mirror, and the object on which the tube is directed is in consequence on a level with the observer's eye.

At first the management of the contouring-glass is rather tedious ; but after a little practice it can be used without pausing for a single step.

INVALUABLE as is the Spirit-level, with its various modifications, it is nothing but an adaptation of that natural law which causes the bubbles to float on the surface of a stream instead of

T

being submerged below it. We have all seen the multitudinous bubbles of soda-water, or of any effervescing liquid, and have noticed how they are very small when generated, but enlarge quickly, and rise to the surface with a rapidity equal to their enlargement. The same phenomena may be observed in any water-fall, or even in the very familiar and unpoetical operation of pouring beer from a jug into a glass.

The reader will see that in the plumb-rule, the level, and the spirit-level one single principle is employed, namely, the attraction of matter towards the centre of the earth. In the two former instruments this attraction gives a vertical line, and in the latter it gives a horizontal line, but the principle is the same in both.

CALLIPERS.

WE conclude the history of measuring tools with the Callipers. For ordinary purposes, and upon a plane surface, the Compasses answer every purpose. But there are various arts, espe-

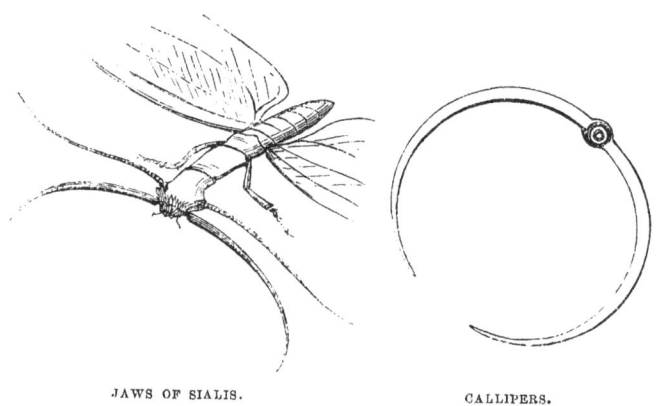

JAWS OF SIALIS.   CALLIPERS.

cially sculpture, in which the compasses, with their straight legs, are absolutely valueless, and their place must be supplied by a differently shaped instrument. For example, no ordinary compasses could measure the exact distance from the nostril to the back of the head, or even touch two points at opposite sides of a limb, and it is therefore necessary to have compasses with curved legs. These are termed Callipers, and can be used on a plane as well as on a rounded surface.

NATURAL Callipers are plentiful enough, and may be found extensively among the insect tribes. There are, for example, the pincers of the Earwig, which have already been described on page 259, and which are, in the common species, formed exactly like the Callipers of the sculptor.

Then we have various insect jaws, especially those of the carnivorous species, one of the most curious being the large insect which is shown in the illustration, upon a very reduced scale. In the male the jaws are exceedingly long and curved, as may be seen by reference to the illustration. I have now before me a pair of sculptor's callipers, and the resemblance between them and the jaws of the Sialis is strangely close, the curve being almost exactly the same in both cases.

The scientific name of this insect is *Sialis armata*, and it is a native of Columbia.

OPTICS.

CHAPTER I.

THE MISSIONS OF HISTORY.—THE CAMERA OBSCURA.—LONG AND SHORT SIGHT.—STEREOSCOPE AND PSEUDOSCOPE.—MULTIPLYING-GLASSES.

The Camera Obscura.—Telescopes, Microscopes, and Spectroscopes, and their separate Objects.—Structure of the Camera Obscura.—The Double Convex Lens.—Its Use as a Burning-glass.—The Meridian Gun in Paris.—Signification of the Word "Focus."—The Human Eye and its Analogies to the Camera Obscura.—Forms of various Lenses.—Long and Short Sight.—Their Causes and Means of Remedy.—Alteration of Sight in the Diver.—Long and Short sighted Spectacles.—The Eye of Birds.—Its beautiful Structure.—Washing-glasses and the "Nictitating" Membrane.—Combination of Images.—Natural Stereoscopes.—The Pseudoscope and its Effects on an Object.—The Multiplying-glass.—The Eight Eyes of the Spider and their Arrangement.—The Seventy Thousand Eyes of the Butterfly.—Form of the Facets.

HISTORY seems to fall into natural divisions, and to write the records of time in successive epochs, recording the advance of the human race. Some of them have apparently disappeared except by the strange relics which they have left behind, but though nothing is known of the men who worked in these ancient times, they stamped their mark upon the earth, and evidently left the world better than they found it.

A very admirable treatise on this subject has been written by the late Rev. J. Smith, called the "Divine Drama of Creation." In this work he divides the progress of the human race into five acts, like those of a drama. The first act is the Hebrew Mission, the second the Greek Mission, the third the Roman Mission and the Middle Ages, the fourth the National Mission, and the fifth the Universal Mission.

Certainly a scene of the last act is now in progress, and may be entitled the Scientific Mission. The last hundred years have been indeed the age of discovery, and, during that time, the life of civilised man has been quite altered, so that practically his sojourn upon earth has been doubled. Steam, with all its various applications, electricity, and other kindred arts have become so intermingled with our lives, that it is difficult to imagine what our state would be if we were suddenly and utterly deprived of them. The loss to all would be incalculable, and not the least of the losses would be that of ready communion with our fellow-creatures.

Of these arts we will now take that which is named at the head of this division of the book, and see how far it is a development of natural facts.

The Camera Obscura and the Eye.

I have already spoken of arts as being akin to each other. They are more than this, and every day of the world's progress teaches us that Art, Science, and Manufacture are sisters, all born of one family, and all depending mutually on each other.

Take, for example, our present theme—namely, Optics— and see how dependent it is upon Manufacture and Art. Without the former, man could not construct those beautiful telescopes, microscopes, spectroscopes, of the present day, which are evidently but the precursors of instruments which will work still greater marvels.

The first enables us to see solar systems without number, to which our own, vast as it seems to us, is but as a grain of sand in the desert. The next instrument makes revelations as marvellous of the infinitely minute as does the telescope of the infinitely great, enabling us to see living organizations so small that thirty-two millions could swim in a cubic inch of water. The third, a comparatively modern instrument, reveals the composition of objects, and can detect and register the materials of which the sun and fixed stars are made, or detect an adulteration in wine. It can adapt itself equally to the telescope and microscope, and the very same instrument which will reveal the character of an invisible gas in the Pole-star, when attached to the telescope, can, when connected with the microscope, point out the presence of half a corpuscle of blood where no other instrument could discover any trace of it.

All these instruments, together with many others, will be described in the present division of the work, and their analogies with Nature shown.

WE will now take the subject of the Camera Obscura, an instrument with which the photographic apparatus of the present day has made most of us familiar. As its action depends chiefly upon the glass, or lens, through which the rays of light pass into the instrument, we will first explain that.

A "lens" is a glass formed in such a manner that the rays of light which pass through it either converge to a focus, or are dispersed, by means of the law of refraction. Every one who has been photographed—and who has not?—will remember that when the sitter has taken his position, the photographer brings to bear upon him a circular glass fixed into a short tube, and then looks through the instrument as if he were taking aim with some species of firearm. It is no matter of wonder that when savages see the photographic camera for the first time they are horribly frightened, for there is really something weird-like in the appearance of the lens thus presented.

Now, this lens is of the shape called "double convex," both sides being equally rounded, so that a section of it would be shaped very much like a parenthesis (). The effect of this form of lens is to bring the rays of light to a point at a given distance from the centre. This point is called the "focus," and is well known by means of the common burning-glass, which will set fire to objects placed in its focus, while itself remains quite cool.

I have seen lead pour down like water when placed in the focus of a large burning-glass, and even the harder metals will yield to the power of the sun's rays when thus concentrated.

There is nothing which gives a more vivid idea of the amount of heat thrown on the earth by the rays of the sun than the effects of a moderately large burning-glass—say one of six inches in diameter. Taking a circle of this size as the surface of the earth, it does not seem as if any very great amount of heat can be received, but when we catch the rays of that circle in our glass, and bring them together upon the focus, the amount of heat can be appreciated. The well-known meridian gun in the Palais Royal is fired by the sun. A burning-glass of no very great size is placed over the touch-hole of the gun, with which its focus coincides. The lens is turned in such a manner that, as the sun attains the meridian, its rays are thrown upon the touch-hole, and consequently fire the gun.

The word *focus* is the Latin term for a domestic hearth, and is used in allusion to the heat which is manifested at the point on which the rays of the sun converge.

It is evident that, after reaching the focus, the rays, if they be not intercepted by some object, will cross each other, and form a large image, but reversed. This part of the subject will presently be explained.

THE accompanying illustration shows two figures, one representing the section of a double convex lens made by the hands of man, and the other that of a double convex lens as seen in Nature.

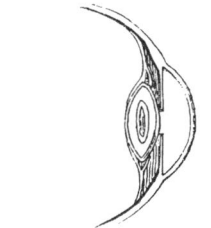

CRYSTALLINE LENS OF HUMAN EYE.   DOUBLE CONVEX LENS.

The former has already been explained. The latter is the double convex lens of the human eye, by means of which the images of external objects are conveyed to the brain. Whenever this lens becomes thickened by disease, the sight is gradually dimmed, and at last total blindness is the result. This disease is popularly called " cataract," and until late days was incurable. Now, however, any good oculist will attack a cataract, and either partially or entirely restore the sight. This operation is performed by carefully removing the convex lens, and supplying its place with a glass lens, which throws the rays of light on the same focus.

The figure shows the double convex lens of the human eye in its place.

HAVING now seen something of the properties of the double convex lens, we will examine its application to the Camera Obscura.

The lens is placed on one side of the camera, and is so made that it can be slid backwards and forwards, and the focus altered at will. The camera itself is a box completely closed, so that no light can enter it except that which passes through the lens. The latter is so arranged that the rays which pass through it are crossed, and throw their image on

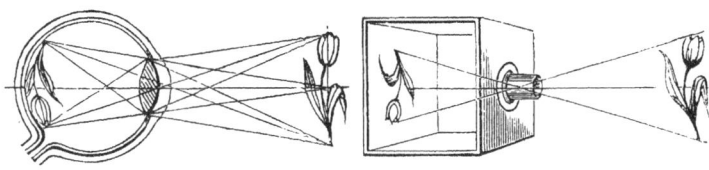

EYE AND IMAGE.   CAMERA OBSCURA AND IMAGE.

the opposite side of the camera. In the photographic camera a piece of ground glass is placed at the end, so that the rays fall upon it, and the operator can see whether the image is a good one. Of course the figures are reversed, so that the sitter seems to be on his head, but that is a matter of no consequence. Exactly the same effect is produced by the marine telescope.

The general structure of the camera is shown in the illustration, all needless details being omitted.

I may here remark that the term "camera obscura," or dark chamber, alludes to the fact that the box is completely closed, and, but for the rays which pass through the lens, would be absolutely dark.

THE opposite illustration shows the most perfect camera obscura that can be imagined, namely, the human eye. Here we have a dark chamber, a double convex lens, and an image falling upon the back. Here the optic nerve comes into play, takes cognisance of the image, and conveys the idea to the brain. With a little trouble, a real eye, say that of an ox, can be dissected out, and employed as a camera obscura, the operator seeing in the back of the eye, or " retina," the same image which the ox would have seen if it had been alive.

In photography, the operator, when he has found that a perfect image is thrown upon the ground glass, which represents the retina of the eye, substitutes for it a sensitive surface, on which the rays are projected, and which, by chemical means, produce a permanent instead of a fleeting object.

EXAMPLES of other lenses may be found in Nature. She, moreover, can perform a task which man has never even attempted, namely, the change of form in a lens according to the duty which it has to do. How this wonderful object is attained we shall presently see.

There is a form of lens extremely useful in Optics, namely, the " Plano-convex " lens. This is, in fact, one half of a double

HUMAN EYE : SECTION OF CORNEA, &C.   PLANO-CONVEX LENS.

convex lens, the section being made through its edges, and the plane sides polished as well as the convex. As, however, this is only a half of the double convex lens, it does not need further explanation. Its natural counterpart may be seen in the annexed illustration.

A somewhat more complicated form of lens is called the " Meniscus," one side of which is convex and the other concave. A good example of the meniscus may be found in the old-fashioned watch-glass, before watchmakers took to flattening them, and watch-wearers were not ashamed to carry a " turnip," in which there was room to spare for the works. If a section of such a glass were taken, it would assume the form of a half-moon. This, in fact, is the meaning of the term "meniscus," which is a Greek word, signifying a little moon. If the same glass were solid, or even filled with water, it would form a "plano-convex" lens.

Of course the outer curve of the meniscus must be larger than the inner curve, but in some cases the disproportion is very strongly marked, the outer curve being very large, and the inner curve very small. An example of such a meniscus may be seen in the human eye. If the reader will refer to the illustration on page 280, in which the structure of the eye is shown, he will see the meniscus lens in combination with the double convex. The former has already been explained, and the latter is formed by the vitreous humour which fills nearly the entire globe of the eye. Its larger curve is due to the form of the eyeball, and the smaller to the convex lens.

LONG AND SHORT SIGHT.

It has already been mentioned that the focus of a convex lens is shorter in proportion to its convexity, and that in consequence its magnifying power is increased. For example, the large glasses through which pictures are viewed are comparatively thin in proportion to their diameter, while the lenses

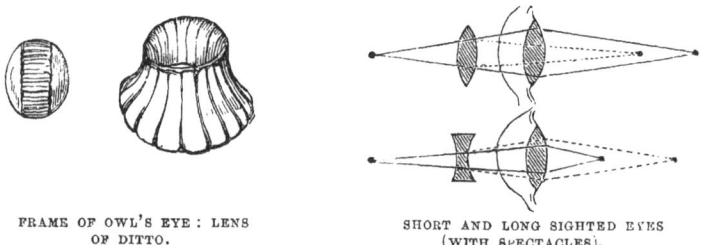

FRAME OF OWL'S EYE : LENS OF DITTO.

SHORT AND LONG SIGHTED EYES (WITH SPECTACLES).

employed for the highest powers of the microscope are scarcely larger than small shot, and nearly as globular. It naturally follows that any instrument to which a lens is adapted, whether it be microscope or telescope, must depend for its focus on the greater or less convexity of the lens in question.

Again taking as our example the human eye, we find that there are very few persons who from youth to age possess or preserve eyes which can read small type at a moderate distance, and can clearly define the outlines of distant objects. Nearly all people, even if in their youth they possess good sight, lose it as they grow older. They can discern distant objects well enough, but, when they come to reading, they are obliged to hold the book at arm's length before they can distinguish the letters.

This defect is caused by the insufficient convexity of the lens, so that the focus is thrown too far back, and it is corrected by wearing spectacles sufficiently convex to supply the deficiency in the lens of the eye.

An admirable example of temporary long-sightedness is familiar to every diver, though he may be unconscious of its cause. Suppose that into very clear water of some twelve feet in depth, a white object, say a common jam-pot, is thrown, it can be clearly discerned from the shore, unaltered in shape or size. But, when the diver searches for it, he sees at first only something white, large, undefined, and wavering, and only finds it resume its proportions as he approaches it. This phenomenon is due to the pressure of the water upon the eyeball, which flattens it, and so throws the focus too far back for a clear image. Nowadays this defect is remedied by the use of very convex spectacles, so convex, indeed, that, if worn in the air, they would render the wearer incapable of seeing anything at more than an inch or so away from him. But, when worn in the water, they only supply the deficiency of the compressed eyeball, and so restore the focus to its proper position.

THOSE who suffer from short-sightedness can see with great distinctness objects which are close at hand, but those at a little distance seem to have no particular outline, and appear as if they were viewed through a fog, thus causing a constant and almost painful strain on the eyes. The cause of this defect is the too great convexity of the lens, which therefore throws its focus short of the required spot. The means of remedy are exactly opposite to those which are used for long-sighted persons, a concave lens being placed in front of the eye, so as to throw the focus farther back, and relieve the organ from the strain.

Although we have not yet invented a machine that can alter the focus at will, we may take a hint from Nature. We have already seen how the pressure of water upon the front of the eye lessens its convexity, and makes it long-sighted.   Consequently, if we could apply pressure round it, we could make it more convex, and so neutralise the weight of the water.

There is a wonderful piece of machinery in Nature which really does perform this office, the eye, at the will of its owner, becoming either telescopic or microscopic. This quality is very desirable in birds, especially those which are predacious and of rapid flight, as they might either fail to see their prey at a distance, or might dash themselves against some obstacle when they were close upon it.

The eye of the Owl affords a beautiful example of machinery which produces this effect, and the means which are used may be understood by inspecting the accompanying illustration.

It will be seen that the eyeball is set in a framework composed of thin bony plates, just like a glass in a telescope. When these plates are relaxed, the whole eyeball is flattened, so as to enable the bird to see an object at a very great distance. But, when they are contracted, they render the whole eye globular in proportion to their pressure, and enable the bird to see objects which are very close to it.   In fact, the eye becomes a telescope or microscope as needed.

Many reptiles possess this arrangement of bones, but the birds have even a more delicate mode of obtaining the focus of the eye. This is by means of a curious organ called, from its shape, the " pecten," or comb, which is placed in the vitreous humour at the back of the eye, and connected with the optic nerve.   It is a congeries of arteries and veins, so that it can be rapidly enlarged by forcing blood into it, or diminished by allowing the blood to withdraw.

As the liquid in which it rests is practically incompressible, it follows that when the comb expands, it causes the chamber of the vitreous fluid to expand, and so forces the lens forward. When, however, the blood retires from the comb, the lens returns to its original place. This, as the reader may have noticed, is the same principle as that which is followed in altering the focus of a telescope in order to suit the sight of different individuals. Perhaps a still better illustration may be found in the coarse and fine adjustment of the microscope, the former of which moves the whole tube, and may be compared to the bony ring ; while the latter causes one part to slide over the other, and is analogous to the comb.

The movements of this organ are believed to be as involuntary as the dilatation and contraction of the iris ; but, whatever may be the case, it is one of the most beautiful examples of natural mechanics, and far surpasses the most delicate machine that can be made by man.

In the illustration of the microscope, which is to be found on page 286, both these movements are given, the double vertical wheel being the coarse movement, and the fine movement being supplied by the single vertical wheel just above them.

WHILE we are on this subject, we may see how Art unintentionally copies Nature, even in trivial details. Every one who is in the habit of using optical instruments, more especially those who are forced to wear spectacles, are aware of the necessity of keeping the glasses as clean as possible, and, where the instruments are delicate, always have by them a piece of clean wash-leather for the express purpose of wiping the glasses.

Here, again, Nature has anticipated Art. In our own case, we have in the human eye a good example of such natural mechanism, the eyelids being formed quite as much for the purpose of washing the surface of the eyeball as of excluding light.

Many animals are provided with a special apparatus for the purpose, called the "nictitating membrane." It is, in fact, a sort of inner or supplementary eyelid, which can be drawn over the eye while the external lids remain comparatively unmoved. It is very conspicuous in the owls, and gives to those birds that almost comical look of perpetual blinking with which we are so familiar.

THE STEREOSCOPE AND PSEUDOSCOPE.

MANY persons have wondered how it happens that, as we have two eyes, we do not see two images instead of one. Practically, this is always the case, for the eyes, especially when they look on solid bodies, see two different images, because they contemplate the object from different points of sight.

This may be easily ascertained by looking at a given object first with one eye, and then with the other, when it will be seen that the image presented to the right eye is slightly different from that of the left eye, but that the two can be combined into one by a very slight inward movement of both eyes, and thus the effect of a solid body be produced. Sometimes, when people are weak, and cannot control the united movement of the eyes, not only two, but five or six images are at once presented to the mind, and produce a strange sense of bewilderment and confusion.

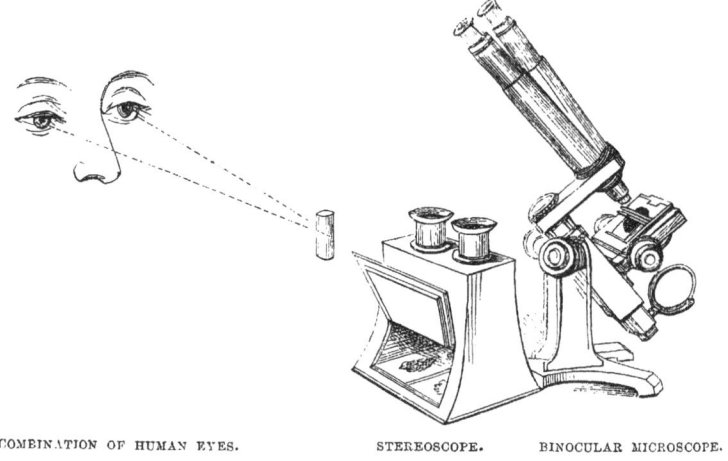

COMBINATION OF HUMAN EYES.   STEREOSCOPE.   BINOCULAR MICROSCOPE.

Painters are obliged to avail themselves of this peculiarity, and to make allowances for the double vision. If they do not, the effect of the painting is flat, and it appears as if the artist had only used one eye.

A good proof of this fact may be seen in Stereoscopic photographs, especially of scenery. If each be viewed separately, it often appears quite unintelligible, but, when they are combined by the instrument, they seem to spring into life as it were, and appear solid enough to be grasped.

Now, the Stereoscope is avowedly constructed on the same principle as the double vision of the eye, so that when it applies itself to two photographs of the same object which have been taken from different points of view, it combines them, and gives them as solid an appearance as if they were realities.

So wonderfully close is the representation, that the idea of a place obtained by means of the combination of the photograph and Stereoscope is quite as vivid and correct as if it had been gained by actual observation.

The principle of the Stereoscope is now applied to the best microscopes, and its value is incalculable, especially when low powers are used, *i.e.* those of not less than half an inch focus. The real beauty of many objects could never have been appreciated but for this discovery, nor their true form defined.

On the left hand of the illustration is shown the combining power of the eyes. Supposing the right eye only to be brought to bear upon the little cylinder, only one side of it will be seen, and it looks nearly flat. The same is the case with the left eye. But, when both eyes are used together, both sides of the cylinder are presented to the mind, and thus we get the effect of solidity.

The Stereoscope is so formed, by means of lenses, that the two figures become combined into one, the rays of light being turned out of their course by the arrangement of the glasses.

The Stereoscope, however, although a useful assistant to the vision, is not necessary. It is perfectly possible to combine the two figures without any stereoscope, and to do so merely by squinting, if we may so call it, at the figures. The power of combination is gained with a very little practice, and in a short time the observer will be capable of producing stereoscopic effects without needing a Stereoscope. This ability is very useful when inspecting photographs in a shop-window. Of course the figures are not so much enlarged as they are with the stereoscope, but they are nevertheless quite as clear and well defined.

THERE is an instrument called the Pseudoscope, which, as its name imports, gives a false idea as to the nature of the object which is viewed through it, converting hollow objects into solid, and *vice versâ*. The following description of its effect is given by Wheatstone :—

" When an observer looks with the pseudoscope at the interior of a cup or basin, he not unfrequently sees it at first in its real form ; but by prolonging his gaze he will perceive the conversion within a few minutes ; and it is curious that, while this seems to take place quite suddenly with some individuals, as if the basin were flexible, and were suddenly turned inside out, it occurs more gradually with others, the concavity slowly giving way to flatness, and the flatness progressively rising into con-vexity.

" Not unfrequently, after the conversion has taken place, the natural aspect of the object continues to intrude itself, some-times suddenly, sometimes gradually, and for a longer or shorter interval, when the converse will again succeed it—as if the new visual impression could not at once counteract the previous results of recent experience. At last, however, the mind seems to accept the conversion without further hesitation ; and after this process has once been completely gone through, the observer, on recurring to the same object, will not find it possible to see it in any other than its converted form, unless the interval should be long enough to have allowed him to forget its aspect.

" Vagaries, however, sometimes occur in these experiments of which it is difficult to give any certain explanation, but which would be probably found referable to the same general principle, if we were acquainted with all the conditions of its operation."

The Multiplying-glass.

Still more extraordinary examples of the combining power of vision are to be found in the eyes of spiders and insects, more especially when we compare them with the work of man.

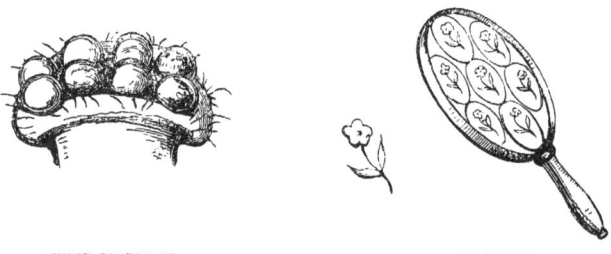

EYES OF SPIDER.   MULTIPLYING-GLASS.

If we take a common Multiplying-glass, such as is shown in the figure, and look at a flower or other object through it, we see the object repeated as many times as there are different foci of vision in the instrument.

Now, taking for example the eyes of a Spider, it would be natural to suppose that the same result would occur, especially as the foci of the eyes point in different directions. The left-hand figure in the illustration represents the eight eyes of one of our common Spiders, belonging to the genus *Clubiona*, which may be found in almost any outhouse, sitting in its curious web, and ready in a moment to run for safety into its silken tunnel.

It will be seen that the foci of all the eyes are in different directions, and so placed as to command a large radius. Observers have remarked that the eyes are placed in Spiders so as to suit their habits. " Those spiders," writes Professor Owen, in his " Comparative Anatomy," "which hide in tubes, or lurk in obscure retreats, either underground or in the holes or fissures of walls or rocks, from which they emerge only to seize a passing prey, have their eyes aggregated in a close group in the middle of the forehead, as in the Bird-spider, the *Clotho*, &c.

" The spiders which inhabit short tubes, terminated by a large web, exposed to the open air, have the eyes separated and more spread upon the front of the cephalothorax.

" Those spiders which rest in the centre of a free web, along which they frequently traverse, have the eyes supported on slight prominences, which permit a greater divergence of their axis ; this structure is well remarked in the genus *Thomisa*, the species of which live in ambuscade in flowers.

" Lastly, the spiders called *Errantes*, or Wanderers, have their eyes still more scattered, the lateral ones being placed at the margin of the cephalothorax."

Yet, although each eye produces a separate image, it is clear that upon the mind of the Spider only a single idea can be impressed, for that otherwise all would be confusion. There must, therefore, be some mechanism in the structure of the eye, the nature of which we are not as yet able to understand.

A STILL more remarkable instance of a natural Multiplying-glass may be found in the eyes of many insects.

The form of multiplying-glass shown in the accompanying illustration is probably familiar to most of my readers. It consists of a convex piece of glass, cut into a number of facets, and showing in each facet a distinct and separate image of the object to which it is directed. Now, the compound eyes of insects are constructed on much the same principle, except that the number of facets is infinitely more. Taking, for example, the eyes of the Tortoise-shell Butterfly, we find that there are about seventy thousand lenses or facets. Now, it is possible, with care, to remove the eye from the insect, cleanse it, and arrange it in a microscope in such a way that objects can be seen through it. When this is done, a separate image is seen in each facet, just as is the case with the Multipying-glass, only, as the facets are very much more numerous, the effect is proportionately more striking.

The reader may notice that the facets of the insect eye appear to be hexagons as perfect as those of the honey-

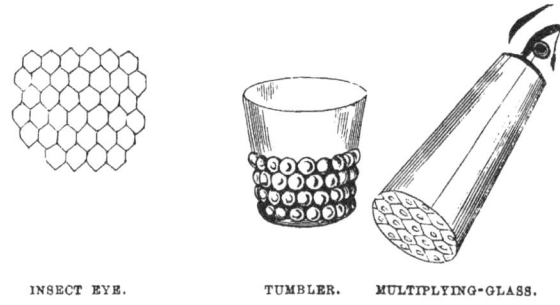

INSECT EYE.   TUMBLER.   MULTIPLYING-GLASS.

comb. This appearance is probably due to the fact that each eye is covered with a convex plate of glassy brightness and transparency, and that, when such objects are viewed from the front, they appear to have hexagonal instead of rounded outlines. A familiar example of this fact may be found in the glass tumblers which are ornamented with rounded projections on their surface. If a photograph of one of these tumblers be taken, the resemblance to the hexagonal markings of the insect eye is so close that the tumbler might easily be taken for the eye.

OPTICS.

CHAPTER II.

THE WATER TELESCOPE.—IRIS OF THE EYE.—MAGIC LANTERN.
—THE SPECTROSCOPE.—THE THAUMATROPE.

Limits to Sight in the Water.—Effect of a Ripple.—The Eyes under Water.—The
Water Telescope, its Structure and Mode of Use.—Gyrinus, or Whirlwig-
beetle, and its Double Set of Eyes.—The Iris of the Eye, and its Double Set of
Contractile Fibres.—Cotterill's Lock and its Structure.—The Magic Lantern
and its Principle.—Chinese Shadows.—Spectre of the Brocken.—An Adventure in Wiltshire.—Effect of the Halo.—The Spectroscope.—Its Structure
explained.—A Star on fire.—Motes in the Sunbeams.—Bessemer Steel made
by aid of the Spectroscope.—Absorption Bands.—Detection of Blood.—A
Man's Life saved by the Spectroscope.—The Pocket Spectroscope.—The
Rainbow, Dewdrop, Soap-bubble, Opal, and Pearl.—The Thaumatrope.—
Structure of the Retina.—Complementary Colours.—The Zoetrope and
Chromatrope.—Wheel Animalcules and their Structure.—An Optical Delusion.

THE WATER TELESCOPE.

EVERY one who has watched the movements of the various
creatures which live below the surface of the water is
aware how entirely dependent he is on the unruffled character
of that surface. No matter how clear the water may be, the
least ruffling of the surface will effectually shut out all sight :—

> " But if a stone the gentle sea divide,
> Swift rippling circles rush on every side,
> And glimmering fragments of a broken sun,
> Banks, trees, and skies in thick disorder run."

And there is an end of the observations. If, however, the eyes
can penetrate below the surface, the ruffling is of little consequence, so long as the water is clear. Consequently, whenever the top of the bank is sufficiently near the water, it is
possible to continue the observations by lying down, and
immersing the head above the eyes. This plan, however, is

not a very comfortable one, although I have often followed it on a windy day when the surface was too ruffled to permit of vision in any other way.

Still, there is an instrument by which it is possible to counteract the ruffle of the surface, and to see objects with tolerable plainness. This is called the Water Telescope, and it is of very simple construction. Like the ordinary telescope, it consists of a tube, but, instead of the convex and concave lenses of that instrument, it has only a single glass at one end, and that glass is perfectly plane.

When used, the eye is applied to the open end, and the glazed end lowered into the water. The sight is then undisturbed by

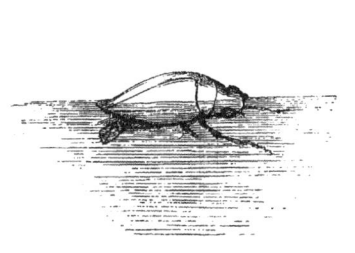

WHIRLWIG-BEETLE.

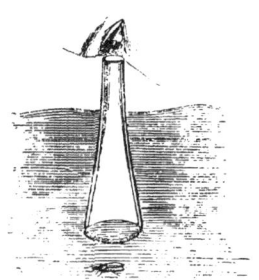

WATER TELESCOPE.

the ripple, and the effect is the same as if the eyes themselves were lowered beneath the surface.

It is much used in looking for shells, sea-urchins, and other creatures which live in the bed of the sea.

In the insect world we have an example of a natural Water Telescope. I do not say that the inventor of the Water Telescope took his idea from the insect, but the reader will see that he might very well have done so.

There are sundry little beetles popularly called Whirlwigs or Whirligigs, and scientifically known by the name of *Gyrinus*. All these names allude to the insect's habit of whirling about on the surface of the water, with a movement which seems ceaseless and untiring. Allusion has already been made to the Whirlwigs on page 22.

Their object in their perpetual waltz is not so much amusement as food, which chiefly consists of the tiny insects which fall into the water. Now, in order to enable it to see both above and below the water, a peculiar structure is required. Generally the insects possess one pair of compound eyes, each group being set on the sides of the head. In the Gyrinus, however, there are two sets of these eyes, one pair being on the upper surface of the head, and the other on the lower surface. Thus, while it can use the upper pair for seeing objects which are out of the water, the lower pair of eyes, which are submerged, act the part of the Water Telescope, and enable it to see objects that are below the surface. Were it not for this precaution, even the ripples which it makes by its own rapid progress would prevent it from seeing.

The Iris of the Eye.

I HAVE often wondered, when contemplating the astonishing mechanism by which the Iris of the Eye is able to contract or enlarge the pupil according to the amount of light, whether any

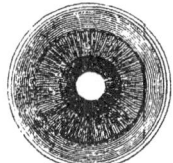

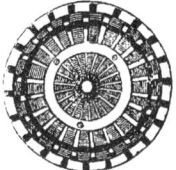

IRIS OF HUMAN EYE.   COTTERILL'S LOCK.

similar mechanism would be used in Art. As anatomists know, the Iris is composed of two layers. One consists of radiating fibres, which serve to enlarge the pupil, while the other layer surrounds the latter, and by its elasticity serves to contract it. As any one may see by looking in a mirror and shifting the light, the pupil is perpetually changing its diameter, but always retaining its circular shape. A glance at the illustration will show the two layers, and aid the reader in understanding the mode in which they work.

Some years ago, while looking at the account given by Mr. J. Price of a lock invented by Mr. Cotterill, I saw at once that the inventor, whether consciously or not, had followed the mechanism of the eye, as far as metal could be expected to imitate animal fibre.

In the very centre of the lock there is a small circular opening, resembling the pupil of the eye, and serving to admit the key, just as the pupil admits light. Around this pupil, if we may so call it, are ranged some twenty thin steel slides which move in channels, up and down which they slide. Round the circumference of the lock are a corresponding number of spiral springs, each of which presses on the base of a slide, and forces it towards the centre.

The reader will now see that the radiating slides of the lock represent the radiating fibres of the iris, and that the spiral springs represent the circular fibres. Both perform the same office, the steel slides regulating the size of the aperture, and the spiral springs pressing them all towards the centre. The key of the lock answers the same purpose as does light in the eye, which by its mysterious pressure enlarges or contracts the pupil.

This is not the place to describe this very ingenious lock in detail, but I may state that it has never been picked. Even Mr. Hobbs, who tried it for twenty-four hours, gave it up, and, when he saw the interior mechanism, said that if he had tried for a month he should have made no progress. This is an unconscious testimony to the wisdom of following Nature in Art.

The Magic Lantern.

WE are all familiar with the Magic Lantern, whether it may take the form of the mere child's toy, be developed into Dissolving Views, or throw black shadows on a curtain, in which case it is called by the name of Chinese Shadows. In all these cases the principle is the same. First we have a light behind the object whose reflection is to be seen. Next we have the object itself, and lastly the surface upon which it is reflected. As to the variety of mirrors, lamps, and lenses which are used to produce different effects, we may put them aside as foreign to our present purpose.

Generally the object is reflected upon a white curtain or sheet, but sometimes, when a specially weird-like effect is needed, a cloud of thick smoke takes the place of the sheet, and

MAGIC LANTERN.

upon it the reflection is shown, as seen in the accompanying illustration.

NATURE has her Magic Lanterns as well as Art, and wonderful things they are sometimes, the well-known Brocken Spectre being an excellent example. It is not, however, necessary to visit the Brocken in order to see this apparition, for I have seen it in perfection in England.

Many years ago, when living in Wiltshire, I went before daybreak to the top of a very high conical hill. The morning mist was so thick that I could scarcely see my way up the hill. When I reached the summit, I stood there for some time, trying to see the landscape, but the mist was so thick that I could barely tell the points of the horizon by the brighter look cast by the coming Day in the east.

I was looking westward, when suddenly the sun rose behind me, and I saw the Brocken Spectre as I have sketched it in the accompanying illustration. It was a gigantic shadow of myself, projected on the mist, just as a Magic Lantern projects the image on a sheet or a smoke-cloud. Of course my gestures were repeated, and it really looked almost awful to see this gigantic spectral figure set in the mist.

Perhaps the most extraordinary part of it was the enormous halo of rainbow colours round the head. No matter where I moved, the halo surrounded the head of the image, its colours being comparatively bright near the centre, and becoming gradually paler towards the circumference.

Another point about this natural Magic Lantern ought to be mentioned.

Wishing to show a friend the extraordinary sight of a Brocken Spectre, I took him up the hill on a misty day like that which has been briefly described. According to surmise, two spectres appeared instead of one, but the halo was not doubled as well as the shadow. I could see my friend's shadow,

BROCKEN SPECTRE.

and he could see mine. But, although the halo was as bright as before, each of us could only see it encircling his own head. We stood as close to each other as we could, we moved apart as far as the nearly conical top of the hill would allow, and in both cases each of us could only see his own halo.

Perhaps the reader may remember the wonderful spectre-scene drawn by Mr. Whymper, and viewed from the Matterhorn just after the accident which had killed several of his companions in the ascent of the hitherto impregnable peak. In the mist there suddenly appeared three vast dark crosses enclosed in an oval. Considering the highly-strung nerves of the survivors, it was no wonder that they were all shaken by such an appearance, and that the guides were for a time too frightened to proceed.

THE SPECTROSCOPE.

NEXT we come to one of the most astonishing and beautiful optical instruments ever made by the hand of man. It is called the Spectroscope, because it deals with a certain arrangement of rays which is called a " spectrum." Many years ago Newton discovered the cause of the lovely colours which deck the rainbow, and the fact that, by passing a ray of white light through a prism, it was decomposed into seven colours, which invariably came in the following order—Red, Orange, Yellow, Green, Blue, Indigo, and Violet. He also discovered that, by looking at that coloured band through another prism arranged in a different manner, the decomposed rays were again brought together, and white light was the result.

Newton had thrown the light on the prism through a round hole, but some time afterwards Dr. Wollaston employed a narrow slit for the purpose, and then found that the spectrum was traversed by dark lines which never changed their places. On these lines depend all the discoveries that have been made by the aid of the Spectroscope. The chief of them are designated by the letters of the alphabet. (See page 300.)

It was soon found out that if burning gases were viewed with the Spectroscope, lines were still seen, but they were bright instead of dark, and that they invariably occupied the place of one or more of the dark lines shown by the spectrum of sunlight. Then it was discovered that these burning gases absorbed or stopped out the light in the solar spectrum, and from that moment the science rapidly advanced.

At the present day the Spectroscope not only determines the metals which exist in the sun, but also those of the fixed stars. It even analyzes the constitution of double stars, and shows the reason why one star should be red and the other green.

One of the most astonishing discoveries in astronomy was due to the Spectroscope.

During the month of May, 1866, one of the stars in the Northern Crown (*Corona Borealis*) was seen to undergo a rapid change. It was originally one of the tenth magnitude, but in

a short time increased in size and brilliancy until it nearly
equalled Sirius, Capella, or Vega. It remained bright for some
time, and then rapidly faded until it resumed its former size.

How this change was effected we never should have known
but for the Spectroscope. No sooner, however, was this instrument pointed at the star than there appeared in the spectrum
the three well-known lines—red, green, and violet—which
denote burning hydrogen. There was no doubt on the matter,
and the Spectroscope showed us that we were witnessing a
conflagration the like of which was never seen or scarcely
imagined.

Supposing our sun, which is known to be one of the stars,
and about which there are vast volumes of hydrogen gas, were

RAINBOW.

to blaze out in a similar manner, the result would be that the
whole of the planets would be consumed in a few seconds, and
converted into gases. In an instant every living thing would
be swept off the surface of the earth by this fearful heat, and,
as Mr. Roscoe says, "our solid globe would be dissipated in
vapour almost as soon as drops of water in a furnace."

So, as Mr. Huggins observes, the old nursery rhyme,—

> " Twinkle, twinkle, little star,
> How I wonder what you are,"—

is no longer tenable, for we really do know the composition of the stars.

The Spectroscope not only tells us the substance of which the sun and the most distant stars are made, but gives us the same information about the " gay motes that people the sunbeam." It tells us that they are common salt in very minute particles. They have been dashed into the air by the winds as spray, and then dispersed over the whole globe. This is one reason why we have so much salt in our bodies, and why the blood and the tears are so salt.

It is also applied to the arts. The well-known Bessemer process consists in pouring melted iron into a peculiarly shaped vessel called a " converter," and blowing air through it for the purpose of burning out the carbon. From the mouth of the converter issues a volume of magnificent flames, and at a certain moment the skilled workman who directs the process inverts the vessel and pours out the steel. A very few seconds too soon or too late would spoil the whole of the metal, in the former case it being simply brittle cast-iron ; and, in the second, becoming so thick that it could not be poured out.

Only a few workmen could judge rightly the exact point at which to shut off the air-blast. They watched the flame, and by some change in it, too slight to be noticed by any except experienced eyes, knew the moment when the iron was converted into steel.

Such men could, of course, demand any wages they liked, and, by striking, stop the whole works. The Spectroscope, however, performed this delicate discrimination far better than the best workman. When directed to the flame, the bright lines indicating carbon are seen in the spectrum. When the blast has continued for some twenty minutes, the carbon lines suddenly disappear, showing that the carbon has been burned out, and giving to the workman the signal to shut off the air-blast.

Another discovery was, that liquids gave dark lines, technically termed absorption bands, of different widths and in different parts of the spectrum. Even liquids which had no perceptible colour threw bands as bold as those which were coloured, while coloured liquids threw totally different bands, irrespectively of their own colour.

For example, the green colouring matter of leaves, called chlorophyll, throws a single broad band on the extreme left— *i.e.* across the red part of the spectrum—so far back, indeed, that it is not easily seen at first.

Then, suppose that we make some pale solutions of red substances, such as carmine, magenta dye, port wine, logwood, permanganate of potash, and blood, it is possible to have them so exactly resembling each other that not even the microscope can discriminate between them ; yet the Spectroscope instantly detects the colouring matter of each solution.

The instrument is, therefore, invaluable in detecting adulterations of wine. For example, supposing that red wine is

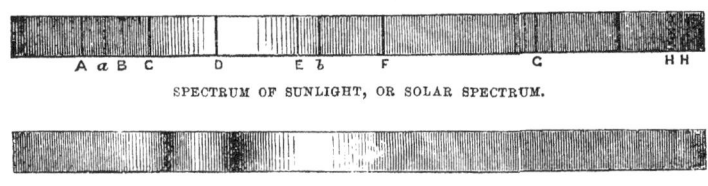

SPECTRUM OF SUNLIGHT, OR SOLAR SPECTRUM.

SPECTRUM OF BLOOD.

suspected of owing its redness to logwood, and not to the genuine grape, a drop is mixed with water and viewed through the Spectroscope, which instantly tells whether the colouring matter is grape or logwood. And as, by photography, the spectrum can be exactly copied, an indelible record is procured of the true nature of the object.

So marvellously delicate is the instrument with regard to blood, that it detects the thousandth part of a grain of colouring matter in a blood-stain.

If upon the spectrum were printed the word BLOOD in the largest and blackest of capitals, it could not be more legible to an ordinary reader than are the two blood-bands to the eye of a spectroscopist. There is nothing like them in nature, and whether it be by association of ideas, or by absolute fact, these two bars have a strangely menacing look about them. Not only that, but if the blood should be that of a person suffocated with carbonic acid gas, the Spectroscope will say so.

Some years ago a man owed his life to the Spectroscope. A

mysterious murder had been committed, and the police had arrested a man who was found near the spot. He could give no intelligible account of himself, and the sleeves of his coat and a part of his waistcoat were deeply stained with a red substance just like clotted blood. A piece of each garment was cut off and given to a well-known spectroscopist, who tried the red matter in the instrument, and at once declared it not to be blood. What it was he had not time to ascertain, so he sent it to a brother in science, who, after examination, pronounced it to be red gum.

By degrees, the man, who had been intoxicated when arrested, stated that he had been to see a friend who was a journeyman hatter. It was then found that he had been leaning on the workman's board, and so had carried off some of the gum-mastic with which hats are stiffened. Had it not been for the infallible Spectroscope, the man might have lost his life.

Thus we see that the Spectroscope is the elephant's trunk of optics, equally fitted for the greatest and smallest, the farthest and nearest, of objects. It is equally at home in earth and sky. When attached to the telescope, it reveals the constituents of the stars, and, when affixed to the microscope, it shows us the colouring matter of a green leaf. It produces the best steel, and detects adulteration in wine. And, lastly, as we have seen, it turns lawyer, and settles the evidence by which the life of a man is lost or saved. It can determine the purity of the smallest coinage, and tell us why a star changes in magnitude.

Yet all these wondrous revelations are made by a few prisms and a magnifying-glass. I possess a Spectroscope, made and presented to me by Mr. J. Browning, the celebrated optician. This astonishing instrument is only three inches long, and half an inch in diameter, so that it can be carried in the waist-coat pocket. I always keep mine in a finger of a white kid glove, which is amply sufficient for it. Yet it gives the spectrum of the sun with its principal lines, will detect the fraudu-lent wine merchant, and could have decided whether the accused man should be acquitted or hanged.

MARVELLOUS and mighty as is this engine, it lay concealed in Nature ever since the sun's rays shone upon earth and a drop of water existed. The Rainbow is nothing but a vast spectrum, a transverse slice of which would be a good representation of the coloured band which is shown in the instrument. It is prefigured in the ever-shifting rainbows of the water-fall and fountain, which latter may even be seen in the fountains of Trafalgar Square, while at the Crystal Palace their beauty has long been noticed.

There is not a dewdrop which is not a miniature Spectroscope, as it glitters with its wondrous iridescence in the rays of the rising sun ; there is not an opal with its shifting hues, nor the splendour of the soap-bubble, nor the nacre of the common river mussel or the ormer shell, which does not owe its beauty to the same principles which govern the Spectroscope. Every green leaf, and blue or pink or yellow petal, every varying tint of the mackerel sky, every blaze of sunset and blue-grey of sunrise, owes its beauty to those wondrous laws of light which had been hidden for so many centuries, until they were unveiled by the simple prism of the Spectroscope. As in so many instances, the revelation lay concealed until the coming of the revealer, whose inspired hand raised the dark veil of centuries.

THE THAUMATROPE.

MIDDLE-AGED persons will recollect that since the days of their childhood a great variety of optical apparatus has been invented ending in the word " trope." This is a Greek word, signifying to turn, and is given to the instruments because they revolve.

All these toys—and they may some day become more than toys—depend on a curious property of the human eye. The reader will remember that in the description of the human eye, as compared with the camera obscura as applied to photography, it was mentioned that the image was thrown from the front to the back, and in the one case was received on a naturally sensitive membrane, and in the other on a film rendered artificially sensitive by chemical means. This membrane is called the " retina," because it not only receives the impression, but retains it for some little time after the object is removed. It has been calculated that the duration of the image is about the eighth part of a second.

Thus the eyelids are perpetually and unconsciously closing and opening with a rapid movement, popularly called "winking." This movement is for the purpose of cleansing the eyeball, and, were it not for the image-retaining power of the retina, we should pass a considerable part of our time in absolute darkness. As it is, the impression of external objects on the retina lasts longer than the time occupied in winking, and, in consequence, we are not conscious that any interval of darkness has elapsed.

Again, when we have been looking steadfastly at an object, and then move our eyes, the image of that object is seen in the new focus; and it is worthy of notice that such object is always seen in its "complementary" colour. For example, if we have been looking at a scarlet spot, and suddenly move our eyes, we shall see a spot exactly similar in size and shape, but of green.

I well remember that when I was a boy I was reading with almost feverish anxiety the green handbill of a travelling circus, to which I hoped that I might be allowed to attend. Having finished it, I asked for some note-paper, for the purpose of putting my request in writing, but, to my astonishment, mixed, perhaps, with a little irritation, all the paper supplied to me was of a bright pink. For a time no arguments could convince me that the paper was really white, until by degrees the pink hue became paler and paler, and the paper assumed its normal whiteness.

The fact was, that the eye had become saturated with the green—*i.e.* the blue and yellow rays—and could see nothing but their complementary colour, which was pink.

A good example of this property may be found in a lighted stick, which, if rapidly whirled round, appears to form a continuous circle of fire. The reason of this is, that the impression made on the retina by the fiery point does not cease until the stick has again come round in its course.

Then there are those well-known chromatic tops, in which are inserted pieces of bent wire. When the top is spun these pieces of wire assume exactly the appearance of transparent jugs, vases, glasses, and similar articles. A very pretty illustration of this principle is given by a little machine, which is made to revolve rapidly by means of a multiplying wheel.

Upon its surface are fixed little pins, with polished globular steel heads, and, when the handle is turned, these heads form the most beautiful and intricate figures with exact accuracy.

Another toy, called the Thaumatrope, or Wonder-turner, is equally ingenious and beautiful, and is sufficiently simple to be made by any one with a slight knowledge of drawing. A disc of white cardboard is cut, and upon each side of it is portrayed some object. If the disc be caused to revolve rapidly, these two subjects will be seen at the same time, the image of each being held on the retina long enough to allow the other to take its place.

Some very beautiful combinations may be made by means of this instrument. For example, a horse may be on one side, and a man on the other, and, by spinning the disc, the man will be seen mounted on the horse. Then we may have a boat on one side, and a rower with his oars on the other. Similarly a mouse can be put into a trap, or a bird into a cage.

The reader must remember that these subjects must be drawn as if they were upside down with regard to each other, so that the man who is to ride the horse is drawn as if he were standing on his head, and the mouse which is to enter the trap looks as if it were lying on its back.

The most simple manner of spinning the disc is by means of two threads, each being inserted near the edge of the disc, and exactly opposite each other.

A very ingenious modification of the Thaumatrope is made by inserting at one side of the disc two strings, of which one is elastic. It is evident, then, that by lengthening or shortening the elastic string, the axis can be changed, and the objects on the opposite sides placed in positions relatively different from each other. Thus the jockey may be made to jump on and off his horse, the bird to go in and out of its cage, the mouse to enter the trap, and so on. This simple invention allows of infinite combinations, so that a tree may be made to sprout, a man to move his limbs, and a bird to flap its wings. It was invented, I believe, by Dr. Paris, author of " Philosophy in Sport made Science in Earnest."

On the right hand of the illustration are seen three figures, each representing a means of obtaining an ocular delusion through the principle of which we are now treating.

The lower figure is called the Zoetrope, or Wheel of Life. As the reader may see, it consists of a hollow cylinder, revolving on a centre, and having within it a series of figures. When the wheel revolves, and the figures are viewed through the slits, each figure seems to be in lifelike motion, whence the name of Zoetrope. In the present case the figures are those of boys jumping over posts.

The mode in which this effect is produced is as follows :— Suppose that a boy were really to jump over a post, he would go through a series of motions, and his body be placed in a certain series of positions, before he cleared the post. Supposing, then, that several points were chosen in his course, and his

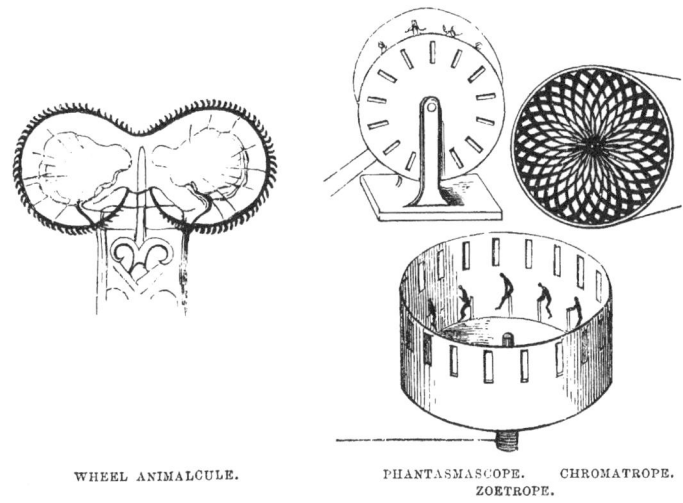

WHEEL ANIMALCULE.   PHANTASMASCOPE.   CHROMATROPE.
ZOETROPE.

body drawn as it would appear at these points, and the drawings placed in their proper order in the Zoetrope, it is evident that the figures must appear in movement. Before the retina loses the image of the boy standing in front of the post, it takes in that of the boy stooping, with his hands on the top of the post, and so on until he has reached the ground on the opposite side.

Another mode of producing the same effect, called the Phantasmascope, is seen above the zoetrope. In this case the images are placed on the inside of the disc, which is held opposite a mirror, and the figures viewed through the slits.

The last of these figures is the rather complicated one, like

the back of an "engine-turned" watch. This is called the
Chromatrope, or Wheel of Colour, and is always a favourite
object in a magic lantern. It consists of two circular plates of
glass, one upon the other, and painted in variously coloured
curved lines, as seen in the illustration. When the image is
thrown upon a screen, and the glass plates turned in opposite
directions, a most singular and beautiful effect is produced. The
lines, unless the eye follows them very closely, disappear, and
torrents of coloured spots seem to pour from the centre to the
circumference, or *vice versâ*, according to the direction in
which the glass wheels are turned. So perfect is the illusion,
that it is almost impossible to believe that the movement is
only circular, and not spiral.

Now we will pass from Art to Nature. The figure on the
left hand of the same illustration represents part of one of the
Wheel Animalcules, so called because they look exactly as if
the fore-part of their bodies were furnished with two delicate
wheels, running rapidly round, and evidently moving or stopping at the pleasure of the owner.

Soon after the powers of the microscope became known,
these Wheel-bearers were discovered, and for a long time they
were thought to have a pair of veritable revolving wheels upon
their heads. They were naturally held in high estimation, as,
although almost every kind of lever can be found in the animal
world, a revolving wheel had never been seen. However, as
the defining powers of the microscope improved, the so-called
wheels were found not to be wheels at all, but stationary
organs, and that their apparent revolution was nothing but an
optical delusion.

The wheels are, in fact, two discs, around the edges of which
are set certain hair-like appendages, called "cilia," from a
Latin word signifying the eyelashes. Each of the cilia has an
independent motion of its own, and, as they bend in rapid and
regular succession, they produce an effect on the eye similar to
that of a revolving body. As for the animal itself, they
produce a double effect, either acting as paddles, and forcing the
animal through the water, or, when it is affixed to some object,
causing a current which drives into its mouth the minute beings
on which it feeds.

The particular species of Wheel-bearer whose mouth is here shown is called scientifically *Limnias ceratophylli*. It derives the latter name from the fact that it is mostly found on the submerged stems and leaves of the Hornwort (*Ceratophyllum*), which is very common in ponds and slow streams. The creature is, however, to be found on the water-growing plants, and Mr. Gosse, in his "Evenings with the Microscope," gives a very full and graphic account of itself and its habits.

He specially mentions the use of the wheels, and, by dissolving a little carmine in the water, had the pleasure of seeing the coloured granules swept into the mouth by the current caused by the cilia, through the jaws, and so into the stomach.

USEFUL ARTS.

CHAPTER I.

PRIMITIVE MAN AND HIS NEEDS. — EARTHENWARE. — BALL-
AND-SOCKET JOINT.—TOGGLE OR KNEE JOINT.

Contrast between Savagery and Civilisation.—Manufacture of Weapons.—
Earthenware of Art.—Sun-baked Vessels.—Earthenware of Nature.—Nest
of Pied Grallina.—Analogy with the Babylonish Brick.—Nest of the Oven-
bird.—A partitioned Vessel.—Necked earthenware Vessels.—Nests of
Eumenes, Trypoxylon, and Pelopœus.—Proof of Reason in Insects.—The
Ball-and-socket Joint.—" Bull's-eye " of Microscope.—The human Thigh-
bone.—Vertebræ of the Serpents and their Structure.—The Sea-urchin and
its Spines.—Legs and Antennæ of Insects.—The Toggle or Knee Joint, and
its Use in the Arts.—The hand Printing-press and the Toggle-joint.—The
human Leg and Arm.—Power of the natural Toggle-joint.—Fencing and
Boxing.—Heads of Carriages.—" Bowsing " of Ropes.—Leaf-rolling Caterpillars.

IN the primitive ages of Man the aids to civilisation were
very few and very rude. Some of them, especially those
which relate to hunting and war, have already been mentioned,
and we now have to deal with some of those which bear upon
domestic life.

Here we are in some little difficulty, for it is not very easy
to draw the line where domestic life begins, or the mode in
which it shall be defined. We may at all events connect
domestic life with a residence of some sort, and may, in consequence, neglect all such primitive savages as need no domestic
implements.

Such, for example, are the few surviving Bosjesmans of
Southern Africa, not one of whom ever made a tool or an
implement, or looked beyond the present day. The genuine
Bosjesman can make a bow and poison his arrows, and he can
light a fire ; but there his civilisation ends. He cannot look
beyond the present hour, he has not the faintest notion of

making a provision for the future, nor did his wildest imagination ever compass the idea of a pot or a pan.

He kills his prey, and, if hunger be very pressing, he will eat it at once without waiting for the tedious ceremony of cooking; or at the best will just throw the meat upon the fire, tear it to pieces with his teeth, and swallow it when it is nothing but a mass of bleeding flesh, charred on the outside, and absolutely raw within. The Bosjesman has not even a tent which he can call his own, any bush or hole in the ground answering for a house as long as he wants it, and then being exchanged for another.

As far as we know, the only trace of civilisation in the Bosjesman is his manufacture of weapons, and even his bow and arrows are of the rudest and clumsiest forms. Nor is it likely that he will ever advance any further; for, as is the wont of all savage tribes, he is disappearing fast before the presence of superior races, and will shortly be as extinct as the Tasmanians, the last of whom died only a few years ago.

EARTHENWARE.

THE advent of real civilisation seems to depend largely upon the construction, not of weapons, but utensils, and the most useful of these are intended either for the preparation or the preservation of food. That such vessels should be made of earth is evident enough, and it is worthy of remark that the rude earthenware pot of the naked savage and the delicate china of Sèvres should both be products of the earth, and yet be examples of the opposite ends of civilisation.

The most primitive earthenware vessels were simply baked in the rays of the sun, the use of fire for hardening them being of later date. Rude and simple as they are, some of these vessels possess tolerable strength, and can answer every purpose for which they are intended. I possess several pots made by the aborigines of the Essequibo district. They are very thick and heavy in proportion to their dimensions, and are still so fragile that I have been obliged to bind them with string whenever they are moved.

Simple as they are, however, they are pleasing to the eye, chiefly, I presume, because they are made for a definite office, and fulfil it, and have no pretence about them. Then, as they are moulded by hand alone, without any assistance from machinery of any kind, even a wheel, the individuality of the maker is stamped upon them, and no two are exactly alike either in form, colour, or ornament. A couple of these rude vases are to be seen on the right hand of the accompanying illustration.

On the left hand of the same illustration are shown two examples of earthenware vessels made by birds, which are nearly, if not quite, as good as those made by the hands of civilised man.

The upper figure represents the nest of the Pied Grallina (*Grallina Australis*), a bird which, as its specific name implies, is a native of Australia.

This nest is formed chiefly of clay, but a quantity of dried

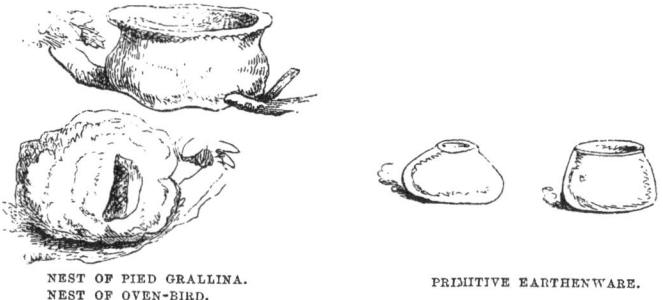

NEST OF PIED GRALLINA.   PRIMITIVE EARTHENWARE.
NEST OF OVEN-BIRD.

grass is always mixed with it, and serves to bind it together. If one of these nests be broken up, and compared with the bricks of which ancient Babylon was built, it will be found that they are almost identical in material, and that both are merely baked in the sun. In form it so closely resembles an Essequibo jar in my possession, that if it were removed from the branch, and similarly coloured, it would not be easy to distinguish the one from the other.

Below this is the nest of the Oven-bird of South America (*Furnarius fuliginosus*), a bird allied to our common creeper. The drawing was taken from a specimen in the British Museum.

Like the nest of the Grallina, it is placed upon some horizontal bough, and fixed so firmly that it cannot fall except by being broken to pieces. Not being afraid of man, the Oven-bird often chooses a beam in some outhouse for a resting-place, and has been known to build even on the top of palings. As may be seen by reference to the illustration, the nest is a very conspicuous one, and concealment is almost impossible.

As in the Grallina nest, the material is remarkably hard and firm, as indeed is necessary, to allow it to withstand the effects of the rain-torrents which fall during the wet seasons of the year.

There is a curious analogy in this nest with many articles of earthenware. Not only among ourselves, but among un-civilised races, earthenware vessels are constructed with partitions, so as to divide one portion from another. If one of these nests be cut open, it will be found to have a sort of partition wall across the interior, rising nearly to the top of the dome, and so dividing it into two parts. The wall also answers another purpose—*i.e.* that of strengthening the entire structure. Within the inner chamber is the real nest, which is lined with a thick layer of feathers, the outer chamber being bare, and, as it is thought, being occupied by the male.

WE now come to pottery of a more elaborate shape. Both in the Grallina nest and the earthen pot of the Essequibo Indian we have a vessel with a mouth nearly as wide as its greatest diameter, and with a lip which is very slightly turned over. There are, however, many varieties of pottery in which the neck is narrow and long, and the lip is boldly formed. Some examples of this form are given on the right hand of the accompanying illustration.

ON the left hand are shown some nests of a solitary wasp belonging to the genus Eumenes. It is a British insect, but seems to have been little noticed, except by professed ento-mologists.

It especially haunts heather, and affixes to the stems of the plant its little globular nests, which are made of mud, and shaped as seen in the illustration. Perhaps some of my readers may have seen the "Napier Coffee Machine," which draws the coffee into a glass globe furnished with a short neck. The globe is shaped exactly like the nest of our Eumenes, and, when I first saw one, I could not remember why its shape was so familiar to me.

As is the case with the birds' nests which have been mentioned, the mud of which the walls are built is of a most tenacious character, and, when dried in the sun, can resist the heaviest rain. The cells are intended as rearing-places for the young, only a single egg being placed in each cell, which is then stocked with small caterpillars by way of food.

THERE is a South American insect also belonging to the solitary wasps, and remarkable for building a round nest exactly similar in material, and nearly identical in shape, with that of

NESTS OF EUMENES.   ANCIENT NECKED POTTERY.

the Eumenes. Its scientific title is *Trypoxylon aurifrons*. The nest of this insect has a much wider mouth than that of the Eumenes, and exactly resembles the upper left-hand jar in the illustration.

ANOTHER South American solitary wasp, belonging to the genus Pelopœus, makes nests of similar material, but nearly cylindrical in shape instead of globular. The nest is built up of successive rings of moistened and well-kneaded clay, exactly as human houses are built by bricklayers. Indeed, the process of making a Pelopœus' nest has been happily compared to that of building a circular chimney.

I may as well mention here that the name Pelopœus is

formed from a Greek word signifying mud, and that the entire word may be translated as "mud-worker."

As a proof that these insects possess reason as well as instinct, Mr. Gosse mentions that one of them, instead of making her nest for herself, utilised an empty bottle, and, after storing it with spiders, stopped up the mouth with clay. Finding, after an absence of a few days, that the nest had been disturbed, she removed the spiders, inserted a fresh supply, and then closed the mouth as before.

BALL-AND-SOCKET JOINT.

WE will now see how some of the most useful mechanical inventions have had their prototypes in Nature.

There is, for example, the well-known " Ball-and-socket joint," without which many of our instruments, especially those devoted to optical purposes, would be impracticable.

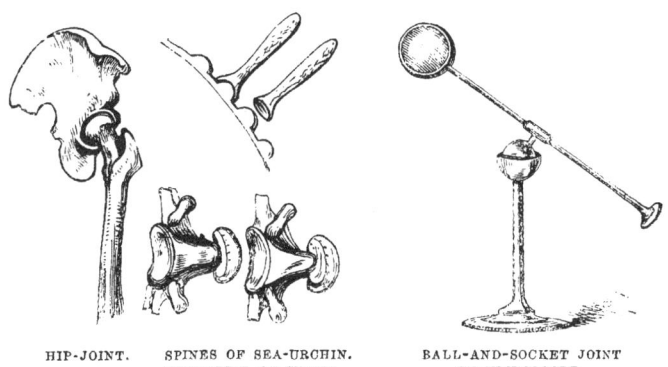

HIP-JOINT.   SPINES OF SEA-URCHIN.   BALL-AND-SOCKET JOINT
VERTEBRÆ OF SNAKE.   OF MICROSCOPE.

The figure on the right hand of the illustration represents the " bull's-eye " of my own microscope. It will be seen that there is a ball half sunk in a cup, so that it can be turned in any direction. In point of fact, the upper part of the ball is nearly concealed by another cup, but, in order to show the structure, the upper cup has been removed. Who was the inventor of the ball-and-socket joint I do not know, but I have little doubt that he must have had in his mind many natural examples of this joint, three of which are represented in the illustration.

On the left hand are seen the upper part of the human thigh-bone and that part of the hip-bone into which it fits.

The reader will see that at its upper end the bone takes rather a sharp turn, and is then modified into a ball. This ball fits into a corresponding socket, technically named the "acetabulum," and is thereby endowed with freedom of motion in almost every direction. Generally we do not practise our limbs sufficiently to develop that full freedom, but those who have seen any good professional acrobats must have been struck with the wonderful mobility of which the human body is capable.

The socket is not a deep one, but dislocation of the hip is exceedingly rare, the bone being held in its place by three powers. The first is due to a short ligament, which, however, does not always exist, but, when it is present, is useful in retaining the bone in its place. Then there is the contractile power of the thigh muscles, which are always forcing the ball into the socket. Lastly, there is the pressure of the atmosphere, a force which is seldom taken into consideration, but which has great influence on many parts of the human frame. This part of the subject will be resumed when we come to treat of Atmospheric Pressure.

The arms are jointed to the shoulder-blades in a very similar manner, the upper arm-bone, or "humerus," being furnished with a rounded end, and fitting into a cup-like cavity in the shoulder-blade, or "scapula." This formation can easily be seen by separating the different bones of a shoulder of mutton.

At the bottom of the illustration are given two vertebræ of a snake, separated in order to show their structure. It will be seen that each joint has a ball in front and a socket behind, thus giving the creature that wonderful flexibility which is quite proverbial, and without which it could not seize its prey.

The following eloquent passage is taken from Professor Owen's work entitled "The Skeleton and the Teeth:"—

"Serpents have been regarded as animals degraded from a higher type, but their whole organization, and especially their bony structure, demonstrate that their parts are as exquisitely adjusted to the form of their whole, and to their habits and sphere of life, as is the organization of any animal which we call superior to them.

" It is true that the serpent has no limbs, yet it can out-climb the monkey, outswim the fish, outleap the Jerboa, and, suddenly loosing the coils of its crouching spiral, it can spring into the air and seize the bird upon the wing : all these creatures have been observed to fall its prey.

"The serpent has neither hands nor talons, yet it can out-wrestle the athlete, and crush the tiger in the embrace of its ponderous overlapping folds.   Instead of licking up its food as it glides along, the serpent uplifts its crushed prey, and presents it, grasped in the death-coil as in hand, to its slimy, gaping mouth.

" It is truly wonderful to see the work of hands, feet, and fins performed by a modification of the vertebral column— by a multiplication of its segments with mobility of its ribs. But the vertebræ are especially modified, as we have seen, to compensate, by the strength of their numerous articulations, for the weakness of their manifold repetition, and the conse-quent elongation of the slender column.

" As serpents move chiefly on the surface of the earth, their danger is greatest from pressure and blows from above ; all the joints are fashioned accordingly to resist yielding, and sustain pressure in a vertical direction ; there is no natural undulation of the body upwards and downwards—it is permitted only from side to side.   So closely and compactly do the ten pairs of joints between each of the two hundred or three hundred vertebræ fit together, that even in the relaxed and dead state the body cannot be twisted except in a series of side coils."

THE upper right-hand figure represents a portion of the shell of an Echinus, or Sea-urchin, together with two of the spikes.

The reader will remember that in the description of the Heart-urchin, and the mode in which it dug its way into the sand, the peculiar mobility of the spines was mentioned.   How that mobility is produced we shall now see.

If a living Sea-urchin can be procured, and placed in a glass vessel filled with sea-water, it will at once be seen that

its surface is thickly covered with spines. In some species
these spines are as thick as ordinary drawing pencils; but in
most of those which are found on our shores they are very slight,
and scarcely longer than darning-needles. They are in almost
perpetual motion, and generally have a sort of revolving movement, the base being the pivot.

Now, if we take a dried shell of the Sea-urchin, we shall
find that the spines will come off with a touch, and, indeed, to
preserve one with all the spines complete is a most difficult
business. Let us, therefore, pull one from its attachment, and
examine its base. This will be found to be swollen into a cup-
like form, as seen in the illustration ; and, if we look at the
spot whence it came, we shall see that there is a little, rounded,
polished prominence, exactly fitting into the cup, just as the
ball of the human thigh-bone fits into the acetabulum. It
has also its ligament to keep it in its place, and its same set of
muscles that move it, and is altogether a most wonderful piece
of mechanism. There are in some species of Echinus about
four thousand of these spines.

THE legs of an insect afford excellent examples of the ball-
and-socket principle, the socket being on the body, and the
ball on the base of the leg. Some of our largest insects—such,
for example, as the common Stag-beetle—exhibit this principle
very well. I have now before me a Stag-beetle which has been
dead for many years, and is quite dry and hard. Yet I can
rotate the legs almost as freely as if the beetle had been
just killed, so easily do the joints work. Even the antennæ,
which are affixed to the head by a similar joint, move about
by their own weight on merely changing the position of the
insect.

These are only a few of the many natural examples of the
Ball-and-socket joint, but they are sufficient for our purpose.

THE TOGGLE OR KNEE JOINT.

ANOTHER most useful invention now comes before us, called
the Toggle-joint, or Knee-joint, the latter name being given
to it on account of its manifest resemblance to the action of
the human knee.

This joint is shown in the illustration. It consists of two levers, jointed together at one end, and having the other ends jointed to the objects which are to be pressed asunder. It will be seen that if the centre of the Toggle be pushed or pulled in the direction of the arrow, so as to straighten the levers, the amount of pressure upon them is enormous. Such an apparatus as this combines simplicity and power in a wonderful manner, and is greatly used in machinery, especially in presses, where the force is required to be great, but not of long duration.

An ordinary two-foot rule, when bent, affords a good example of the Toggle-joint, and will exert a wonderful amount of force.

The illustration represents one of the common printing-presses that are worked by hand. When the workman draws

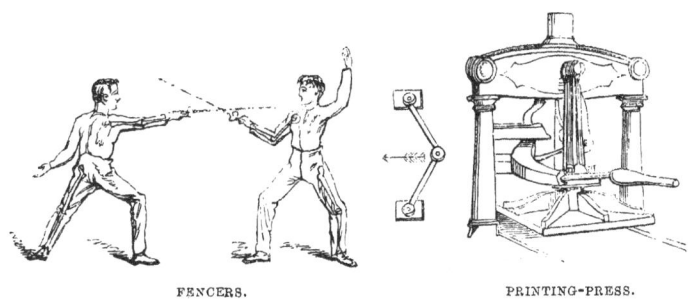

FENCERS.   PRINTING-PRESS.

the handle horizontally, he causes the two portions of the Toggle to approach a straight line. The upper half of the Toggle being jointed to the fixed beam above, and the other half to the movable plate or " platen " below, it is evident that the latter will be pressed downwards with enormous force. Indeed, so great is the power of this instrument, that a man of moderate strength can exert a pressure of many tons.

WE now proceed from Art to Nature, and take first the human knee, being the joint from which this piece of mechanism has derived one of its names.

If the reader will look at the figure of the fencers, he will see that the arm and leg are both Toggle-joints. In the one who is standing on the defence they are bent, and in the other, who has just made a longe, the Toggles of the right arm and left leg are straightened. It is by the straightening of these joints, and not by the action of stabbing, that the rapidity and force of a thrust are achieved.

It is just the same in boxing. No one who has the least knowledge of sparring strikes a round-handed blow, for, putting aside the ease with which it is parried or avoided, it has scarcely any force in it. When a boxer hits " straight from the shoulder," he not only straightens the Toggle-joint of his left arm, but that of his right knee also, so that the force of the blow comes quite as much from the leg as the arm.

It is by the right use of this joint that a small man, provided he be an expert boxer, will easily conquer an ignorant opponent who far surpasses him in size and weight. I have seen in a sparring-match a man not only knocked down, but fairly lifted off his feet, by a blow from a smaller opponent. The blow took effect under the chin, and, as the boxer hit exactly the right moment in straightening both limbs, a very great force was exerted with little apparent effort. I do not know which of the two combatants was the more astonished, the one to find himself on his back without exactly knowing how he got there, and the other to see his antagonist prostrate without exactly knowing how the thing was done.

The jointed apparatus by which the heads of carriages are raised or lowered is a good example of the Toggle, and exemplifies the force which a comparatively slight piece of machinery can exercise.

ANOTHER form of the Toggle-joint is the process called by sailors " bowsing " of rope. If a rope be fastened at both ends, and then pulled in the middle, the ends are drawn forcibly towards each other. This plan is mostly adopted in getting up sails. When a sail, say the mainsail of a cutter, has to be hoisted as far as it will go, the last few inches are always very obstinate. The word is then given to "bowse." The rope, or haulyard, is no longer pulled at the end, but a turn is taken round the cleat, so that it does not give way. The rope is then forcibly pulled away from the mast, when up goes the gaff a little higher. In this way, by repeated bowsings, the gaff is coaxed, so to speak, up the mast, and forced into its place.

Some of the leaf-rolling caterpillars act in a similar manner, by alternately bowsing and shortening their lines. As, however, their mode of working will be described under another heading, we will say no more of them at present.

USEFUL ARTS.

CHAPTER II.

CRUSHING INSTRUMENTS. — THE NUT - CRACKERS, ROLLING-
MILL, AND GRINDSTONE. — PRESSURE OF ATMOSPHERE. —
SEED DIBBLES AND DRILLS.

Importance of Leverage in Crushing Power.—Nut-crackers a Lever of the Second
Order.—The Chaff-cutting and Tobacconists' Machines.—Jaws of various
Animals.—The Wolf-fish or Sea-wolf.—The Rolling-mill and its Action.—
Gunpowder-mills and Granulating Machine.—The "Jacob's Ladder."—
The Mangle and its various Adaptations.—The Grindstone.—Primitive
Grindstones of the Savage Races.—The Kafirs and the Inhabitants of Pales-
tine.—Ceasing of the Millstone.—"Facing" of Millstones.—Tusk of the
Elephant and its Structure.—Its Facings always preserved.—Power of
Self-renewal.—Pressure of Atmosphere.—The Napier Coffee Machine.—
The Cupping Instrument.—The Pneumatic Peg.—The Magdeburg Hemi-
spheres.—Plane Surfaces of Glass or Metal.—Suckers of the Cuttle-fish.—
Foot of the Water-beetle.—The Limpet.—The Star-fish and its Mode of
Progression.—The Sucking-fish and the Fables connected with it.—Its real
Structure.—Modification of the Dorsal Fin.—The Gobies and Lump-fish.—
The Gecko and Tree-frog.—The Lampern and the Medicinal Leech.—Seed
Dibbles and Drills.—Labourers *versus* Machinery.—Natural Dibble of the
Grasshopper.—The Daddy Long-legs.—Drills and Dibbles of the Ichneumon-
flies.—A wonderful Specimen from Bogotá.—The Pelecinus and its Mode
of laying Eggs.

CRUSHING INSTRUMENTS.

AS we are on the subject of leverage, we will take some
examples of levers in Art and Nature, without, however,
even attempting to exhaust the topic.

On the right hand of the illustration is shown a very familiar
example of a lever, namely, nut-crackers, with a nut between
them. This useful implement is simply an adaptation of levers
of the second kind, the power being represented by the human
hand, the weight by the nut, and the fulcrum being the joint of
the instrument.

The common chaff-cutter, which is worked by hand, is
another familiar example of this kind of lever, and so is the knife
used by tobacconists in cutting cake Cavendish into threads,

and by druggists for similar purposes. In these instruments the point of the knife is jointed to some fixed object, and becomes the fulcrum ; the hand of the cutter supplies the power, and the weight is the object which is being cut. It will be seen that, by increasing the length of the handle, very great power can be obtained.

Exchanging the power for weight, we have in the common tongs, whether used for the coals or for sugar, a leverage of a similar character, the weight moving over a greater space than the power. A good example of this is to be found in the deltoid muscle of the human arm. The muscle, which furnishes the power, contracts about an inch, and, so doing, moves the

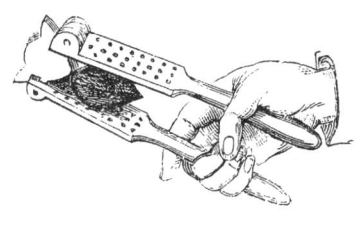

JAWS OF WOLF-FISH.

NUT-CRACKERS.

hand over some forty inches of space. It has been well stated that if a man is able to hold in his hand, and with extended arm, a weight of twenty-five pounds, the muscle must be exerting a power of forty times as great, *i.e.* about a thousand pounds.

THERE is little doubt that, in such Crushing Instruments as have been mentioned, the idea has been taken from the jaws of sundry animals. We know, for example, that with ourselves, if we desire to crack a walnut or a filbert in our teeth, we always put it as far back as possible, so as to make the leverage as powerful as possible. No one would ever dream of cracking a nut with his front teeth, an act which would be very much like that of trying to break a piece of coal by pinching it with the tongs.

The left-hand figure of the illustration represents part of the jaws of the Wolf-fish, or Sea-wolf, as it is sometimes called, and a very wonderful crushing machine it is. The Sea-wolf (*Anarrhicas lupus*), sometimes called the Sea-cat, or Swine-fish, is tolerably common on our coasts, and, as it sometimes attains a length of seven feet, and is proportionately stout and muscular, the power of its bite may be estimated. The fish in question feeds chiefly on crustacea and hard-shelled molluscs, and is therefore furnished with an apparatus which can crush their shells. Extremes meet. The Sea-anemones, which are mere films of animal matter, and can be torn in pieces with the finger and thumb, can seize, swallow, and digest a crab or an oyster in spite of the thick and strong shells in which they are enclosed. So can the Sea-wolf, and fishes of a similar character. But nothing intermediate can touch them, and it is curious to reflect that such opposite means should produce a similar effect.

On reference to the illustration, the reader will see how exact is the parallel between the Nut-crackers and the Sea-wolf's jaws, both being worked on the same principle, and both being furnished with a series of projecting points, which are used for the purpose of preventing the escape of the object which is to be crushed. The terrible grasping power of the crocodile, the dolphin, and other predacious creatures can be explained on the same principle.

THE ROLLING-MILL.

WE now come to another variation of the Crushing Machine, *i.e.* that in which the motion is constant, and not intermittent, as is the case with those machines which have just been mentioned.

Perhaps some of my readers may have visited those great iron-works in which huge masses of iron are rolled into plates of greater or less thickness, or are cut up into strips as easily as if they were butter.

The mechanism is in its principle simple enough. The cylindrical rollers are placed nearly in contact, and forced towards each other by mechanical means, such as levers, screws, or springs, or all three combined. These cylinders revolve in opposite directions, and, if any object be placed between them, they draw it through them, and present it on the other side in a flattened condition.

Many years ago, one of my schoolfellows, who had been brought up entirely under the care of some maiden ladies, was visiting a workshop, and must needs put his finger between two revolving rollers. Of course the hand was drawn between them, and simply squeezed flat. The machine was instantly stopped, and the hand extricated; and the strange thing was, that the crushed and shapeless hand afterwards recovered its full

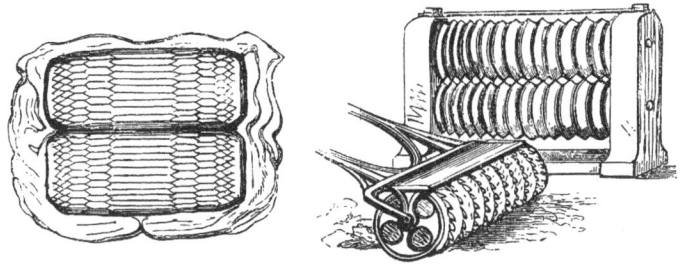

JAWS OF SKATE.   CRUSHING-MILL AND ROLLER.

power, though not its shape, and was able to touch the keys of the piano.

The whole process of the Rolling-mill is singularly interesting, whether it be used for large or small objects.

Supposing that the grooved rollers of the illustration were cut across so as to present a number of points, it is evident that anything which got between them would be bitten to pieces, each piece being of a tolerably uniform shape.

This plan is now adopted in the granulation of gunpowder. After the future powder has emerged from the hydraulic press in the form called "press-cake," it was formerly broken to bits with wooden or copper mallets, and then placed in a very peculiar kind of sieve. This was shaped like an ordinary sieve, but the bottom was made of cowhide, pierced with innumerable holes. A round pebble was placed in the sieve, and, when the latter was violently shaken backwards and for-

wards, the powder was driven through the holes by the pressure
of the stone, and was afterwards separated into its various
degrees of fineness.

I have only twice seen this process, and confess to have been
in a very nervous state on both occasions. The sieve is
whirled about with enormous velocity, and the pebble flies
round as if it were a thing alive. Let but a broken needle or a
fragment of stone get into the sieve, or even let the stone itself
break asunder, and there will be an instantaneous explosion,
which will hurl the house, the machinery, and the workmen
into unknown regions.

Now, however, the mode of granulating powder is radically
altered. There is a series of double cylinders, such as shown in
the illustration, and each of them has the ridges cut into teeth
in regular order. Thus the first set of rollers or cylinders
merely bites the press-cake into convenient pieces, though
seldom of the same weight.

The press-cake, thus bitten to pieces, is passed through a
series of cylindrical sieves, each graduated with the utmost
accuracy, and being turned by means of machinery. Being
set on a slope, the powder runs by its own weight down them,
and all those particles which cannot pass through the meshes
are poured out untouched at the lower end.

The portions which are too large to pass the openings of the
first sieve are then handed onwards by means of a machine called
a "Jacob's Ladder," which consists of a series of little vessels or
buckets strung on a tape, and revolving over a couple of wheels.
The first set of buckets takes the coarsely bitten press-cake to
the second set of rollers, the teeth of which are comparatively
small. Thence it is passed over to a third set, and so forth, until
it is delivered in any quality of grain which may be required.

The modern Mangle, again, affords a good example of this
principle. The old obtrusive, costly, and cumbrous Mangle,
which was nothing more than a heavy box of stones upon
rollers, has given place to the modern system of duplex action
in rollers, and one of the old Mangles is not easily to be seen,
unless it be worked as a curiosity. In fact, it is nearly as obso-
lete as the spinning-wheel, which yet may be seen in some of our
country villages, where scarcely one per cent. of the population
has ever been in a town, and many of them, the women espe-

cially, make it their boast that they have never been beyond the outskirts of their village.

This clumsy machine is now replaced by the very simple invention which has been in vogue for some years, and which can not only release, but regulate, the pressure at any moment, by means of springs, levers, and weights. This machine is, in fact, exactly the same as that which is represented in the illustration, except that the rollers are quite smooth. They can be adjusted to almost any amount of pressure by levers and weights which are attached to the upper roller, and, when the linen has passed through them, it has undergone the double operation of wringing and mangling. It does not occupy one-quarter of the space of the old machine, and is light enough to be moved easily from place to place.

THE GRINDSTONE.

BEING on the subject of jaws and teeth as a mode of breaking to pieces objects which are placed between them, we will take those implements which grind to powder, or "triturate," instead of breaking or flattening.

From the very earliest ages, and as soon as man had begun to discover the "staff of life," the art of grinding naturally assumed an ever-increasing importance.

The first and most primitive mode of grinding corn and converting it into meal was that which was followed by Sarah, when she welcomed her husband's guests, which we know, from internal evidence, was followed by the uncivilised races who formerly inhabited this island, and by many semi-savages of the present day.

Nothing could be simpler than the machinery used, and nothing could cause a greater waste of muscular power. Two stones were employed, a large one upon which the grain was placed, and a smaller which was held in the hands, and used for grinding the corn to powder, just as the painters of the last century used to grind their colours. The Kafirs of Southern Africa use this simple mill, and so exactly do they keep unconsciously to the customs of long-perished natives, that if one of their mills were buried for a few years and dug up again, it might be mistaken for one of the ancient "querns."

As the stone held in the hand was rounded, it naturally wore a rounded hollow in the lower stone, and this made the process of trituration easier. Perhaps some of my readers may have noticed that when a chemist makes up a prescription, and is obliged to reduce one of the ingredients to powder, he always does so by rubbing, and not by pounding, as is generally believed. He works the pestle round and round the mortar with a kind of twisting motion, and thus obtains a powder much too fine to have been produced by any amount of pounding.

The labour of this operation is necessarily very severe, and therefore the Kafir of the present day, as did his predecessors

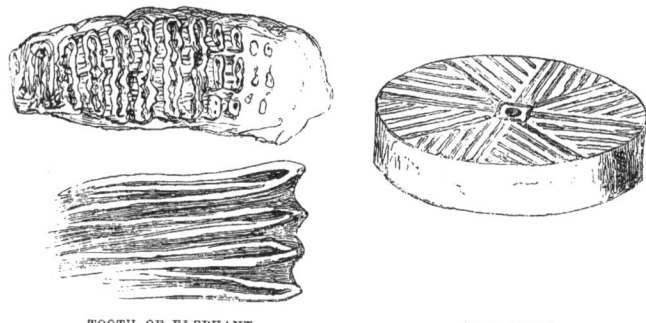

TOOTH OF ELEPHANT.   GRINDSTONE.

of the long-lost races, declines to do it himself, but hands it over to the women. In Palestine, as in other parts of the world, a simple mill has been invented, which takes away much of the labour, and, above all, releases the grinder from the obligation of leaning with her full weight upon the upper stone. In this mill the stones are similar. The upper is moved backwards and forwards round a pivot, and the grain is passed between them by means of a conical aperture in the upper stone, which answers the purpose of our "hopper."

In order to work this mill, two women are required, sitting opposite each other, with the mill between them, holding the same handle, and assisting each other in turning the stone backwards and forwards. No one who has not seen this operation can fully appreciate the force of the saying that "two women shall be grinding at the mill; the one shall be taken, and the other left."

It is worthy of remark that, even at the present day, the custom of grinding corn is carried out in Palestine as it was so many centuries ago, and that it is repeated in Southern Africa among the Kafir tribes. In both parts of the earth the first sound of early morning is caused by the millstones of the grinding women, and the amount and duration of the noise afford a sure test of prosperity. Cessation of the millstones signifies adversity and a thin population, as has been said by a writer who lived not very far from three thousand years ago. Speaking of tribulation, he mentions that "the grinders cease because they be few, and that the doors shall be shut in the streets when the sound of the grinding is low."

After awhile improvements were gradually introduced into the business of grinding, not the least of which was covering its surface with ridges, instead of leaving it entirely smooth, as it had been formerly. Millers of the present time know the value of these ridges, and the additional grinding power which this "facing" gives to a stone. One of these stones is represented in the illustration, so as to show the system on which the ridges and grooves are constructed.

Now, passing from Art to Nature, we find that the whole system of the millstone, its movement and its ridged surface, existed in the times when man had not yet come upon earth.

The reader is probably aware that among the tooth-bearing animals there are three types of teeth. First come the incisors, or cutting teeth, which occupy the front of the jaw, and find their fullest development in the rodent animals, such as the beaver, the squirrel, the rabbit, and the rat. Next them come the canine or piercing teeth, which are so highly developed in all the cat tribe. Lastly, there are the molar or masticating teeth, so called from a Latin word signifying a millstone, because their office is to grind food.

As it is with these last that we have now to treat, we will say nothing about the others.

The molar teeth find their greatest development in the Elephant, the structure of whose molars is exactly like that of our modern millstones. There is certainly one very great difference. When the surface of a millstone is rubbed away, the stone must be re-faced, and sooner or later is worn out altogether, and must be replaced with a new one. This, however, is not the case with the Elephant's molar teeth, which not only keep their facing perfectly sharp, but have the faculty of renewing themselves as fast as they are worn away.

How these important objects are attained we shall now see.

If the reader will refer to the upper left-hand figure of the illustration, he will see that its surface is for the most part round, with irregularly oval figures, close and thick at one end, and almost disappearing at the other. These are the "facings" of the Elephant's tooth, and they are formed as follows :—

The tooth, which is of enormous size, is not solid, but is composed of a number of plates laid side by side, like a pack of cards when set on their edge. Each of these plates is composed of a hard external layer of enamel, and an internal layer of comparatively soft bony matter. A slice of badly made toast affords a familiar parallel, the half-charred outside representing the enamel, and the soft, sodden interior being analogous to the bony matter. In order to show the arrangement of these plates, a side view of part of the tooth is given on the same illustration. Sometimes, when the teeth of fossil elephants are discovered, these plates all fall asunder, the material which connected them having been dissolved away in the earth.

When, however, we look upon the upper surface of a recent tooth, we see it present the appearance which is shown in the illustration. The elongated oval marks are the edges of the hard enamel plates, while the spaces between them are filled with the soft bony matter. It will be evident, then, that if two teeth such as these be in opposite jaws, and perform the task of grinding food, their surface will always be well "faced." Owing to the different hardness and density of the enamel and bony substance, the latter will wear away with comparative rapidity, leaving the former to project slightly, and thus to preserve the facing of the natural mill.

This is, indeed, but a modification of the beautiful animal mechanism which keeps the teeth of a rodent animal always sharp, and always bevelled off at the proper angle. If we could invent some plan whereby, in our millstones, we could make the facing of much harder material than the stone, we should make an advance in the miller's art that would render the millstones of the future as far superior to those of the present as are our present millstones to the hand "quern" of the Kafir women.

Yet another improvement has to be made. Would it be possible to construct a millstone which should not only retain its facing, but possess the power of renewing itself in proportion as it is worn out? This property is found in the Elephant's tooth, and the illustration will give a tolerably good idea of the simple and beautiful mechanism by which it is brought into operation.

The tooth, instead of being one solid mass, consists, as I have already stated, of a series of plates set side by side. These plates are so constructed that they are more worn away in front than behind. In proportion as they are worn, a new tooth is built up behind the old one, and gradually pushes off the old one. Now, if we could only construct millstones with such properties, we should possess an absolutely perfect instrument.

PRESSURE OF ATMOSPHERE.

THERE are many useful inventions which depend on the weight of the atmosphere and the creation of a more or less perfect vacuum. There is, for example, the common Pump, which raises water simply by the action of the atmosphere. A pipe passes into the water, and in that pipe an air-tight piston is inserted. When the piston is drawn upwards a vacuum is formed, and the water is at once forced into it by the pressure of the atmosphere.

Then there is the graceful and useful Napier Coffee-making Machine, consisting of a glass globe, and vase of the same material.

Coffee and boiling water are put into the vase, and some hot water into the globe. The two are then connected with the tube, and under the globe is placed a spirit-lamp. Presently the water in the globe boils, expelling the air and filling the globe with steam. The lamp is then removed, and the steam in the globe is condensed, leaving a vacuum. The pressure of the atmosphere then comes to bear upon the coffee in the vase, which is forced through the tube into the globe, producing beautifully clear and well-flavoured coffee.

SURGERY employs the weight of the atmosphere in the operation called "Cupping," now rarely employed, but formerly in such constant use that scarcely any man who had attained middle age had not undergone it. The operation was intended for the purpose of removing the blood from some definite spot. Persons, for example, who appeared to have a tendency to apoplexy were regularly cupped between the shoulders twice a year, i.e. in the spring and autumn.

The mode of performing the operation is as follows :—A vase-shaped glass vessel called a cupping-glass is placed close to the skin. The flame of a spirit-lamp is then introduced for a moment in the glass so as to expel the air, and the glass is rapidly placed with its mouth downwards on the skin. If this be done with sufficient rapidity, the partial vacuum in the cupping-glass causes it to adhere to the skin, which is forced into it by atmospheric pressure, as shown in the illustration. The blood is, of course, drawn towards the surface by the same means.

The glass is then quickly removed, and a little brass instrument applied, which, at the touching of a spring, sends out a number of small lancet-blades so formed as to make very slight cuts. The glass is again applied, and rapidly becomes filled with blood from the cuts, the air having forced it in exactly as it forces the coffee in Napier's machine.

IN the upper right-hand corner of the illustration is shown the Pneumatic Peg, a comparatively recent invention, and useful in cases where much strength is not required. The base of the peg is fitted with a sort of cup made of india-rubber. When this base is pressed against a smooth and flat surface, such as a pane of glass, the air is forced out of the cup, and a vacuum formed. The pressure of the atmosphere then causes the cup to adhere to the glass with sufficient force to enable objects to be suspended from it.

The boy's well-known toy, the Sucker, is made on exactly the same principle. A piece of leather, generally circular, though the shape is not of much consequence, has a hole bored through its centre, so as to allow a string to be attached. The leather is then soaked in water until it is quite soft. If it be firmly pressed on any smooth object, such as a stone, the air is forced from under it, and it becomes capable of sustaining a weight in proportion to its dimensions. As the air has a pressure of about fifteen pounds on every square inch, it is easy

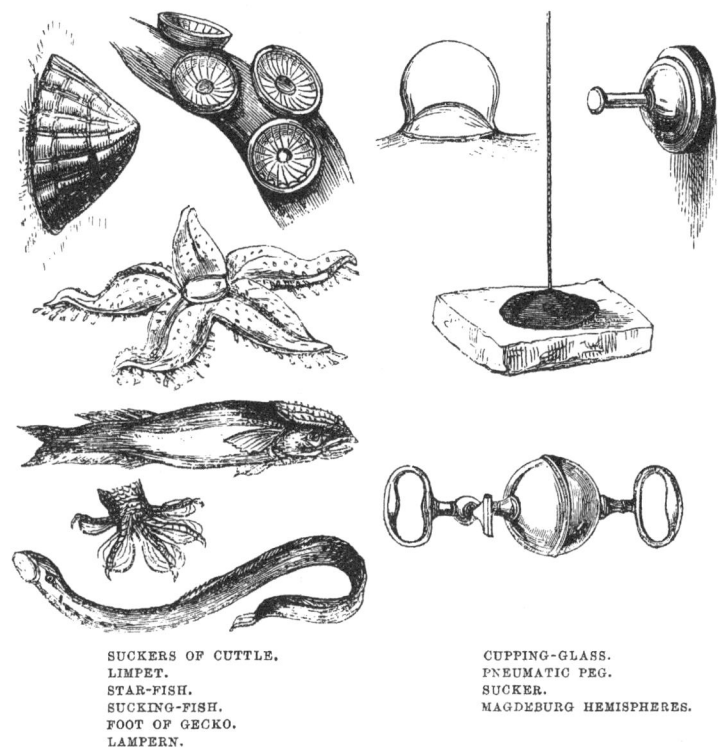

SUCKERS OF CUTTLE.
LIMPET.
STAR-FISH.
SUCKING-FISH.
FOOT OF GECKO.
LAMPERN.

CUPPING-GLASS.
PNEUMATIC PEG.
SUCKER.
MAGDEBURG HEMISPHERES.

to calculate the weight which it will uphold, a margin being left for imperfection of vacuum.

The lower figure represents the instrument called the Magdeburg Hemispheres, which are made for the purpose of showing the enormous power of air-pressure. They are two hollowed hemispheres, having their edges very accurately ground together. When used, a little lard is rubbed on the edges in order to insure their exact fit, and they are then pressed tightly together.   The air is removed by means of the common exhausting syringe, and it is found that the two adhere together with such force that two strong men cannot pull them asunder.   But, if the tap be turned, and air admitted, they come apart without the least difficulty.

Similarly, if two plates of glass or metal be ground to exactly plane surfaces, and pressed together, they adhere nearly as strongly as if they were one solid piece.

WE will now turn from Art to Nature, and examine some natural producers of vacuum.

One of the most celebrated is that series of suckers which may be found upon the arms of the various Cuttles.   At the upper part of the illustration a figure is given of part of an arm, on which are four suckers.   When the animal wishes to attach itself to any object, it presses the disc of the sucker against it, and simultaneously withdraws the centre, exactly as the boy does with his toy sucker.   And, as each arm contains a great number of suckers, it is evident that the holding power must be very great.   Indeed, on one occasion when a comparatively small specimen had fastened on a man's arm, he could not remove it, but was obliged to have it cut away piecemeal by an assistant.

The common Water-beetle has similar suckers upon its first pair of feet, and can adhere to smooth surfaces with great tenacity.

ON the left of the cuttle-arm is the common Limpet, shown as it appears when adhering to the rocks.   Every visitor to the seaside who has attempted to remove the Limpets may remember how difficult it is to stir them when they have once taken their hold.   If they can be taken by surprise, they come away with a touch ; but if they become alarmed, they press the edges of the foot firmly against the rock, withdraw the centre, and thus create the necessary vacuum.

NEXT follows a Star-fish, shown as it appears when in the act of walking, or rather, gliding along.

This movement is obtained by the use of a vast number of long suckers, exactly resembling the pneumatic peg, except that they are flexible, and can be curved in any direction. It is really beautiful to see the manner in which a Star-fish will glide along by means of its suckers, its arms accommodating themselves to the irregularities of the ground, and its multitudinous suckers protruded and withdrawn with a never-ceasing movement.

And, as the Star-fish is apparently blind, not having any organs which can even be conjectured to serve the purpose of vision, this mode of directing its course is not easily understood. Yet, blind though it may be, it guides itself with as much accuracy as if it possessed eyes, and evidently does so with a definite purpose, using its suckers with as much decision as a centipede uses its legs.

These suckers can be seen very well by placing a Star-fish in a shallow vessel of sea-water, and laying it on its back. The suckers immediately protrude themselves from their little apertures, and the arms slowly curve themselves so as to find something to which the suckers can adhere. Presently one or two of the suckers will take hold of the bottom of the vessel. Others soon follow, and in a very short time the Star-fish is on its legs, if we may so call them, and is quietly gliding on its way.

BELOW the Star-fish is seen the celebrated Sucking-fish (*Echeneis remora*) about which so many strange tales have been told, and which is possessed of a structure remarkable enough to need no aid from invention. The dorsal fin of this fish is modified in a most singular manner. The spines of which it is so largely composed are metamorphosed into flattened plates very much resembling the laths of a Venetian blind, and form an instrument of suction identical in principle, though not in form, with those which have already been described. When the sucker is pressed against a smooth surface, a vacuum is formed, and the fish in consequence adheres firmly to the object.

The fact has been known for centuries, though it has only been lately discovered, that the sucker was not a separate apparatus, but merely one of the fins modified in a simple though effective manner. Indeed, any one who has some slight notion of the structure of a fin can easily see, by looking at the Sucking-fish from above, that the apparatus is nothing more than the dorsal fin laid flat.

I may mention here that the name of Echeneis is taken from two words signifying " ship-holder." It was given to the fish on account of a curious notion which was fully believed until quite modern times, that the Sucking-fish had the power of attaching itself to ships, and holding them so firmly that they could not proceed in spite of sails and oars. The word Echeneis is used by Aristotle in his " History of Animals." The specific name *remora*, or " delay," is Latin, and is given to the fish for the same reason.

The little Gobies, which are so plentiful along our coasts, have the ventral fins formed into a sucker, with which they can cling firmly to any object, such as a leaf of seaweed or a smooth rock or stone. A similar modification of the ventral fins is also found in the beautifully coloured Lump-fish, or Lump-sucker, sometimes called the Cock-paidle. One of these fishes, when placed in a bucket of water, adhered so strongly to the bottom, that, when lifted by the tail, it bore the whole weight of the pail and water.

JUST below the Sucking-fish is drawn a foot of the curious little lizard, the Gecko, so called from its peculiar cry. It is common in the West Indies, and haunts houses, traversing their walls just as flies run up panes of glass. It is enabled to perform this movement by means of the structure of the feet. As the reader may see by reference to the illustration, the toes are greatly widened and flattened. If the lower surface be examined, it will be found to be furnished with a number of plates very much resembling those of the sucking-fish, and performing the same office.

So rapid is the operation of these plates, that the animal can even leap upon a perpendicular flat surface, and stick there. Perhaps the reader may remember that the beautiful Tree-frogs, which cling so tightly to leaves, are furnished with suckers on their toes, whereby they can hold on even to an upright pane of glass. In fact, the smooth surface of the glass seems to please them, and when they adhere to it they give an excellent opportunity of examining the structure of the feet with a magnifying-glass.

Another example of the pressure of the atmosphere has been slightly mentioned, when treating of the ball-and-socket joint. This is the joint by which the thigh-bone is attached to the hip. As the rounded head of the thigh-bone fits exactly into the cavity of the hip, and is, moreover, well lubricated with the animal oil called synovia, no air can obtain admission between the two. Consequently, they are held together so firmly by the pressure of the atmosphere, that they retain their places even after the whole of the muscular attachments have been removed. Not without very great force can the thigh-bone be dislodged from the shallow socket in which it lies ; but, if a hole be bored so as to admit the air, it comes out at once.

Similarly, however firmly a limpet may cling to the rock, if the finest needle were introduced so as to admit air, the creature could not retain its hold for a moment.

THE last figure on the illustration represents the common Lampern (*Lampetra fluviatilis*).

The mouth of this little fish is formed on the principle of the sucker, and very firmly it can adhere, as I can state from much personal experience. Indeed, it is rather alarming, to those who are unacquainted with the character of the fish, to have it turn round and fasten upon the hand. However, it is quite harmless, and those who are accustomed to them will have half-a-dozen hanging on their hand at a time, and take no notice of them.

ALREADY has it been mentioned that Surgery has pressed into its service the weight of the atmosphere by means of cupping. She also makes use of Nature in a similar manner by employing the Leech for local and surface bleeding.

The mouth of the Medicinal Leech forms an exact parallel with the cupping-glass and lancets, only that it is very far superior in its powers. To make the analogy perfect, the lancets ought to be within the cupping-glass, and the latter ought to be able to exhaust the air from itself, and to be attached to a reservoir into which the blood could be passed.

I need hardly mention that the action of sucking as practised by the young of all mammalian beings, from man downwards, is due to the same principle. By the action of sucking a partial vacuum is formed, and the pressure of the atmosphere upon the breasts forces the milk into the mouth of the young.

We might multiply examples *ad infinitum*, and we will therefore pass to another subject.

SEED-DRILLS.

AMONG the modern improvements in agriculture we may reckon the invention of the Seed-drill as one of the most important. By means of this invention, seed is greatly

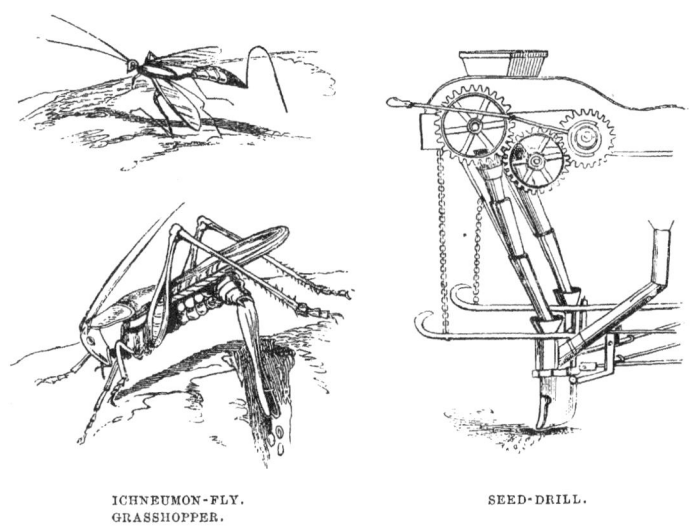

ICHNEUMON-FLY.
GRASSHOPPER.

SEED-DRILL.

economized, the supply can be regulated, and the sower knows exactly where every grain of seed goes. There is no scattering, as in the wasteful broadcast plan, by which the seeds are flung almost at random over the field, and may or may not fall into the furrows. The Seed-drill, on the contrary, either stamps holes or ploughs narrow furrows, measures the seed into them, and in some machines replaces the earth. The former kind of machine rather deserves the name of a dibble, and was invented for the purpose of superseding the use of the hand-dibble.

It is really a pitiful thing to see human beings endowed with reason and aspirations performing such a task as dibbling by hand, one going backwards with a dibble in each hand, and the other following and putting seed into the holes. Yet the field labourers have the greatest objection to the machine dibble, as, indeed, they have to any sort of labour-saving machine, thinking that it will lessen the demand for labour, and prevent them from earning a livelihood.

I well remember how a country clergyman, pitying the hard toil of the hand-dibblers, took occasion when he visited town to purchase a machine dibble wherewith one man could set eight rows of beans at once. It was a very simple affair, comprehensible even by the dull brain of a Wiltshire labourer. His trouble was all in vain, for no one would use it, and there was such a disturbance about it in the village, that for the sake of peace its owner laid it up in a loft and abandoned its use. There might be some semblance of reason in thinking that it would deprive them of their field labour, but no cottager would even use it in his own garden, though it was freely offered to any one who wished to borrow it.

THESE machines have their parallels in Nature, two of which are represented in the illustration.

The lower left-hand figure represents the female Grasshopper depositing her eggs. She is furnished with a sharply pointed ovipositor, composed of two blades. When she is about to lay her eggs, she searches for a suitable piece of ground, where the earth is tolerably soft, and with the closed ovipositor bores a hole. She then separates the blades slightly, and an egg glides between them into the ground, precisely as is done by the machine dibble with its beans. When I first saw and used the instrument, some twenty-five years ago, the parallel struck me at once.

THE female of the familiar Daddy Long-legs (*Tipula*) acts in a similar manner. She is furnished with an ovipositor too short to be used like that of the grasshopper, and so she attains her object in a rather different manner. Making use of her long stilt-like legs, she sets herself nearly upright, with the point of the ovipositor in the ground. She then twists herself

from side to side, just after the principle of the bradawl, and
so proceeds until she has made a hole large enough for her
purpose.   The blades of the ovipositor are then separated, and
the egg placed in the hole, as has been described of the grasshopper.

THE upper figure represents one of the large Ichneumon-flies
depositing the egg in the grub of some wood-inhabiting larva.
How she bores the hole has already been described when treating
of Boring Tools, and the process need not again be discussed.
The principal point at present is, that after the hole is bored, an
egg can pass between the blades of the ovipositor, though they
are but little thicker than human hairs.

One of the most extraordinary instances of this kind of ovipositor is found in an Ichneumon-fly brought from Bogotá. The
body, from the head to the end of the tail, is not quite an inch
long, while the ovipositor is six inches and a half in length, and
scarcely thicker than that of the insect whose portrait is given
in the illustration.   Nothing is as yet known of its habits, so
that the object of this wonderfully long ovipositor is a mystery.
But that it should be used like other ovipositors is evident
enough, and the chief wonder is, what are the mechanical means
whereby an egg can be propelled between blades so long and
slender.

There is a genus of Ichneumon-flies called Pelecinus. They
deposit their eggs in wood-boring larvæ, and we might imagine
that the ovipositor would be a long one.   It is, however, extremely short, and the requisite length is obtained by the form
of the abdomen, the joints of which are so long and narrow that
they almost look as if they had passed through a wire-drawing
machine, the length of the head and throat being three-eighths
of an inch, and that of the abdomen an inch and a half.   This
long abdomen belongs only to the female, that of the male
being short and club-shaped.

USEFUL ARTS.

CHAPTER III.

CLOTH-DRESSING. — BRUSHES AND COMBS. — BUTTONS, HOOKS
AND EYES, AND CLASP.

The Teazle and its Structure.—Its Use in raising the "Nap" on Cloth.—Its Value
in Commerce.—Artificial Teazles.—The modern Cloth-dressing Machine.—
The Brush an Article of Luxury.—Definition of the Brush, and its various
Uses.—Brushes in Nature.—The Foot of the Fly and the Tail-brush of the
Glow-worm Larva.—Mode in which they are used.—The Comb.—Varieties
of the Comb as made in different Countries.—Combs in Nature.—Foot of the
Spider and its Uses.—Beak of the Toucan.—Comb of the Scorpion.—Buttons,
Hooks and Eyes.—Use of the Button.—The Egyptian Garment.—The
Buckle and the Shoe-tie.—The Clasp.—Wing-hooks of various Insects.—
The Saddle-back Oyster.

CLOTH-DRESSING MACHINE.

IN former days, when so much was done by hand that is now
done by machinery, the thistle called the Teazle (*Dipsacus
fullonum*) was of great value in British commerce, being used
by countless thousands in the manufacture of broadcloth.

When the woollen threads are woven so as to form the fabric
of the cloth, there is no nap upon them, this having to be produced by a subsequent process. The plan of former days was,
to procure a quantity of the seed-vessels of the Teazle, and dry
them. They were then fastened to an instrument something
like a wooden battledore, and swept over the surface of the cloth.
By degrees the delicate hooklets which terminate the many
scales of the seed-vessel tore up the fibres of the cloth, and produced the desired nap without impairing the strength of the
thread. When this nap is worn off, the threads are again
visible, producing the effect called "threadbare."

As the art of weaving continued to progress, the demand for
Teazles increased in due proportion, and vast quantities were

imported from abroad. Instead of being used by hand, they were then fastened to the circumference of wooden wheels as broad as the width of the cloth, and made to revolve rapidly, while the cloth was pressed against them.

For many years attempts had been made to construct artificial Teazles which would not wear out so rapidly as did the dry seed-vessels, but nothing could be constructed that was not too stiff or too strong, and which did not injure the threads while producing the nap. At last, however, this difficult problem has been solved, and the Teazle is no longer an important article of

TEAZLE.

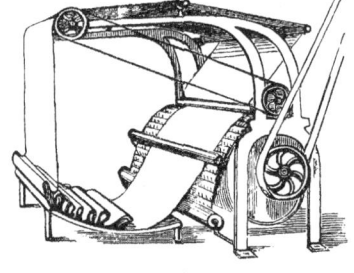

CLOTH-DRESSING.

commerce, its place being supplied by delicately made cards of the finest and most elastic wire.

In the illustration a head of Teazle is given on the left hand, and on the right is seen the mode in which the wire cards are placed in the machine, and the cloth drawn over them so as to produce the required nap.

BRUSHES.

IT is worthy of notice that there are many articles of comparative luxury which could not be used until man had attained some degree of civilisation. Among these we may class the Brush and the Comb, no true savage ever troubling himself about either article. The Brush, indeed, belongs to a much more advanced stage of civilisation than the Comb, for whereas we find combs, however rude they may be, used in semi-savage, or rather, barbarian countries, the Brush is, as far as I know, an adjunct of a high state of civilisation.

Brushes may be defined to be instruments formed of fibres set more or less parallel to each other. The vast variety of brushes used in different parts of Europe is indicative of the civilisation of the nations who use them. Take, for example, the brushes used in household management, such as the hearth-brush, the housemaid's brush, the Turk's-head brush, the crumb-brush, the stair-brush, the carpet-brush, the dusting-brush, and many others.

Then we have those which are applied to our garments, such as the ordinary clothes-brush, the velvet-backed hat-brush, and the three kinds of boot-brushes.

In architecture, again, we should be very badly off without the painting-brushes, the whitewasher's brush, and the paper-hanger's brush; not to mention the exceeding variety of brushes used by artists both in oil and water colours.

As to brushes applied to our persons, we have an infinite number of them. There is, of course, the hair-brush, without a pair of which, one for each hand, no one with a respectable head of hair could be expected to be happy.

We may add to this the revolving brush worked by machinery, which is to be found in the rooms of any respectable hairdresser, and which is a sort of an apotheosis of the Hair-brush, especially when it is worked, as in some places, by the electrical engine.

Then there is the shaving-brush, once an absolutely necessary article in a gentleman's dressing-case, and above all requisite if the owner should happen to be a clergyman. Nowadays, shaving is rapidly decreasing, and of all the professions, those who are most largely bearded, both in number of beard-wearers and dimensions of the beard, are to be found among the clergy.

Then there are any number of tooth-brushes for the interior of the mouth, and of flesh-brushes, with or without handles, for the service of the bath. There are even gardeners' brushes, for the purpose of clearing the plants of the aphides, or green-blight, as these insects are popularly called by gardeners. So it will be seen that—absurd as the proposition may appear at first sight—we may really accept the use of the brush as a safe test of the progress of civilisation.

We will now glance at the illustrations of this subject.

On the right hand is depicted the once honoured Shaving-brush, the terror of all stiff-bearded men on frosty mornings, and yet clung to with a strange inconsistency. Many years ago a military member of the House of Commons was sensible enough to wear his beard, and was, in consequence, the butt for interminable jokes. At the present time, if the House were counted, a great majority of the younger, and not a few of the older, members will be found to wear either the beard or moustache, or both.

Perhaps some of my readers may object that many nations in a state of very partial civilisation are accustomed to shaving. So they are, but they do not use the shaving-brush. Most of

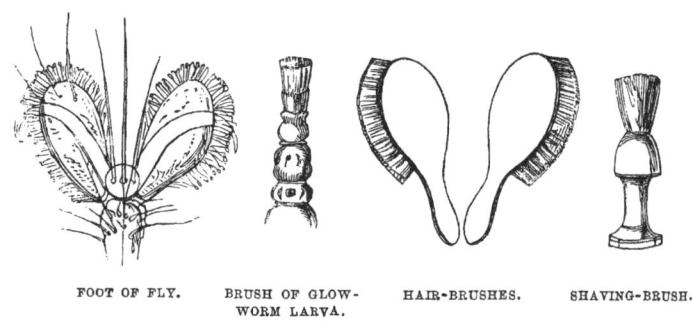

FOOT OF FLY.   BRUSH OF GLOW-   HAIR-BRUSHES.   SHAVING-BRUSH.
                 WORM LARVA.

them content themselves with pulling out the hairs by the roots, while others merely saturate the hair with hot water, and so need no brush.

Next to the shaving-brush is drawn a pair of ordinary Hair-brushes, such as have been mentioned.

PASSING to the left, we find an object which bears a curious resemblance to the shaving-brush. This is an apparatus belonging to the larva or grub of the Glow-worm. This creature feeds upon snails, and, in consequence, gets itself covered with the tenacious slime. In order to enable it to rid itself of this inconvenience, the larva is furnished near the end of its tail with the curious apparatus which is here shown. It consists of some seven or eight soft white radii, arranged so as to produce a brush-like outline, and being capable of extension or withdrawal at will.

It had long been known that this "houppe nerveuse," as it is called, was employed as an assistant in locomotion ; but until comparatively late years—I believe about 1826—no one seemed to be aware that it was used as a brush. Its functions as a brush may be compared with the somewhat similar offices fulfilled by the pincers of the Earwig, as mentioned on page 259.

Next to the brush of the glow-worm larva is shown one of the fore-feet of the ordinary house-fly, much magnified. Passing, as irrelevant to the present subject, the use of the feet as organs of locomotion, we may take them as being used for the purpose of cleansing the body of the insect.

I suppose that none of my readers has been sufficiently inobservant not to have noticed the way in which a fly cleanses itself, behaving almost exactly like a cat under similar circumstances. The fore-feet are repeatedly passed over the head, which is bowed down to meet them, while a similar office is performed for the rest of the body by the hind-legs. The feet are then rubbed against each other, so as to free them from all accumulations, just as the housemaid cleanses the hair-brush with the comb before washing it. So mechanical is this process, that a fly has been known to go through it even after it had been deprived of its head.

The reader will see, on reference to the illustration, that the two sharp and curved claws are capable of answering the purpose of combs, and, indeed, are so employed.

COMBS.

WE will now proceed to the COMB, and see how Art has been anticipated by Nature.

As long as human beings possess hair upon their heads, whether it be the short, frizzed, woolly pile of the negro, the thick, coarse crop of the Fijian, the coarse, straight hair of the Mongolian, or the long and fine hair of the Georgian races, they must, as soon as they attempt any kind of civilisation, form some instruments by which the hair can be dressed. The simplest machine for this purpose is the Comb, and I possess many varieties of this article, suitable to the different races for whom it was made.

Putting aside the ordinary Combs of our European civilisation, such as are given in the illustration, there are many
others which are modified according to the use which they have
to fulfil.

The simplest is the Comb of the celebrated Amazon regiment of Dahomey. This is nothing but a slight skewer of
ivory, some ten inches in length, and amply sufficient for
arranging the short woolly lumps which do duty for hair on
the head of a true negro. One of these very primitive combs
is in my collection, together with an undress costume of the
Amazon in question, and both being very much suited to each

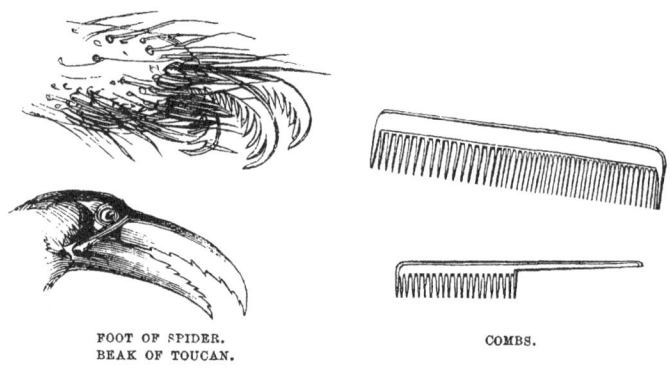

FOOT OF SPIDER.
BEAK OF TOUCAN.

COMBS.

other. The comb being a simple skewer, the dress is only a
few thongs of leather, but they are both equal to the requirements of their wearers.

As much time would be lost in combing the hair with a
single skewer, especially when that hair belonged to any but
the pure negro races, a simple but obvious improvement was
introduced. A number of skewers were lashed together side
by side, with their ends a little diverging, and thus was formed
the germ of our present Combs.

As to the varieties of the Comb, they are simply endless;
and whether they are intended, in the form of the Currycomb,
to smooth the harsh coat of a horse, or, as a small-tooth Comb,
to search the hair of the young, they are all based on one
principle.

It is really curious to see how often two men, who cannot

possibly have seen each other, will hit upon the same idea, not only simultaneously, but often in the very same words. So it is with regard to the Comb. In no two parts of the world can the natives be more opposed to each other than is the case with Fiji and Western Africa; yet I possess specimens of combs from both countries, made on the same principles, and so exactly in the same manner, that, except for the coarseness of the African Comb, it would be almost impossible to distinguish between them. There is but a slight difference in the size and shape of the two combs, and yet nothing can be more distinct than the characters of the two nations.

I have also a Japanese Comb of the most ingenious construction. It is made of wood, and cut exactly like our double ivory small-tooth comb; but it is furnished with a curious kind of handle, consisting of a flat piece of wood with a deep longitudinal slit, into which either side of the comb fits; and so beautifully is it made, that when it is fitted upon either side of the comb it looks as if handle and comb had been cut out of the same piece of wood.

The Fijian Combs are much after the same fashion as those of Western Africa, except that, with the artistic nature of their kind, the Fijians, instead of merely lashing together the numerous spikes of which the comb is made, employ a variety of patterns, and seem to luxuriate in the exuberance of artistic spirit which can make hundreds of combs, and no two of them alike.

On the left hand of the illustration are two examples of Natural Combs which are well worthy of notice. The upper one is a foot of the common Garden Spider (*Epeira diadema*), which has been several times mentioned in this work in connection with different subjects.

Every one who has watched the life of one of these creatures must have noticed how often its hairy body becomes clogged with little bits of its own web, and how dexterously it releases itself from such encumbrances. The figure in the illustration shows how this can be done, the strangely formed foot acting at the same time the part of comb and brush. It will be seen that the curved spikes of the claws act as a comb, while the bristle-like hairs discharge the duty of a brush.

NOT only are these projections used as Combs, but as appendages which insure the security of footing along the lines of the web. The reader will easily remember that when a Spider rushes along its web to secure its prey, it always runs along one of the radiating lines, which have no viscid drops, and that it never misses its hold. The latter point is secured by the structure of its claws, which are so made that if one projection misses the line, another is sure to fasten upon it. Some years ago, while watching "Blondin" go through his wonderful performances, I was especially struck with the pattern on which he had constructed the stilts upon which he traversed the rope. They were made in the most exact imitation of the Spider's foot, and though it is not probable that he borrowed them from that object, the resemblance was so close that he might readily have done so.

BELOW the spider's foot is given the head of a Toucan, one of those beautifully coloured and large-billed birds that inhabit tropical America. These birds are very particular about their plumage, and even when in captivity dress their feathers with the utmost care. When they do so, the saw-like notches of the beak act the part of a comb, and the fibrils of the feathers are by their action dressed parallel to each other, and give to the whole bird its proper appearance of health.

I MAY here mention that there is one comb in Nature, the use of which has never been clearly ascertained. This is the remarkable organ found in the Scorpion, and simply known as the "comb." There are two of them, one on each side of the under surface. Their colour differs slightly according to the species, but is generally a light yellow brown. The number of teeth also differs extremely, for in the Rock Scorpion there are only thirteen teeth, while in the Red Scorpion there are twenty-eight.

BUTTONS, HOOKS AND EYES, AND CLASP.

HAVING now treated of brushes and combs as articles belonging to the toilet, we will proceed to those which belong to the dress rather than the person. It is a curious fact that, as far as is known, buttons and hooks belong only to advanced civilisation. The simplest garment is, of course, a cloth of some material wrapped round the waist, and, as we see in the wonderful Egyptian paintings which have survived their painters some three thousand years, the simple fold can retain its grasp round the loins, even through the exertions of a long day's work.

I was always at a loss, when looking at these drawings, to understand how a single fold could retain so simple a garment in its place, but when I made my first visit to the Hammam Turkish Bath in Jermyn Street the mystery was at once solved. The "check," as it is there called, is long enough to pass about once and a half round the waist of an ordinary man. One end of it is placed on the left side, so as to bring the lower edge on a level with the knee. It is held by the left hand until the right hand passes it round the waist. It is then turned over in a broad single fold, and will remain in position for hours, the left leg having free scope between the two ends, and yet not being needlessly exposed.

Next to the simple fold comes the tie, which is in use all over the world. The chief object of a good Tie is that it should retain its hold as long as needed, be loosened with a touch in necessity, and, as a matter of consequence, should never "jam."

Still, even the best of ties are liable to objection. I once heard an argument on the subject of ties and buckles with regard to shoes. The speakers were both Derbyshire men, and their phraseology was somewhat obscure. However, both stuck to his own principles, one saying that "when a shee-uew is boo-oo-oockled, it's boo-oo-ookled;" and the other asserting, in equally strong terms, that "when it's tee-ee-eed, it's tee-ee-eed."

The buckle was here asserting its supremacy in civilisation over the tie, and was palpably right. Any one, so rose the argument, can tie two strings together, but the structure of the buckle is too complicated to be understood, much less invented, by any uncivilised being.

NEXT come, in natural order, the Button and the Clasp, each being identical in principle. In the case of the former the "eye" is placed over the button, while in the latter the clasp or hook is passed through the eye. Several examples of the Button and the Clasp are given on the right hand of the illustration, and are too familiar to need description.

As to the corresponding articles in Nature, they are very numerous. We will take, for example, the Saddle-back or Crow Oyster of our own shores. It is a most remarkable being. It deposits upon the object to which it adheres a sort of button of shelly matter, and the lower valve, which is nearly flat, has in it an aperture which is placed over the knob, just

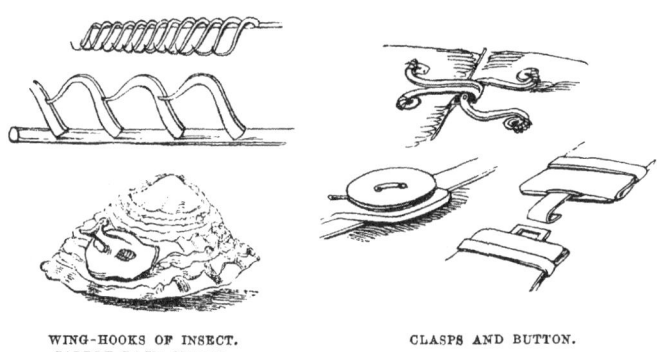

WING-HOOKS OF INSECT.
SADDLE-BACK OYSTER.

CLASPS AND BUTTON.

as a button-hole goes over the button. As this arrangement is confined to the lower valve, and cannot be seen unless the upper valve be removed, the lower valve only is shown in the illustration, as it appears when fastened to the side of a large limpet.

OF the Hooks and Eyes in Nature I have only taken two examples, though there are many others.

We all know the Bees, Wasps, Hornets, and other similar insects, and that they possess four wings. I may here mention that no insect which does not possess four transparent wings is capable of stinging.

When the insect is at rest the four wings may be easily distinguished, but when it is in flight they coalesce, so that practically the insect has two wings instead of four. This object is attained in the following way :—

The lower edge of the first pair of wings is turned over in a rather stiff fold. The upper edge of the second pair of wings has a row of small, but strong and elastic hooks. When the insect is about to fly, the hooks are hitched into the fold, and so the wings are fastened together. These hooks are shown in the illustration, and the reader will easily see how effective they must be in their operation. An almost exactly similar structure is found in the feathers of birds, and it is by means of these tiny hooks that wings are enabled to present a continuous, light, and elastic surface in the air.

USEFUL ARTS.

CHAPTER IV.

THE STOPPER, OR CORK.—THE FILTER.

Vessels and their Covers.—Corks.—Mode of bottling Wine.—Conical Corks and Stoppers.—Self-fitting Candles.—Candle-fixers.—The Vent-peg.—The Blow-guns and their Missiles.—The Serpula and its Conical Stopper.—The Filter.—The Bosjesman procuring Water.—How to make a simple Filter.—The Earth as a Filter.—The Sea-mouse, or Aphrodite, and its filtering Apparatus.—The Duck's Beak, and its beautiful Structure.—The Jaw of the Greenland Whale.—Fork-grinder's Respirator.—How Insects breathe.—Spiracles, and their general Structure.—Spiracle of the Fly.—Experiment upon a Cockroach, and its Result.

THE STOPPER, OR CORK.

THIS object, as depicted in the illustration, is a product of civilised life, though, as soon as a savage could make a vessel, he seems to have made a Cover for it if it were of large diameter, or a Stopper if the opening were small. Even the very Bosjesman, who is quite unable to make a clay vessel, and uses empty ostrich eggs by way of water-bottles, is yet capable of making plugs with which he can stop up the apertures. Then the Kafir, with his gourd vessels, whether they be for water or snuff, makes a plug that fits tightly enough to exclude the air, as well as to retain the contents.

The invention of glass bottles necessarily brought with it the introduction of a new kind of plug, and a material for such a plug was found in the bark of the cork-tree, a species of oak. This bark possesses the capability of compression to a very great extent, and, being highly elastic, it expands as soon as the pressure is removed.

Thus, in bottling wine, the corks are always made much too large to go into the mouths of the bottles. They are first dipped in a cup containing the same wine, and are then compressed violently by a machine worked by a handle, and which, being practically a powerful pair of nut-crackers with a rounded gripe, must suit the shape of the cork. It is then taken out of the machine, and, before it has had time to expand, is rapidly fitted to the neck of the bottle, and driven home with a wooden mallet. Expansion then takes place, and the bottle is rendered air-tight, so that no damage is done to the wine.

If the whole of the wine were to be drunk when the cork was removed, this plan would be amply sufficient. But there are many cases where the bottle is opened, and only part of the wine consumed. To re-cork the bottle would be too troublesome, and to leave it uncorked would spoil the wine. So the Conical Stopper was invented, which fits the neck of any ordinary wine-bottle, according to the depth to which it is introduced, and, by a slight screwing movement, sufficient compression is obtained to render the bottle air-tight. One of these Conical Stoppers is shown in the illustration on page 352. Sometimes they are made of cork, and sometimes of india-rubber ; but the principle is the same in either case.

Perhaps some of my readers may have seen the Self-fitting Candles, which require no paper to make them fit the candlestick. These are enlarged at the base, which is made in a conical form, and slightly grooved. The " Candle-fixers " that are so much in use at the present day are made exactly on the same principle, being hollow cones of paper, which take the place of the solid cone.

The Vent-peg of casks is another instance of the cone used as a stopper.

Another example is to be found in the Blow-guns and Arrows of tropical America. In some districts the base of the arrow is fitted with a conical appendage of light cotton, rather larger than the tube, but capable of compression, so that it exactly fits the tube when pressed into it. In other districts the cone is hollow, and made of some thin and elastic bark.

Some years ago one of our most eminent gun-makers hit upon the same idea while making improved missiles for the game of " Puff and Dart," and very much surprised he was when I showed him the South American arrow, not only with the same hollow cone at the base, but having also spiral wings along the shaft, so as to give it a rotatory motion as it passed through the air. The hollow cones of his darts were made of india-rubber, but the shape of the two was identical.

If the reader will refer to the left-hand figure of the illustration, he will see a beautiful example of the Conical Stopper as existing in Nature.

This is the "Stopper," as it is popularly called, and, scientifically, the "infundibuliform operculum." I prefer the former term myself, as being less liable to misapprehension.

The Serpula lives in a shelly tube of its own construction, and has the power of protruding itself when it desires to obtain food, and of withdrawing itself within the tube when alarmed.

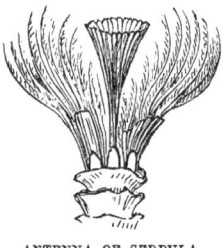
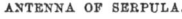
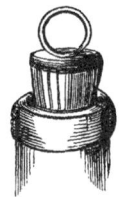

ANTENNA OF SERPULA.   CONICAL STOPPER.

This movement is performed so rapidly, that the eye can scarcely follow it, and the mechanism by which it is done has already been described when treating of War and Hunting.

When it withdraws itself, the Stopper closes the mouth of the tube with perfect exactness, so as to leave the inhabitant in safety. The reader will see, on referring to the illustration, how exactly similar is the Conical Stopper of Art to that of Nature, and how the inventor of that article, as well as of the self-fitting candle, the candle-fixer, the blow-gun arrow, and the vent-peg, might have found prototypes of their inventions in Nature, if they had only known where to look for them.

THE FILTER.

Even in a state of uncivilisation man has been driven to invent a Filter of some kind.

The simplest kind of Filter is that which is used by the

Bosjesman women when procuring water for the use of their families. When, as often happens, the only water to be obtained is to be found in muddy pools which have been trampled and perturbed by thirsty animals, the women have recourse to a simple, though rather repulsive, expedient.

Each woman is furnished with empty ostrich egg-shells by way of water-vessels, and she also takes a couple of hollow reeds. Over the end of one of these reeds she ties a bundle of grass, and then plunges it as deeply as she can into the mud. After a little while she sucks up the water through the tube, the grass acting as a filter, and she then discharges it by the second tube into the egg-shells. In this way the women will obtain water, where none but themselves could have procured it. As to the repulsive mode of obtaining it, no one can be fastidious when dying of thirst. Sir S. Baker mentions that when he was on his travels he managed in a halt to save up enough water for a bath for himself and his wife. He was about to throw away the soapy water, when the vessel was snatched from his hands by two of his attendants, and the contents eagerly drunk.

The different varieties of the Filter which we use at the present day are too familiar to need description. Whether they be made principally of charcoal, which is a powerful disinfectant, or of merely stones, gravel, and sand, they are all constructed on the same principle, namely, the straining out solid substances, and allowing only the pure water to pass through the interstices.

As to the Filters of Nature, they are almost innumerable. In the first place, the Earth itself is the primary filter of all, taking into itself all kinds of decomposing substances, separating them for the use of vegetation, and delivering the pure, bright, and sparkling spring water which we so highly and rightly value. The whole human body, again, is practically a collection of the most elaborate and effective filters that the mind of man can conceive. But we will pass to the more obvious examples of filters as seen in animal life.

On the upper left-hand portion of the illustration may be seen a long, fat, hairy creature, called popularly the Sea-mouse, and known to zoologists as *Aphrodite aculeata*. Although it inhabits the mud—and sea-mud is about as

noisome a substance as can be imagined—it is clothed with a
garment of such beauty that the rainbow itself can scarcely
rival, and not surpass it.   The hairs with which it is so pro-
fusely covered glitter and sparkle with every imaginable hue,
among which red and green seem to be predominant.

These hairs occupy the sides of the body, but in the upper
surface there is a thick coating of felted hairs, interwoven with
each other so closely that they can with difficulty be separated.
These hairs form a natural filter, strain away the mud from the
water, and allow the latter to pour itself upon the organs of
respiration.   If, therefore, a specimen be examined when it is

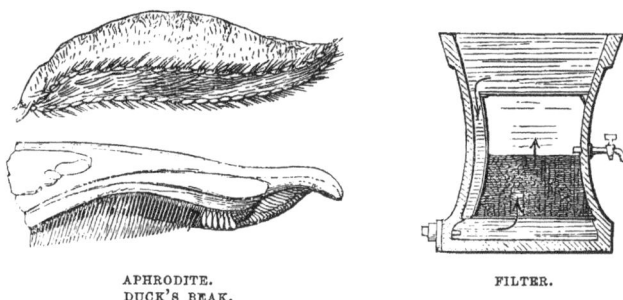

APHRODITE.
DUCK'S BEAK.

FILTER.

first brought up by the dredge, the felted hair will always be
found to contain a considerable amount of mud, and much
washing is needed before the creature can be introduced into
an aquarium where the water is intended to be transparent.

I may here mention that the name of Aphrodite is a singularly happy one.   It signifies something that arises from the
foam of the sea, and was given to the goddess of beauty,
because in the ancient myths she was said to have sprung from
the foam of the sea.   Unpoetical as it may appear, the German
word Meerschaum, which is so familiar to us in connection with
pipes, is the exact equivalent of Aphrodite.

BELOW the Aphrodite is a figure representing the filtering
apparatus which is found in the beak of the duck.   This singularly beautiful apparatus is well worthy of examination, and
the more important details of its structure can easily be made
out by the unassisted eye.

In the first place, the upper half of the beak, or upper

mandible, as it is scientifically called, is furnished along its edges with a row of curved horny projections, very like the teeth of a comb, and each of them coming to a point. There are some fifty or sixty of these teeth on each side, and they are regularly graduated in size, being longest in the middle of the beak, and becoming very short at either end. They are set diagonally, with the tips pointing backwards. The edges of the lower mandible are turned up in a sort of fold, on the outside of which is a row of grooves corresponding with the teeth of the upper mandible, and, like them, being set diagonally.

These teeth and grooves would of themselves make a very efficient filter, but they are further aided by the tongue. This is thick, fleshy, and very mobile ; so much so, indeed, that when the mouth is opened the tongue is automatically thrust forward. The edges of the tongue are, like those of the mandibles, furnished with a filtering apparatus. Instead, however, of being horny and stiff like those of the mandibles, they are membranous and exceedingly delicate. Indeed, in order to see them properly, it is necessary to place the tongue under water, so that the membranous filaments shall be floated apart instead of clinging together by their own weight.

The whole of this apparatus is abundantly supplied with nerves, and is evidently a most exquisite instrument of touch. The reader will now understand the peculiar movements of a duck's beak while feeding. Although the bird can and does eat solid food, such as barley, and, by reason of its superior width of beak, will very much defraud the poultry in a yard where ducks and hens are kept together, it is chiefly fitted for extracting nourishment from water, and will find abundant subsistence where a hen would die of starvation.

When the beak is plunged into the water, the mandibles are rapidly opened and shut, the tongue incessantly working backwards and forwards between them. Consequently, not only are the solid parts of the water strained between the comb of the upper beak and the grooves of the lower, but they undergo a further sifting or filtering from the delicate fibrils which fringe the edge of the tongue.

ANOTHER familiar example of the Filter is to be found in the jaw of the Greenland Whale. In this animal, as well as in its

congeners, the "whalebone," or "baleen," as it is more properly called, is so formed that it allows liquids to pass through
it, while it retains solids.   Feeding as it does upon small
marine matters, it would starve but for the filtering power of
the baleen, which enables the animal to take into its vast
mouth the sea-water with its inhabitants, and to expel the
water through the plates and fibres of the baleen, while retaining the animals.

The process of filtering, as well as the structure of the
baleen, is so familiar that it does not need further description.

WE will now proceed to another filter, which is used in the
air, and not in water, namely, the Mouth-guard or Respirator
of the fork-grinder.

There is, perhaps, no trade which is more destructive of
human life than that of the fork-grinder was until the peculiar
respirator was made obligatory.   The minute particles of steel
thrown off by the grindstone fills the air, and were necessarily
inhaled.   Now, the human lungs are capable of enduring very
bad treatment, but the introduction of steel-dust into them is
more than they can bear.   Consequently the duration of human
life was very short, consumption almost invariably setting in
at an early age, and carrying off the men before they had
achieved middle age.

Nor did the mischief end there.   It was bad enough that life
should be shortened, but far worse that it should be wasted, as
was mostly the case.   The men, knowing what their fate must
be, were simply reckless, and plunged into all kinds of
debauchery, under the plea of "a short life and a merry one."
They knew no better, and could scarcely be blamed for their
mode of living.   And, as a matter of course, each succeeding
generation was worse, smaller, and feebler than the preceding.

Then .there came the invention of the Magnetic Respirator,
by which the fork-grinder's trade was rendered as healthy as
any other.   It was made of steel-wire gauze, and magnetised,
so that the floating particles of steel were not only stopped in
their progress to the lungs, but arrested by the magnetism, and,
so to speak, taken prisoners by it.

Even a well-made respirator of several layers, like those
which are used by persons suffering from weak lungs, would

have been useful, but the addition of magnetism doubled the efficacy while greatly diminishing the cost, a single layer of wire being quite adequate to the office, and was, in fact, quite a stroke of genius.

The value of this invention is at once shown by the many complaints which the workmen made when the Respirator was first introduced. They complained that the apertures of the Respirator became so choked that they could not breathe. This was perfectly true, but the complaint showed the real value of the instrument.

It was necessary for the workmen, every now and then, to clear off the innumerable particles of steel which adhered to the

SPIRACLE OF FLY.   RESPIRATOR OF FORK-GRINDER.

magnetised wires, and impeded respiration. But they never seemed to realise the fact that, if it had not been for these wires, all the particles would have been drawn into the lungs, and gradually choked them up, brought on inflammation, and extinguished their life altogether. And, with the usual repugnance to new ideas which is inherent in undeveloped minds, the men stoutly resisted the introduction of the Respirator, and did their best to reject an invention which doubled the length of their lives, and enabled them to find long happiness in the world instead of brief pleasure ended by sure and painful death.

Now, we will see how the principle of the Respirator is carried out in Nature.

On the left hand of the illustration is drawn one of the most perfect Respirators, or air-filters, if we may use the term, that can be imagined. Perhaps some of my readers may know that insects do not breathe as we do. They have no lungs, but their entire system is permeated by air-vessels, just as is our system with blood-vessels, and therefore the air, instead of being restricted to the lungs, is conveyed to every part of the insect, the air-vessels extending to the very tips of the wings and antennæ, and to the claws of the feet.

Neither does the insect receive the air through mouth or nostrils as we do. Along the sides of the body are certain oval apertures called "spiracles," from the Latin word *spiro*, which signifies breathing. These spiracles can easily be seen by examining an ordinary silkworm. They are situated in the soft and flexible skin which connects the rings or segments of which all insects are composed, and pass directly into two large air-tubes which run on either side of the body.

It is evident that since an insect is so thoroughly permeated with air, it must be furnished with means to render that air as pure as possible, and at all events to preserve the respiratory system from being choked with dust or other adventitious substances.

How important the air is to an insect can easily be seen by dipping it in oil, or even brushing an oiled feather on its sides so as to fill up the spiracles. A man under the hands of the hangman or garotter could not die more swiftly, so much does an insect depend on air. In fact, an insect is almost wholly composed of air-tubes, but for which the great thick-bodied dor-beetles could never use their organs of flight.

Of course, although the spiracles can act as filters as far as the air is concerned, they cannot be analysts, and consequently insects are peculiarly obnoxious to a bad atmosphere. There is, for example, the well-known "laurel-bottle" of entomologists. A few young laurel-leaves are crushed and placed in a bottle. As soon as an insect is introduced, it breathes the prussic acid which is exhaled from the leaves, and at once dies.

So it is with the more delicate "death-bottle," into which a little cyanide of potassium is introduced, and covered with plaster of Paris. The plaster prevents the poison from touching the insects and damaging their beautiful colours. It permits the deadly vapour to roll through its interstices; consequently, even the large-bodied moths, which are tenacious of life almost beyond credibility, can barely run round the bottle, when they roll over, and expire almost without a struggle, the venomous atmosphere having saturated the entire body.

All entomologists know that the spiracles act as sieves, preventing any extraneous objects from gaining admission into the breathing-tubes. But, unless they have had personal experience, they cannot appreciate the efficacy of the spiracle when acting as a respirator. Even the microscope, though it may magnify the object to any extent, does not show the wonderful filtering power of the spiracle. The figure in the illustration represents a spiracle of the common "blue-bottle" fly, and any one who wishes to examine such an object for himself can have but little difficulty in doing so, especially in the warm season of the year.

How effectual is the barrier thus interposed by Nature between the external world and the interior of the insect may be inferred from the following narrative :—

Many years ago, while absorbed in the comparative anatomy of insect structure, I believed myself to have hit upon a plan for injecting the minutest of tubes with mercury. So I took a male cockroach, placed a vessel of mercury in the receiver of an air-pump, and suspended the cockroach exactly over it. As the reader will fully have surmised, my idea was, first to exhaust the air from the inside of the insect, then to plunge it into the mercury, and then to admit the air, which, at a pressure of fifteen pounds to the square inch, was likely to drive the mercury into the smallest of tubes. Such a plan was very successful with ordinary tissues, and might succeed with insects.

Accordingly, I exhausted the air from the vessel in which the cockroach was placed, and kept it in a state of exhaustion for a whole day, so as to prove that every particle of air was withdrawn from the insect. I then plunged the cockroach deeply beneath the mercury, and admitted the air, hoping that the severe pressure would drive the mercury into the respiratory vessels. But not one particle of the mercury could pass through the wonderful filter with which the cockroach had been provided, and, except that I had learned the power of the spiracle, I might have saved both the time and trouble.

It is worthy of notice that, almost countless as are the species of insects, no two of them possess exactly the same structure of the spiracles, the individuality being marked as clearly in these tiny organs as in the entire insect.

USEFUL ARTS.

CHAPTER V.

THE PRINCIPLE OF THE SPRING. — THE ELASTIC SPRING.— ACCUMULATORS.—THE SPIRAL SPRING.

Springs and their various Structure.—The Elastic Spring.—The Boy's Catapult and its Powers.—The Pistolograph, its Principle, and Uses to which it can be put.—Leaf-rolling Caterpillars, and their Way of Work.—The Carriage Spring.—The Horse's Hoof and its complex Structure.—Fungi and their united Power.—The Chinese Cross-bow.—The ancient Balista.—Skull of the Crocodile.—Bones of young Children.—The Spiral Spring and its many Uses.—The Toy-gun.—The Needle-gun.—Valved Brass Instruments.—Watch and Clock Springs.—The Bed Spring.—Parallels in Nature and Art.—Buffers of Railway Carriages.—Spring Solitaires.—The Bell Spring.—Spiral Springs in Vegetable Tissues.—Poison Cells of various Marine Animals.—Effects of the Spiral Springs.

ELASTIC SPRINGS.

HERE we come upon a subject so large, that it is difficult to define its exact requisite limits. The principle of the elastic spring pervades all Nature, and the numerous adaptations in Art are closely, though perhaps not directly, attributable to the wide distribution of the spring in Nature.

There is, for example, the simple elasticity which enables a tree, when bowed by the wind, to spring back so soon as the pressure is removed, and which, indeed, is the power which enables a bow to propel an arrow. Then there are spiral springs innumerable, many of them so minute that they can only be seen by the aid of the microscope, and there are many springs which exhibit their elasticity by their power of extension and shortening, just as is done with the elastic fabrics which are so much in vogue at the present day, and which seem so necessary to ordinary comfort that we feel disposed to wonder how our forefathers managed without them.

We will now proceed to examine some of these springs in detail.

THERE is one form of elastic spring which has of late years become more familiar than agreeable, namely, the toy which is learnedly called a "catapult," though it has little in common with the ancient weapon whose name it bears.

As may be seen by reference to the illustration, it consists of one or more india-rubber straps attached to a fork-like handle, and carrying a small pouch in which is contained the missile. Although it is not remarkable for accuracy, it can throw a stone or a bullet a considerable distance, and its power can be very quickly increased by adding to the number of the straps. Thus a catapult has been made which was capable of sending a small pistol bullet through a wooden board, so that the child's toy might really become a dangerous weapon.

Indeed, cases are known where the catapult has hurled a stone with fatal effect upon human beings. In my own neighbourhood there are many examples of glass being pierced by stones thrown from catapults just as if they had been subjected to bullets shot from firearms, the holes being quite small and round.

The power of accumulating force by increasing the number of springs was utilised by Mr. Scaife, when he invented his wonderful photographic machine which he termed the "Pistolograph," on account of the sound which was produced when the portrait was taken.

The idea was simple enough, though the practice of it was not so easy. He wished to be able to take a photograph with an exposure of the least possible time, and thus to attain freedom and action, instead of the dull stiffness which generally characterizes photograph portraits. The mode which he adopted was by introducing a peculiarly sensitive film, which would take an impression in a mere moment, and then arranging the machine so that an exposure of more than a moment was impossible.

This was done by covering the lens with an exactly fitting door, revolving on a pivot. The axis on which the door revolved was attached to a number of india-rubber bands, exactly like those which are used for confining papers. As the power of the springs increased with their number, it naturally followed that the rapidity of the revolution was in exact ratio with the number of the bands, so that the duration of exposure to light could be measured with tolerable accuracy.

So wonderfully well did this plan succeed that photographs of eclipses were taken with perfect accuracy, a matter of great importance when time has to be considered. Horses were also taken at full gallop, so as to display their action, and the crowning achievement was the photographing of a cannon in the act of firing, and the bursting of a charged shell. So rapid is the action of the instrument, that in several cases where a cannon or mortar had been photographed, even the track of the ball or shell is visible.

It necessarily followed that when the springs caused the circular cover to revolve with such rapidity, they made it close

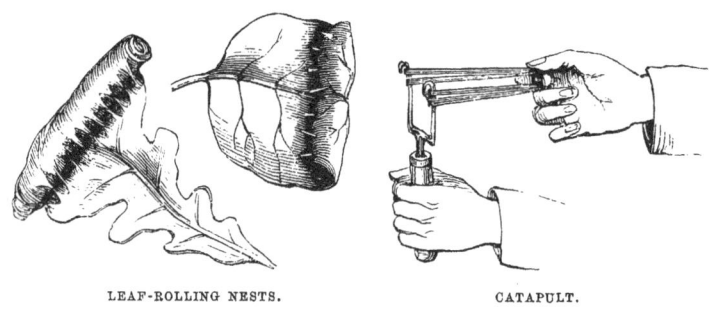

LEAF-ROLLING NESTS.   CATAPULT.

with a sharp report, and so gave rise to the name of the machine. Moreover, as it had to be used for rapidly moving objects, it was not fixed on a pedestal, but was held in the hands, while aim was taken at the object, just as with a pistol. When the observer thought that he had his aim correct, he touched a trigger, round spun the cover, and the photograph was taken.

On the right hand of the illustration is seen the Catapult, made with several springs, and on the left is shown an example of the Accumulator as formed by Nature.

The reader may probably be acquainted with the Leaf-rolling Caterpillars, of which there are so many. I had often inspected these curled leaves, and, on comparing them with the size of the caterpillars, had noticed that the muscular strength of the insect was quite inadequate to the work which was done. That much of it was owing to the "bowsing" system, which has already been described when treating of the Toggle-joint, was very probable, but that some other force must be employed was evident.

On unrolling a leaf, the hidden force was at once explained, and showed itself to be a system of accumulators exactly like those of the pistolograph or the catapult. The caterpillar spins successive belts of silken threads, and affixes them to the leaf, as shown in the illustration. These threads are nearly as elastic as the india-rubber bands of the catapult, and accordingly draw the leaf together. Another set of belts is added above the former, and, as they harden and contract in the air, they roll the leaf still further. The first row is then shortened and tightened, and a third and fourth row are added in the same fashion. So elastic are these belts, that if the leaf be carefully handled it can be almost wholly unrolled, and will spring back again as soon as the force is removed.

ANOTHER form of accumulated force may be seen in the ordinary Carriage Spring, one of which is shown in the illustration. It is made of a number of strips of elastic steel lying upon each other, and suffered to play upon each other by means of slots and rivets. The weight being placed in the centre, it is evident that this very ingenious spring is really an elastic girder, yielding to sudden pressure, and recovering itself when that pressure is removed.

INGENIOUS as is this spring, it has many parallels in Nature, one of which is here given.

It is popularly thought the hoof of the horse is a solid mass of horn destined to protect the feet against hard and rough ground. Such certainly seems to be the opinion of farriers, who, in shoeing horses, act exactly as if the horn of the hoof were structureless; whereas it is a marvel of complicated mechanism. On looking at the exterior of a horse's hoof, it will be seen to be marked with a vast number of very fine, but easily visible longitudinal lines, looking as if they were scratches from a very fine needle. If the hoof be removed rom the foot, and examined upon the interior, it will be seen that each of the apparent scratches signifies the edge of a very thin plate of horn, not so thick as the paper on which this book is printed. The hoof, in fact, is built up of multitudinous plates of horn, set side by side, and each acting as a separate spring. It is this beautiful structure which allows the horse to tread without a jar being sent through its whole system by every step which it takes.

A similar structure is to be found in all hoofed quadrupeds, and is especially noticeable in the case of the Elephant. All those who have watched the walk of an Elephant, no matter what its size may be, must have been struck with the curious noiselessness of its movements. Its weight may be measured by tons, and yet the enormous animal steps as noiselessly as a cat. On examining one of the hoofs, after it is removed from

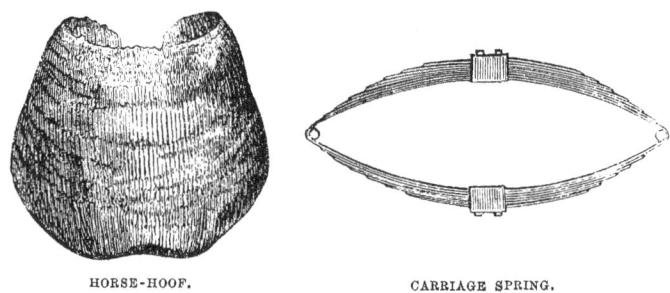

HORSE-HOOF.   CARRIAGE SPRING.

the foot, the cause of this marvellously silent tread is perfectly evident. The whole of the hoof is composed of nearly parallel horny plates, and by their united action they produce the required result.

Each plate in itself is very feeble, but, when united as they are at the ends, they afford mutual support to each other. Similarly the separate feathers in a couch would be crushed by a comparatively slight weight, but when a number are confined together they support each other, and form the soft, yielding couch with which we are so familiar. Horsehair, when used as the stuffing for a couch or chair, acts in the same way, and so do the fine filaments of wool when used under the name of " flock."

Another good example of the power of accumulated force, although it has no direct relation to the spring, is the well-known fact that fungi, which are separately so fragile, are capable of lifting and retaining in the air stones so large that two men could hardly carry them. Were the stones laid down upon the fungi, the latter would be crushed, but, as they grow beneath the stones, they accumulate their powers, and slowly, but certainly, raise the weight from the ground.

THIS very principle of accumulated force has long been used in weapons of war, and I possess several examples of such weapons. One of them is a Chinese repeating Cross-bow, which was taken at the capture of the Peiho Fort, and was really a formidable wall-instrument, carrying a reserve of arrows, and

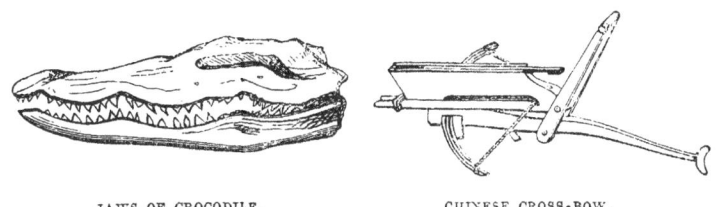

JAWS OF CROCODILE.   CHINESE CROSS-BOW.

delivering them with great rapidity. In point of fact, it consists of three bows, placed upon each other, and playing upon each other just as do the portions of a carriage spring. Such strength is thus obtained, that the bow cannot be drawn by hand, but is worked with a lever, as shown in the illustration. The whole machinery of the weapon, including the self-notching and self-supplying system, is very interesting, but is outside our present object. The very powerful bow of the ancient Balista was made on the same principle, and was strong enough to throw large stones and wooden beams.

I also have bows in my collection which are strengthened on the same principle, though not exactly in the same manner. There are several Indian, Chinese, and Japanese bows which are curved almost like the letter C, and have to be reversed when strung. These bows are of no very great size, but possess wonderful elasticity. They owe the latter quality to sundry layers of sinew which have been affixed to the back when wet, and which add enormously to the power of the bow, while they very little enlarge its dimensions.

Another bow, made by the natives of Vancouver's Island, has the back strengthened by a number of cords spun from sinew fibres, and possessing the strength and elasticity to which we are accustomed in the strings of the harp, guitar, or violin.

WE will now turn to a parallel in Nature. This is to be found in the lower jaw of the Crocodile, as is pointed out by Professor Owen, in his work on the "Skeleton and the Teeth."

All persons who have a smattering of anatomy are aware that even in the human body the most solid bones of the adult were originally composed of several pieces, and that they only become fused together in course of time. The jaw-bones, for example, were once so composed, and in the Crocodile the junction is never completed, the pieces of bone remaining separate, but being pressed firmly against each other during life.

I have now before me the skull of a Gangetic Crocodile, in which, although the animal was an adult when killed, the bones of the long lower jaw are so loose that unless they were tied together the jaw would fall to pieces.

This analogy between Art and Nature is thus described by Professor Owen in the work which has just been mentioned :—

"The purpose of this subdivision of the lower jaw-bone has been well explained by Conybeare and Buckland, by the analogy of its structure to that adopted in binding together several parallel plates of elastic wood or steel to make a cross-bow, and also in setting together thin plates of steel in the carriage spring."

Dr. Buckland also adds: "Those who have witnessed the shock given to the head of a Crocodile by the act of snapping together its thin, long jaws, must have seen how liable to fracture the lower jaw would be were it composed of one bone only. . . . . The splicing and bracing together of thin flat bones of unequal length and of varying thickness afford compensation for the weakness and risk of fracture that would otherwise have attended the elongation of the parts."

A good example of the value of this structure of bone may

be found in young children. Before they are old enough to take care of themselves they are perpetually falling down, and never hurting themselves. I have seen a little girl of five years old roll from top to bottom of a lofty staircase. It looked as if the child must be killed, but she was only giddy with her many revolutions, and a little bruised about the elbows. The reason of this curious immunity from injury is, that the bones, especially those of the skull, are not completely united, and so act on the principle of the compound spring.

The Spiral Spring.

This subject is so large, and there are so many examples, both in Art and Nature, that it is not very easy to make selections which will sufficiently answer the purpose.

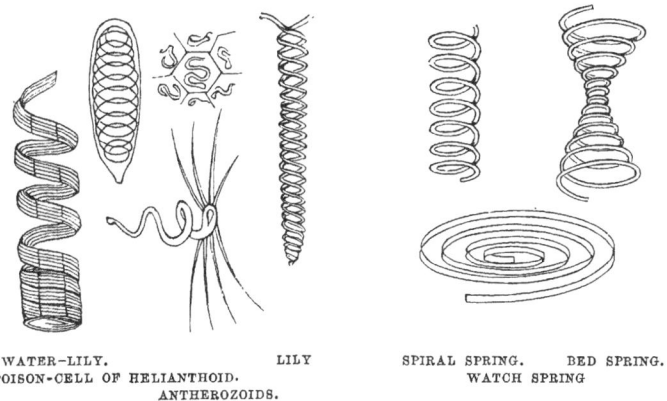

WATER-LILY.   LILY   SPIRAL SPRING.   BED SPRING.
POISON-CELL OF HELIANTHOID.   WATCH SPRING
ANTHEROZOIDS.

The upper left-hand figure of the illustration represents the ordinary Spiral Spring made of wire, and used for its power of resuming its shape when compressed. In early childhood most boys have had practical experience of this spring in the toy guns and cannons with which they are supplied. The spring is compressed by the ramrod, and held in its place by a catch. If a pellet be placed in the gun, and the catch released by pulling the trigger, the spring flies back to its former shape, and drives the pellet.

An exactly similar spring is used in the well-known "Needle-

gun," the spring driving a needle through the explosive mixture, and so igniting the charge.

Our brass instruments would be very badly off without the
spiral spring, which is placed under the pistons. The elasticity
allows the pistons to be pressed down, and when the fingers
are raised the pistons spring up again.

Another form of this instrument is seen on the right of the
ordinary spring. This is used in the manufacture of spring
mattresses and couches, and is made thinner in the centre, so
as to allow of greater elasticity.

Below them is the spring which is used for watches and
clocks, one end being fastened to the rim of the barrel, and the
other to the pivot. When the latter is turned the spring
becomes "wound up," and, when released, keeps the works
going by pressing against them. Of the "pall-and-ratchet"
wheel, by which the movements are retarded, we shall treat in
another place.

On the left hand of the illustration are a few figures of
the Spiral Spring as seen in Nature.

On the extreme left of the group is a spiral cell taken from
the flower-stem of the Water-lily. As the reader will see, it
is composed of a number of fibres laid parallel to each other,
and twisted into a hollow spiral. In order to exhibit its shape
the better, the spiral has been partially uncoiled.

On the extreme right is a corresponding spiral cell from the
common Lily, in which the spring power is given by two fibres
twisted in opposite directions. The reader will now understand and admire the mechanism by which these plants attain
their great strength and elasticity, the stems being made of
myriads of these spiral fibres.

The oval body on the upper part of the illustration is a
poison-cell of a marine polyp, and is given here as an example
of an animal spiral spring, the others all belonging to the
vegetable world.

We shall see more of its structure a little further on, and
will not now examine it in detail.

The two remaining figures represent the remarkable objects
called Antherozoids, i.e. the living creatures of anthers. They
exist in vast numbers in the non-flowering plants, and inhabit

those parts which correspond with the anthers of the flowering plants. When placed in water they have a curious way of coiling and twisting themselves spirally, so as to make their way through the water in a tortuous, but tolerably rapid, course. This movement is effected by the contraction and expansion of the spirally twisted filament. The upper figure represents a group of Antherozoids in their cells, and the lower is a much more magnified figure of a single Antherozoid as it appears when free, and in the act of moving through the water.

On the accompanying illustration are many examples of Spiral Springs, both natural and artificial. We will take these in their order.

The upper left-hand figure represents the "Buffer," by which the carriages of railway trains are prevented from jarring against each other.

Perhaps some of my readers may be old enough to remember the days of the old railway carriages that were connected by short chains, and furnished with buffers that were merely padded. As the train started a separate jerk was given to every carriage by the tightening of the chains, and, as it stopped, all the carriages bumped against each other in a most unpleasant manner. Now, however, the buffers are furnished with powerful springs, and are pressed strongly against each other by means of screw-bolts, so that they form one continuous line.

In fact—and here is another analogy between Art and Nature—a train, when properly made up, bears a close resemblance to a human spine, the carriages being analogous to the vertebræ, and the spring buffers to the elastic cartilages between the vertebræ.

Nowadays, owing to this arrangement, the whole train moves together, and can be started and stopped so gently that the passengers are hardly aware of movement or stoppage. For example, one of my friends was in a train which came into collision with some obstacle. The carriages in front were dashed to pieces, and several of the passengers killed. His carriage, however, which was nearly at the end of the train, and had the benefit of all the springs, was hardly shaken, and

the inmates did not know for some little time that an accident
had occurred.

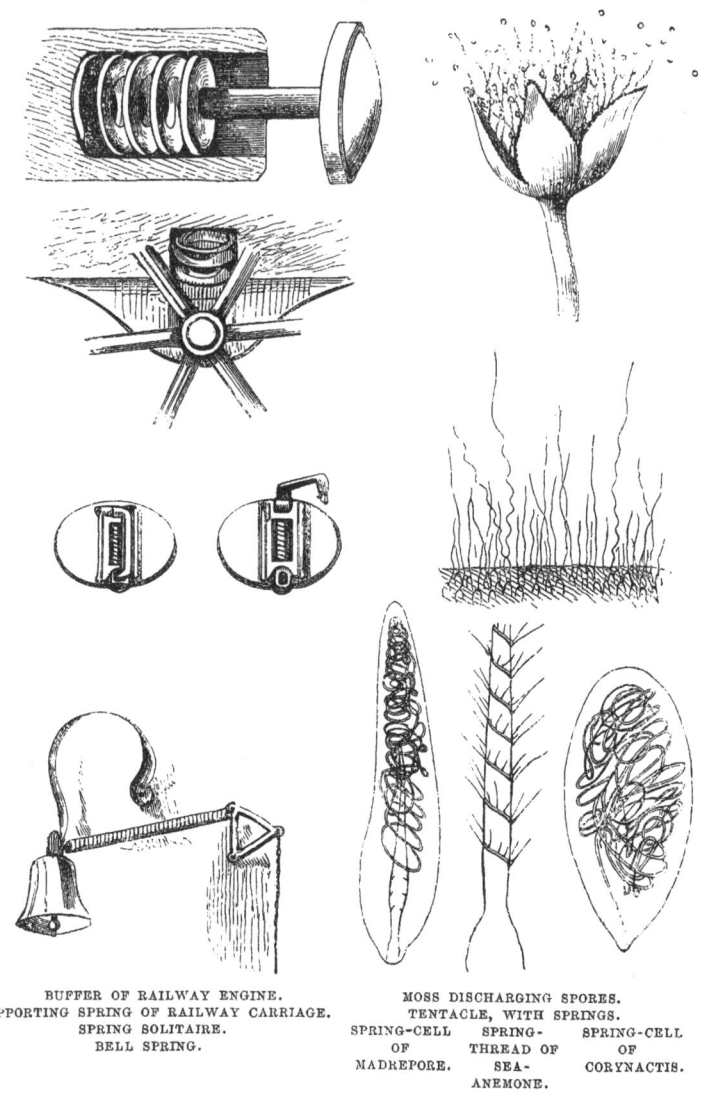

BUFFER OF RAILWAY ENGINE.
SUPPORTING SPRING OF RAILWAY CARRIAGE.
SPRING SOLITAIRE.
BELL SPRING.

MOSS DISCHARGING SPORES.
TENTACLE, WITH SPRINGS.
SPRING-CELL   SPRING-   SPRING-CELL
OF   THREAD OF   OF
MADREPORE.   SEA-   CORYNACTIS.
ANEMONE.

Below the buffer is a Wheel Spring, made exactly on the
same principle, but set perpendicularly instead of horizontally.

THE two figures beneath the wheel spring represent an object very familiar to us, namely, a Spring Solitaire, one figure showing it as open, and the other as closed. In this article the clasp is held in its place by a spring, and is only released by pressure.

BELOW the solitaire is a very prosaic application of the Spiral Spring, namely, that by which a house-bell is kept in vibration after the force of the pull has ceased, and which renders the bell, as Dickens happily remarks, so greedy to ring after it has been pulled.

I made and employed a spring of a similar character in closing the door of my parrot's cage. Polly is a wonderfully clever bird, and a capital talker. First, she had a cage with upright bars, two of which could be slid upwards by way of a door. She soon found out the trick of the bars, and used to escape, carefully replacing the bars afterwards.

When she was transferred to a metal cage, she discovered that the door slid upwards, and began at her old tricks. So I took a piece of galvanised iron wire, coiled it into a spiral spring, fastened one end to the upper part of the door, and the other by a hook to a staple at the bottom of the cage. Consequently, when Polly lifted the door, and loosened her grip for a fresh hold, the door closed itself again. So, after awhile, Polly gave up the door, and now never tries to open it.

PASSING to the upper right-hand corner of the illustration, there is shown a portion of Moss as it appears when magnified, and discharging its spores. When they are ripe a vast number of little spiral springs are let loose, and shoot the sporules into the air.

BELOW the moss are four figures, which are, in fact, the same object differently magnified, and seen from different points of view. These peculiar organs are technically termed " cnidæ," from a Greek word which signifies a nettle. The appropriateness of the name we shall presently see.

I have already mentioned that the tentacles of various marine animals are furnished with poison-cells. The object of

these cells is to capture and kill the prey, and the mode of
doing so is very remarkable.

On the right and left of the illustration are two such
bodies, in which is seen a sort of elastic wire coiled spirally,
apparently without regularity, but really possessing a most
beautiful order. That on the left is the poison-cell of a Madrepore, and the other is the same organ in a Corynactis. No
sooner is the tentacle touched than the poison-cells are mechanically acted upon. They are turned inside out, and the
coiled spring darts forth with wonderful violence.

Slight as is the dart, so fine that it cannot be seen except
with the aid of a tolerably powerful microscope, it is a terrible
weapon. Although it is projected with sufficient force to bury
itself to its base even through so tough an object as the
human skin, it could inflict but little injury, and would, indeed,
scarcely be felt. But it carries with it a most irritant poison,
which is apparently contained in the little capsule. These
cnidæ are very plentiful in the tentacles of the Stinging Jelly-
fish, or Stanger, as it is often called, and are charged with a
terrible poison.

As is the case with all such poisons, its effects differ according to the constitution of the being that is poisoned. There
are some persons, for example, who care no more for the sting
of a bee than for the prick of a needle, and there are those
whom a single bee-sting will bring almost to the gates of
death. So with the tentacles of the Stinging Jelly-fish and
those of the Portuguese Man-of-war, and there are persons who
are scarcely affected with the sting of the scorpion.

So it is with nettles. When I was a boy at school it was
thought necessary to wear an oak-leaf, or at least a portion of
an oak-leaf, on the 29th of May, and all who did not possess
this talisman might be flogged with nettles by those who did.
As the school was situated in the north of England, where the
oak puts forth its leaves late in the season, it was no easy
matter to obtain a veritable oak-leaf, and we used to take any
leaf that we could procure, and cut it round the edges into
the similitude of a suitable oak-leaf.

The effect of the nettles upon the boys was most curiously
diversified. Some cared nothing whatever for them ; others
suffered sharp but brief pangs ; while others, of whom I was

one, endured the most lancinating pain at the time, and for hours afterwards a hot, burning, fevered skin, and a heavy, dull ache, accompanied by throbbings of the brain so violent that it appeared as if the head would burst asunder at every heart-beat.

The fact of this inequality has been throughout life a valuable lesson to me, *i.e.* that a punishment which will nearly, if not quite, kill one man, will be no punishment at all to another.

Of course I cannot answer for the effects of these very minute cnidæ upon others, but I can state that they nearly killed *me*, and that if I had been forced to swim another hundred yards, I should have collapsed, sunk, and had a coroner's jury return a verdict of " Found drowned in consequence of cramp."

On me the effects were as follows :—First a slight, and then a severe, tingling on the parts which had been struck. Then sharp, darting pangs. Then a sudden shock as if a bullet had passed through the breast from one side to the other. Conse-quent collapse, and suspension of the office of both heart and lungs. I once had to walk nearly two miles after being stung by one of these dread animals, and how often I fell before reaching my lodgings I dare not say, but certainly once in every two hundred yards.

Even after partial recovery I should not have known my own face. It was that of an old and wearied man of seventy, grey, wrinkled, and withered ; and many months elapsed before I felt myself sure that the weird-like bullet would not drive through my breast, and leave me lying on the ground gasping and speechless.

These dreaded tentacles can sting as fiercely when separated from the animal as when they are conjoined to it, as I can also testify from personal experience.

I have a natural alacrity in damaging myself, and there is scarcely a representative bone in the body that I have not fractured or dislocated, or both. Fortunately the cerebral vertebræ have hitherto escaped. I have broken the right leg, right arm, two ribs, and right collar-bone ; dislocated the right ankle, and smashed nearly every bone of the right hand. At present, the damage to the left side is restricted to two ribs ; and I hope that the Genius of Ossifraction may now be content with his work.

But I equally seem to have a natural affinity for the tentacles of the Stangers, which deliver their envenomed darts just as fiercely when they are separated from the Medusa as when they are connected with it.

A curious example of this fact befell me in the present year (1875). Seeing that there had been a steady southern gale, which made Lundy Island and Hartland and Baggy Points indiscernible, I dreaded my old foes, and, instead of bathing from the "Pebble Ridge," took to the great "Nassau" Baths at Westward Ho. I sadly missed the roll of the waves, and the placid rapture of lying with outspread arms as the vast Atlantic billows came rolling in, flinging up the great grey boulders as if they were corks, and letting them roll down the ridge again with a thundering, and yet soothing, sound. Three miles or more inland may the thunder of the Pebble Ridge be heard; and at night, even though a storm be raging, tearing the leaves off the trees in whirling showers, flinging great branches into the air like ostrich plumes, and howling so that one person can hardly hear another speak, the dull, low, continuous thunder of the Pebble Ridge is heard over all. I have often remained awake at Bideford, simply on account of the deep roar of the Pebble Ridge, as the rising tide rolled its vast waves along the coast from Baggy Point, through Westward Ho and Clovelly, to Hartland.

When there is a heavy sea, the "undertow" of these waves is so great that even had no such things as Stangers existed, I should not have ventured upon the Pebble Ridge. One of my friends, a strong swimmer, was nearly drowned off that ridge by the undertow; and not long before I visited Westward Ho a promising young man lost his life within a few yards of that treacherous shore.

Much against my will, I went to the new bath, which is always supplied with a running current of sea-water; and I had hardly swum the length of the bath before I felt the familiar nettle-like sting in my foot. Fortunately it was only caused by a small fragment of a Stanger's tentacle, which had been severed from the animal and pumped into the bath, and no harm ensued.

USEFUL ARTS.

CHAPTER VI.

SPIRAL AND RINGED TISSUES.—VARIOUS SPRINGS IN NATURE AND ART.

Spiral Tissues, and their Structure and Uses.—The movable Gas-lamp.—Elastic Tubes.—Breathing-tubes of Insects, and their Spiral Wire.—Ringed Tissues and their varied Structure.—Ringed Tissues applied to modern Dress.—Chinese and Japanese Lanterns.—Proboscis of the House-fly.—Trachea of various Animals.—Mutual Tendency of Rings and Spirals towards each other.—Fibres of the Yew-tree.—Diving and Divers.—Principle of the Diving-bell.—How it is supplied with Air.—Structure of the Air-tubes.—Nests of the Water-spider.—Diving by means of Tubes.—Larva of the Drone-fly, and its Mode of breathing.—How to examine them.—Leaping Springs.—The Skip-jack in Nature and Art.—Skip-jack or Click Beetles.—The Spring-tail, Grasshopper, Kangaroo, Gerboa, and other Jumping Creatures.

SPIRAL AND RINGED TISSUES.

WE have now to consider the Spiral Tissue under another aspect, *i.e.* that of acting as the internal support of an exterior membrane. Ringed tissues are necessarily conjoined with the Spiral, as they both discharge the same office, and in some cases merge almost imperceptibly into each other in the same specimens. This is most beautifully shown in the proboscis of the common House-fly, to which reference will presently be made.

The subject is so large that only a comparatively small selection of examples can be made, the greater number belonging to Nature, and not to Art.

We will first take the common movable Gas-lamp, with its accompanying tube. It is at present the tube of which we have to treat, the gas itself being reserved for a future age.

It is necessary that, in order to enable the lamp to be moved from one spot to another, the tube through which the gas passes must be so constructed that if it be bent, or even coiled, it retains its form, and does not become flattened. In order to obtain this object, a very long thin wire is coiled spirally to a suitable length. Over this wire is sewn the casing of the tube, which is afterwards made waterproof with elastic varnish. A still simpler mode is by enclosing a spiral wire within a tube of vulcanised india-rubber. It will be seen, then, that by the elasticity of the spiral wire the tube must always retain its shape, no matter how much it may be bent.

On the right hand of the illustration are shown the movable Gas-lamp and tube, and a portion of the latter is given

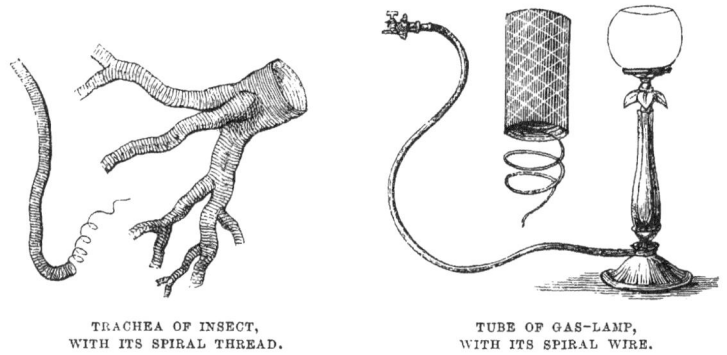

TRACHEA OF INSECT,
WITH ITS SPIRAL THREAD.

TUBE OF GAS-LAMP,
WITH ITS SPIRAL WIRE.

with its spiral wire partially unwound, in order to show its structure.

The large tubes which convey air to divers are made in the same manner, as they would not only succumb to the pressure of the water without the wire, but could not be dragged over obstacles or round corners without collapsing. It often happens that a diver is obliged, when surveying a sunken ship, to traverse the whole of her interior, descending ladder after ladder, and entering every cabin in the ship. This could not be done but for the internal coil of wire within the tube. Reference will presently be made to the subject of diving.

On the left hand is seen an object that looks something like a branch hollowed very thin. It is a magnified view of part of the Trachea or breathing-tube through which air is conveyed into the system of an insect. These breathing-tubes ramify to every portion of the body of an insect, even penetrating to the extremities of the antennæ, the wings, and the legs. It is obvious that as these organs are in tolerably constant movement, and the legs are much bent at every joint by the action of walking, the air-tubes which run through them must possess the same qualities as those of the gas-lamp and diver.

If one of these tracheæ be removed and placed under the microscope, it will be seen to be constructed in a manner exactly similar to that which has been described. Within the membrane which forms the tube proper there is a very fine, but very strong thread, which is coiled exactly like the wire spring. It is not attached to the membrane, and so strong is it that, although it is all but invisible to the naked eye, it can be drawn out as shown in the left-hand figure of the illustration.

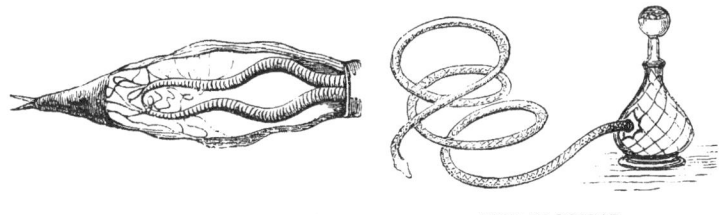

TRACHEA OF DRAGON-FLY LARVA.   TUBE OF HOOKAH

If laid on a piece of glass, it immediately tries to recoil itself, and for some little time will twist and curl about as if it were alive.

On the above illustration are two similar examples of the spiral thread with a flexible tube. The right-hand figure represents one of the many forms of the water-pipe, whether known as Hookah, Narghile, or Hubble-bubble. In the simpler forms of this pipe, such as the latter, the inhaling-tube is quite straight, and the bowl is held in the hands of the smoker. In the more refined pipe, however, the tube is very long, flexible, and made elastic by an inner spiral wire.

Perhaps the reader may remember that the larva of the Dragon-fly is a most remarkable creature in consequence of its methods of propulsion and respiration. The water is taken into the interior of the body through a peculiarly formed aperture, and then ejected with such violence as to drive the body forward on the same principle as that which causes a rocket to ascend.

The figure on the left hand of the illustration is a representation of the abdomen of this larva rather magnified, and opened so as to show the interior. On either side run the two principal breathing-tubes, through the delicate membranes of which the spiral thread can plainly be seen.

These tubes are connected with a smaller set, and they with a still smaller, so that at last they are of such tenuity that they can scarcely be distinguished without the use of a glass. But, however small they may be, they are always fitted with the spiral thread.

WE now come to the cases where the membrane is supported by a series of rings, and not by a single spiral wire.

In the right-hand division of the illustration are two specimens of objects which shall be nameless, but which were drawn per special favour at a milliner's shop. Although the day has now happily gone by when the larger object was in general wear, and seemed to be irrepressively increasing in dimensions, certain modifications of it, under various names, have made their appearance in' almost every book of fashions and every large milliner's shop.

Here we have the external membrane made of linen, calico, merino, or similar material, distended by a number of elastic rings set at tolerably even distances from each other.

The two small objects represent the handy little paper lanterns so common in China and Japan. They are composed of an external coat of tough tissue paper, so thin that it allows the light to pass through it with tolerable freedom, and of an internal series of elastic rings, which not only support it and preserve its cylindrical shape, but allow it to be folded up flat when not wanted.

I possess a singularly ingenious lantern of this kind, made in Japan, and displaying the thoroughness of work which characterizes that nation. It is five inches in diameter, and the lantern itself is affixed at either end to a circular wooden cap, the upper fitting over the lower. Consequently, when the lantern is shut, it is entirely enclosed between these two caps, which effectually preserve it from harm. It is delicately finished, and has no less than thirty rings, made of very narrow strips of bamboo. The upper cap has a little trap-door through which the candle can be admitted and trimmed, and in its centre is a small round hole for the passage of air.

In the left-hand division of the illustration are shown several examples of ringed and spiral tissues belonging to the

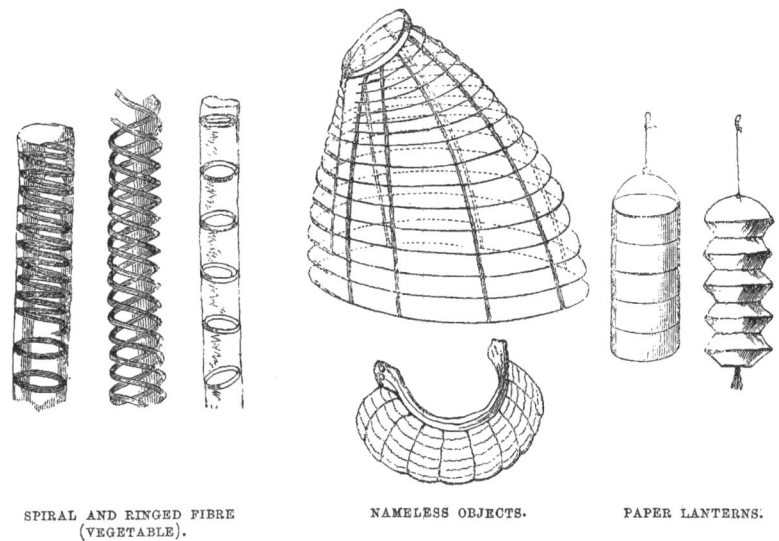

SPIRAL AND RINGED FIBRE   NAMELESS OBJECTS.   PAPER LANTERNS.
(VEGETABLE).

vegetable world, in which the principle is exactly the same as that of the Chinese lantern, &c. That on the right hand is an example of simple rings within a membrane. The central figure shows a double spiral, which produces very much the appearance of a series of rings; and on the extreme left is an interesting example which shows the transition in the internal supports from spirals to rings.

I have already mentioned that the proboscis of the House-fly exhibits this modification. If one of these objects be placed under a moderate power of the microscope—the half-inch is quite enough—and examined, it will be seen that there are some large tracheæ, just like those of the Dragon-fly larva, on each side of the proboscis, and that, where the end is widened and flattened into a sort of disc, their place is taken by a set of very much smaller tracheæ, coming nearly to a point, and each being supported internally by a series of incomplete rings, shaped very much like the letter C. A slide containing this object well mounted can be purchased at any optician's for a shilling.

THE trachea, or windpipe, as we call it, of all vertebrate animals, man included, is formed on exactly the same principle, as any one may see by going to a butcher's shop, and looking

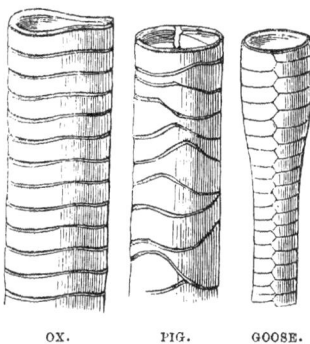

OX.   PIG.   GOOSE.

at the trachea, or windpipe, by which the lungs, or "lights," as they are called, are suspended. Were it not for this structure, we should not be able to bend our necks or turn our heads.

The accompanying illustration shows the tracheæ of three well-known creatures. The left-hand figure is the trachea of an Ox, the central figure that of a Pig, and the right-hand figure that of a Goose. Mr. Tuffen West, who made the drawings, sent with them the following remarks :—

"The tracheæ of animals furnish some very interesting examples of variation in the form and arrangement of the rings. Their purpose, perhaps, one can but guess at in some cases; but doubtless, as being works of the Master Builder, careful study would be repaid.

" In the Ox the rings are very strong and close, and in form
like a horse-shoe with the ends approximated.

" In the Pig the incomplete rings are broad at one part, and
narrow on the opposite side, with a tendency to spiral arrangement. I imagine that this would make a very rigid tube, and,
indeed, it feels so in the hand.

" Then, in the Goose, the narrowed lower part is that which
is figured just before the trachea reaches the sternum. The
(complete) rings are twice as broad in one half as in the other,
and by the alternate disposition of these differing widths, a
tube is formed of great flexibility fore and aft, but almost
absolutely rigid in the lateral direction. This seems to be so

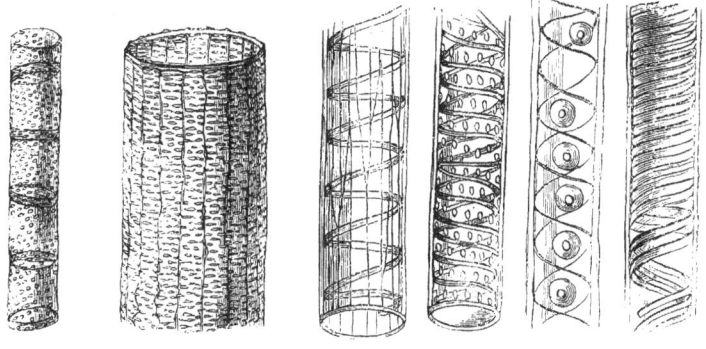

RINGED TISSUES OF SUGAR-CANE.   VEGETABLE SPIRAL TISSUES TENDING TO RINGS.

marked an evidence of design as to be calculated to greatly
raise our admiration."

WE have seen several examples of ringed tissues tending to
the spiral form, and it is but natural that we should expect to
find spiral tissues tending to the ring.

In the accompanying illustration the two left-hand figures
represent the curiously modified ringed tissue which is to be
found in the sugar-cane, the left-hand figure being much more
magnified than the other.

The other figures represent four examples of vegetable spiral
tissues, in which it will be seen that there is a tendency to form
rings, and that if a number of rings were substituted for the

spiral, and the object viewed in a slanting direction, it would be almost impossible to distinguish between the ring and the spiral.

Among the most remarkable of these examples are the two right-hand figures. That on the extreme right represents a spiral vessel taken from the so-called root, or " rhizome," of the Water-lily, and the other is a similar vessel taken from a branch of the Yew-tree. It has been suggested that to this spiral structure is due the proverbial elasticity of the yew-tree, which has from time immemorial rendered it the best wood for the manufacture of bows.

DIVING AND DIVERS.

IT has already been mentioned that the flexible tubes used by modern divers are constructed on the model of several structures belonging to the animal and vegetable kingdoms.

We will now see how they are utilised.

IN the earlier stages of the diver's art the Diving-bell afforded the only means of gaining access to the bed of the sea, even in comparatively shallow waters. The mode in which this result was obtained was simple enough, and though it carried with it the germs of still greater improvements, was but limited and uncertain in its action.

The reader is probably aware that if a vessel be filled with air, no liquid can obtain admittance until a corresponding amount of air be set free. Suppose, for example, that an empty tumbler be inserted over a basin of very clean water, and pressed downwards, it will be found that scarcely any water will enter it, the air having taken up all the available space, and only allowing as much space as may be accounted for by its faculty of compression.

It is evident, therefore, that if an enlarged tumbler could be lowered to the bed of the sea, a man might be enclosed within it, and for a time be able to support life by means of the air contained within the " bell," as this enlarged tumbler was popularly called.

It is equally evident that within a short time the air within the bell must be exhausted, and that, unless a fresh supply could be introduced, the diver within the bell would be as effectively drowned as if there were no bell at all.

The accompanying illustration is a kind of chart, so to speak, of the mode in which air was formerly supplied to the bell.

On the right hand is seen a section of the Diving-bell itself, together with the seat on which the divers can rest. There is also an escape-valve at the top of the bell, by which the vitiated air can pass away ; but, as it is not essential to the subject in hand, and is rather complicated in structure, it has been omitted.

Immediately on the left of the bell is a cask, to which several heavy weights are attached. This cask contained com-

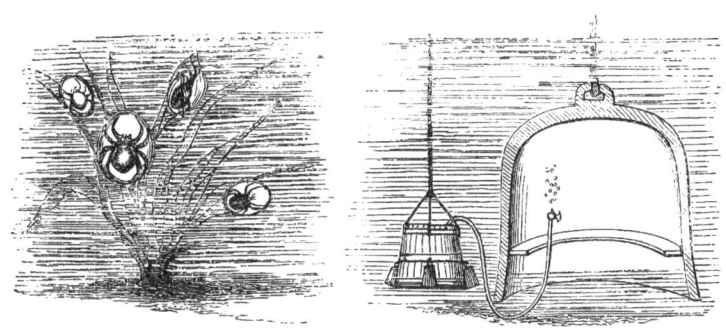

NEST OF WATER-SPIDERS. DIVING-BELL.

pressed air, and, after it was lowered by the side of the bell, the end of the flexible tube was taken into the bell, the tap turned, and the compressed air rushed into the bell, taking the place of that which had been exhausted by respiration, and was allowed to pass through the escape-valve. I may mention that the divers unexpectedly discovered that, when they were breathing compressed air, they could dispense with respiration for a wonderfully long time, the amount of oxygen taken in at a single breath being enough to renovate the blood more than could be done by several ordinary inspirations.

On the left hand of the illustration is seen a sketch of the nest of the now familiar Water-spider (*Argyroneta aquatica*), taken from some specimens in my possession.

The Water-spider is really a remarkable being. Itself a denizen of air, breathing our earthly atmosphere just as we do, and as capable of being drowned as ourselves, it nevertheless passes nearly the whole of its existence under water, and in that strange locality lays its eggs and rears its young. How this wonderful feat is performed we shall now see.

When the female Water-spider wishes to deposit her eggs, she looks out for a suitable locality, and, being a good diver, tests the various aquatic herbage until she has found a favourable spot, and then sets to work on her remarkable nest, which I believe is quite original in zoology.

After stretching a few stout threads by way of a scaffolding, she attaches to the plant a small silken cell, shaped very much like an acorn, but not so large. Ascending to the surface of the water, she contrives to clasp a bubble of air between her last pair of legs, and, laden with this airy treasure, dives below.

As soon as she has reached the entrance to the cell, which is always below, she loosens her hold of the air-bubble. It at once rises into the cell, and expels a proportionate amount of water. Not many of these journeys are required before the nest is filled with air, and then the diminutive architect spends the greater part of its time in holding on to the mouth of the little diving-bell, and supporting life by means of the air within it.

This nest, as the reader will see, is an exact representation of the various diving schemes in which air-bells are the chief portions of the machinery, although the air is conducted into them after a different fashion.

WE now come to another mode of diving, in which the bell is practically superseded by the flexible tube, which allows to the diver far more range than can be obtained by the bell. In this case the diver wears a peculiar dress, the chief part of which is a helmet so constructed that air can be introduced to it from above the surface of the water, and, after respiration, can escape by means of a valve.

Air is pumped into the tube by assistants above water, and, as the tube is long and elastic, the diver can move about with considerable freedom. As is the case with the diving-bell, the diver's tube is strengthened by an internal spiral wire, so that it is always open, however it may be bent or twisted.

The right-hand figure of the illustration represents the diver examining part of a sunken vessel. The tube through which he breathes is seen passing to the surface of the water, and so is the line by which he gives his signals to his comrades above. In his hand he holds a lamp which can burn for a limited time, being connected by a smaller but similarly constructed tube to a vessel of compressed air.

On the left hand of the same illustration are shown the curious Rat-tail Maggots, as they are popularly called. They

RAT-TAILED MAGGOTS.   DIVER WITH AIR-TUBE.

are the larvæ of the common Drone-fly (*Eristalis tenax*), which is so common towards the end of summer, and looks so curiously like a bee.

These creatures pass their larval life buried in the mud and below the surface of the water, and yet are obliged to breathe atmospheric air. This they do by means of the long appendages which have gained for them the name of Rat-tails. These "tails" are very elastic, and are capable of elongation and contraction to a wonderful extent.

When the creature is undisturbed, it lies buried in the mud with its head downwards, and its tail extended so that it reaches the surface of the water. Within this tail are two air-tubes, which are connected with the principal tracheæ, which have already been mentioned. They are wonderfully elastic, and, when the tail is extended to its utmost limit, are nearly straight. When, however, the tail is contracted, the tubes become self-coiled by their own elasticity, and shrink into the base of the tail.

As the tail is very transparent, it is easy to see how these movements are conducted. The larvæ, which may be found in almost any stagnant water, should be placed in a tall and narrow glass. Some mud should be placed at the bottom of the glass, which should then be filled with water to the depth of three inches or so.

When the mud has quite subsided, and the water become clear, the long slender tails of the larvæ will be seen so elongated that their tips reach just above the surface of the water. A magnifying-glass will easily show the two tubes within the tail.

Let the glass be but slightly tapped, and all the tail is withdrawn in a moment, so as to be out of reach of external danger. The magnifying-glass will then show the two tubes lying contracted in the base of the tail, and taking astonishingly little space, considering the amount of elongation which they can sustain. And, on examining the various bends and curves of the tubes, the value and power of the spiral spring will at once be seen. True, they are very small, but in Nature all things go by comparison, and our whole earth itself is as a grain of sand upon the seashore among the grandeurs of the visible universe.

THE LEAPING SPRING.

THE last of the springs which can be mentioned in this work are those which are used for leaping purposes.

The figure on the right hand represents the common Spring-jack or Skip-jack with which children are always so much amused. It consists of a flattened piece of wood called the "tongue," which is inserted into a twisted string, so that it forms a tolerably powerful spring. When twisted round, and then suddenly released, it strikes against the ground with such force that the whole machine is thrown into the air.

Sometimes the Skip-jack is made of a fowl's merrythought, as shown in the illustration; sometimes of the breast-bone of a goose; and sometimes of a piece of wood cut into the semblance of a frog, and painted. In all cases, however, the machinery is practically the same. I may mention *en passant* that these frog Skip-jacks are most acceptable presents to savage chiefs in many parts of the world, and that the most powerful and venerable warriors are as delighted with these toys as any European child of six years old.

Now we will turn to Nature, and see what she has in the way of Skip-jacks.

All entomologists will at once have before their minds the vast groups of Skip-jack Beetles, technically termed *Elateridæ*, and also known as Click-beetles, from the sharp clicking

SKIP-JACK BEETLE.
GRASSHOPPER.

SKIP-JACK.

sound which they produce when in the execution of their curious gymnastics. In this group belong the fire-flies of warm countries, and it may be mentioned that the larvæ of some of our species are too familiar to the agriculturist under the name "wireworm."

All these beetles have very short legs and very long bodies, so that if they should fall on their backs on a smooth surface, they could not recover themselves. Now, as they, when discovered, instinctively try to save themselves by falling to the ground, it is evident that some means must be used to enable them to regain their position. This is found in a most curious apparatus.

Attached to the "prothorax" is a rather long, pointed, and very elastic projection exactly corresponding with the tongue of the Skip-jack.   The end of this tongue fits into a groove in the " sternum."

When the beetle falls on its back, it curves its body as shown in the illustration, the tongue thus being freed from its groove. It then smartly springs the tongue back into its place with the sharp clicking sound already referred to, and does so with such force that it leaps into the air to some height.

Generally it falls on its feet, but if it should fail, it repeats the process.   If one of these beetles be laid on a plate or similar smooth surface, it will skip ten or twelve times without stopping, and after a short rest will begin again.

THERE are some curious little beings, popularly called Spring-tails, which afford excellent examples of the Leaping Spring.   Their exact place in the system of Nature is rather uncertain, some zoologists considering them as insects, while strict entomologists reject them.   They are very small, and mostly of a darkish brown colour.

Plenty of them may be found under stones in damp spots, under bark, and in similar localities, though they are often found in houses, and have frequently traversed the paper on which I have been writing this book.   Cellars are favourite localities of theirs, and a little flour sprinkled on a plate or piece of paper in a cellar is tolerably sure to attract them. Although they are certainly not more than the fifteenth of an inch in length, they may be at once recognised by their peculiar attitude, which very much resembles that of a dog or cat in its usual sitting posture.

As long as they are not disturbed they crawl about in a quiet manner, but if touched, or even alarmed, they suddenly make a tremendous leap, propelling themselves by means of a forked and elastic tail, doubled under their bodies, and acting just like the tongue of a Skip-jack.

BELOW the Skip-jack Beetle is shown the common Grass-hopper, as an example of muscular leaping springs.

We all know what wonderful leaps the Grasshopper, Cricket, and all their kin can make, the leaping movement being evidently intended more as a means of defence than as an ordinary mode of locomotion.   The same may be observed in the Kangaroos and Gerboas, which are content to use an ordinary walking pace when undisturbed, but when alarmed can make tremendous leaps, and outstrip almost any pursuer.

Even in Man, the Horse, the Dog, &c., which are most essentially leaping animals, the same principle is employed, the legs being used as muscular springs acted upon by the will of the owner.

USEFUL ARTS.

CHAPTER VII.

FOOD AND COMFORT.

Parents and their Young.—Milk, and the various Ways of obtaining and using it.
—The Kafir Tribes and Clotted Milk.—The Tonga Islanders.—The Tartars.
—Ants and Aphides.—Honey-dew.—Milch Cows in Insect-land.—Fish-tanks
and Aquaria.—Bill of the Pelican.—Eggs and Chickens.—The Hen-coop.—
Nest of Termite.—Workers and Queen.—Egg-hatching.—The Hen and her
Young.—Artificial Egg-hatching Machine.—The Snake and her Eggs.—
The Gad-fly and Bot-fly.—Preservation of Provisions.—Hanging Meat.—
Eggs of the Lace-wing Fly.—Spider-eggs.—The Butcher's Hook and the
Claws of the Sloth.—Bats and Insects.

THIS subject is necessarily a very large one, and I shall, in
consequence, be obliged to compress it, though it might
well make a separate work by itself. For Food represents the
very existence of Man, considered as one of the animal world ;
and Comfort represents the progress of civilisation, by which
man leaves day by day his savage and solitary nature behind
him, and becomes social, moral, and elevated.

PUTTING aside the instinct which forces the parent to feed
the young without external assistance, we come to those cases
where the parent has to seek food which the offspring could
not have found for itself, and often to prepare it for the use of
the offspring.

In the greater part of the world, the milk of various animals
is the staple of food, not only for children, but adults ; and the
"milk diet," as it is called, is strongly urged by many phy-
sicians of the present day.

The Kafir tribes, for example, a wonderfully powerful race
of men, live almost wholly on sour milk, mixed with maize
flour, never eating such valuable animals as kine except on

great occasions. Yet the natives of the Tonga Islands think that nothing can be more disgusting than for a human being to drink the milk of a cow.

How the operation of milking is conducted we need not say, whether it be performed on the cow as with most nations, or the ass in case of need with ourselves, or the mare as with the Tartars, or the goat and sheep in various parts of the world. The milk of the sheep, by the way, is singularly rich and nourishing.

Suffice it to say that the animals which are to be milked are kept for that purpose, and that the touch of the human hand, rightly applied, induces the animal to part with its milky stores.

In Nature there is an exact parallel.

It has long been known that some species of Ants are in

ANT AND APHIS.   MILKING COW.

the habit of acting in exactly the same manner as ourselves, in not only extracting a nutritious liquid from other insects, but watching and tending those which furnish their daily food just as a good dairyman watches and tends his cows.

The Ants, being insects, would naturally require insect cows, and such are to be found in the Aphides, of which mention has already been made. These insects are furnished with a pair of very small tubercles near the end of the abdomen, and from them flows that sweet liquid which is so familiar to us under the name of "honey-dew." For centuries no one knew the source of the sweet honey-dew which attracted all the bees of the neighbourhood to the tree on whose leaves it was sprinkled, sometimes in patches, and sometimes coating them with a thin shining coat, as if varnished.

At last it was discovered that the honey-dew is, in fact, the liquid exudations from these tubercles upon the backs of the aphides, and that the ants feed regularly upon it. Not only do they lick up the honey-dew that has fallen from the ants, but they milk them, so to speak, exactly as a dairymaid milks a cow. With their antennæ the ants pat and stroke the tubercles of the aphides, and in a few seconds a drop of pellucid liquid appears at the extremity. This is the honey-dew, and is at once lapped up by the ant, which proceeds from one aphis to another until it has obtained its fill of the sweet food.

How the ants carry off the aphides, cherish and guard them for the sake of their honey-dew, is a story too long to be told, but it is well known among entomologists. Our English ants are, however, totally eclipsed by a Mexican species, which not only collects honey, but stores it in the bodies of its kindred.

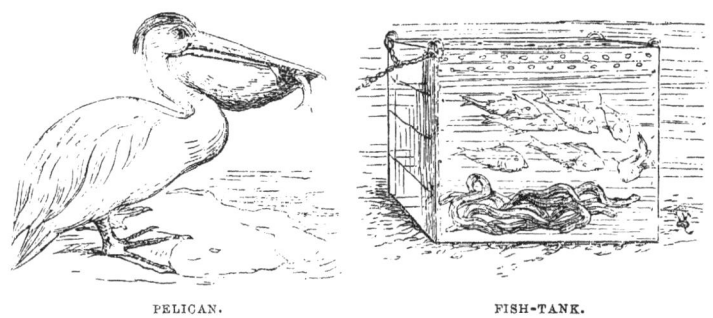

PELICAN.   FISH-TANK.

After taking precautions that no food can escape, the ants feed with their sweet store their companion, who is thus doomed to pass the remainder of life as a mere honey-cell. The abdomen becomes spherical, smooth, and so transparent that the honey can be seen within it. It is quite air-tight, and so preserves the fragance of the honey until it is wanted.

So plentiful are these honey-ants, that they are an article of commerce, and are sold by measure for the purpose of making a sort of mead. There are many of them in the British Museum, with the honey still within their transparent bodies, and they are well worth seeing.

THE accompanying illustration represents the artificial and natural way of preserving food in an uninjured state. The right-hand figure is that of an ordinary glass aquarium, such as was in general use until the properties of air and water were better understood, and it still need not be rejected. It is simply a vessel in which water is contained, so that aquatic or marine animals may be able to live in it for some time.

There are infinite varieties of the "Fish-tank," if we may so call it, the chief of which is the "well," which is so extensively used in bringing fish to market.

Through the bottom of the boat projects a sort of box pierced with holes, so that the water has free access and egress. The sides of the box are so high that there is no fear of the water rising into the boat. When fish are taken, they are thrown into the well, and there can live until they are wanted for sale.

Also, as all know who are acquainted with river-banks or seashores, fishermen have similar wells detached from the boats, and partly or entirely sunk in the water. In them they keep their stock, and, when a customer arrives, they simply draw the box ashore, so that the water runs out, select what fish they choose, and replace the box in the water.

Now, the power of conveying fish to some distance without destroying life has for countless ages been possessed by the Pelican, one of which birds is shown in the accompanying illustration.

As every one knows, the chief peculiarity of this bird is the large and very elastic membrane of the lower jaw. When not in use, it contracts by its own elasticity, and the bill looks quite slender, as well as long. But, when distended with water and fish, it presents the appearance shown in the illustration.

Any one who wishes to see the exercise of this power can do so by attending the Zoological Gardens, and visiting the Pelicans at feeding-time, and an hour or two before it. They hardly seem to be the same birds. Some years ago I made a series of sketches of the same Pelican under different circumstances, and it is scarcely possible to believe that they could be, as they are, truthful representations of the same bird.

THE right-hand figure of the next illustration requires no comment, as it simply represents the ordinary hen-coop.

As everybody is aware, the object of the coop is to keep the
hen within its bars, while the little chicks can run in and out
as they choose, and the coop is made so as to prevent the egress
of the mother, while the offspring find no difficulty in
escaping.

Now, in the world of insects we find an exactly analogous
structure. As is the case with many hymenopterous insects,
there is in the nest of the Termite, or White Ant, as it is
popularly called, a single perfect female, which is the mother of
the nest. A similar arrangement occurs in the common hive-
bee, but there is a notable distinction between the queen Bee

QUEEN TERMITE IN HER CELL.   HEN IN HER COOP.

and the queen Termite, the latter belonging to the neuropterous
order.

The former is unconfined, and moves about from cell to cell,
depositing her eggs within them, and taking the greatest pains
that they occupy exactly their proper place within the cell. The
latter never moves after she has begun to deposit eggs, but
remains motionless in the same spot, and allows her subordi-
nates to dispose of the eggs which she lays.

How this end is achieved will now be seen.

The reader is probably aware that the queen Termite attains
to enormous dimensions, her head, thorax, and legs retaining
their normal size, but the abdomen becoming several inches in
length, and thick in proportion. The legs are necessarily
unable to move so vast a body, and in order that so important
a personage should not receive injury, a large oval cell is built
around her, from which she never moves for the rest of her
life. She has but one duty, namely, to lay eggs, and so is fed

that she may have strength to produce them. She is simply passive, and never even sees her eggs, much less has care of her young.

All the care of guarding and nurturing the eggs and young falls upon the worker Termites. These insects are quite small, about the size of our common Wood-ant.

When they build the clay cell around their queen, they bore a number of holes along the sides, which are just large enough to allow the workers to pass freely, but which effectually exclude the soldier Termites, or any foes larger than themselves.

Through these apertures streams of workers are continually passing—some entering the cell to fetch the eggs, and others coming out with eggs carried carefully in their jaws.

EGGS OF ŒSTRUS.　　　　EGG-HATCHING MACHINE.

Thus, as the reader will see, we have in Nature an exact analogy of Art, the Termite queen being confined within her cell exactly as is the hen within the coop.

BEING on the subject of eggs and egg-hatching, we will take another case in which Art has acknowledgedly followed Nature.

We all know that eggs are developed into life by means of well-regulated heat, and that with birds the general rule is, that the needful heat is supplied by the parent bird, who sits upon them for a certain time, until the young birds make their appearance in the world.

Under ordinary circumstances, the aid of the parent bird is

quite sufficient; but when the progress of civilisation requires that the eggs of poultry should be hatched in numbers too great for the powers of the parent bird, Man has been fain to imitate Nature, and to invent machines whereby eggs can be hatched by artificial heat, regulated to the temperature of the hen's body.

Various as are these machines in detail, they are all alike in principle, and the right-hand figure of the accompanying illustration will give a fair idea of the method which is employed.

A box is fitted up with trays, on which the eggs are arranged. At the bottom of the box there is the heat-producing apparatus, which can be regulated at pleasure. The trays of eggs can be moved from one part of the box to another, so as to insure the right amount of heat, and, if this process be only carefully carried out, the young chicks emerge from the eggs exactly as they would have done if the hen had sat upon them.

This machine is sometimes called the Artificial Mother, and it is worthy of notice that it is no modern invention, the ancient Egyptians having used it more than three thousand years ago.

WITH regard to Nature, it would have been simple enough to give one illustration of a bird sitting on her eggs, but I have preferred to select a different subject, as more relevant to the question of artificial heat.

There is an insect to which we have had several occasions of reference, namely, the Wurble-fly of the ox, scientifically known as *Œstrus bovis*.

The eggs of this insect are deposited in the skin of the ox, and are there hatched by the heat of the animal. In proportion as the larva grows, it raises lumps upon the skin, these being practically the roofs of the artificial home. There are several other species of the same genus, all of which have their eggs hatched by the heat of the animals on which they are placed. There are, for example, the common Bot-fly (*Œstrus equi*), whose eggs are hatched in the interior of the horse, and the Sheep-fly (*Œstrus ovis*), whose eggs are hatched in the head of the sheep. The common Snake leaves her eggs to be hatched in the artificial heat produced by decaying vegetable matter.

WE now come to the preservation of provisions.

In the first place, we have the well-known "cache" of Northern America—*i.e.* a spot wherein provisions are hidden, and their locality only marked by signs intelligible to those for whose use they are intended. It is, perhaps, hardly necessary to mention that many creatures—such as the dog, the squirrel, and most of the crow tribe—are in the habit of concealing provisions for future use.

In those parts of the world, however, where the rights of hunters are acknowledged, any one who kills a deer, or other animal of chase, and is not able to carry off the entire body, can preserve it for his own use. He simply cuts it up in hunter fashion, and hangs the various portions to branches of

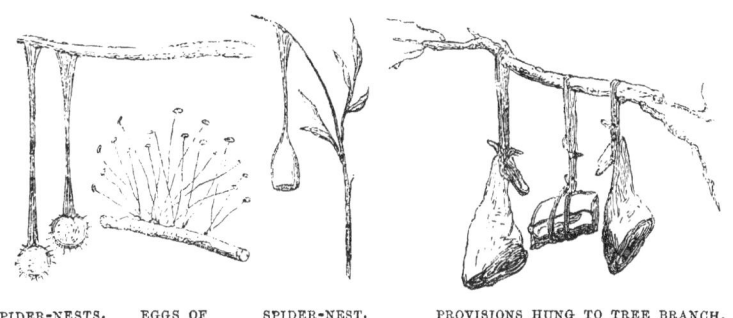

SPIDER-NESTS.   EGGS OF   SPIDER-NEST.   PROVISIONS HUNG TO TREE BRANCH.
LACE-WING FLY.

trees, where they are out of the reach of wild beasts. Stores like these, such as are shown in the illustration, are always respected, and no hunter would dream of helping himself to the game which was killed and dressed by another.

Beasts of prey, however, cannot be expected to be so punctilious, and in consequence the hunters hang their meat to branches which cannot be reached.

IN Nature we find many similar examples, one or two of which are given on the left hand of the illustration.

In the centre is seen a group of eggs of the Lace-wing Fly (*Hemerobius*), so called on account of the delicate, lace-like structure of its beautiful pale green wings.

When the female lays her eggs she always chooses a slight

twig, and upon it deposits a little drop of a slimy consistence.
She then draws out this drop into a thread, which hardens as it
is brought into contact with the air. At the extreme end of
the thread she places an egg, which is thus kept at some height
above the ground, and defies the approach of inimical insects.
The eggs, as well as the stalks, are perfectly white, and have
so singular a resemblance to mosses, that for many years they
were actually classed and figured as such.

These egg-groups are plentiful enough, if the observer only
knows where to look for them. I have several of them in my
collection, and have found that nearly every one who sees them
for the first time takes them for mosses. I never myself saw
the pretty insect lay its eggs, and for the description am
indebted to Mr. A. G. Butler, of the British Museum, who has
kept them and watched their habits.

The objects on either side of the Lace-wing Fly's eggs are
egg-groups of certain spiders, suspended by threads from
branches.

A STILL more remarkable instance of unconscious imitation
may be found in the two objects in the accompanying illustration. It is hardly necessary to say that the right-hand figure
represents a portion of the arrangement by which a butcher
hangs up his meat out of harm's way until it is wanted.

The hooks in question are simply formed into a double curve,
like the letter S, and can be slid along the horizontal bar without any danger of falling.

Now, in the common Sloth we have an exact prototype of
the butcher's hook. The Sloth passes the whole of its life in
the remarkable attitude which is shown in the illustration.
It lives among the branches—not on them, but under them—
its claws being long and curved, just like a butcher's hook.
I have often watched the animal traversing the branches, and
have been greatly struck with the accurately picturesque
description of the late Mr. Waterton, who was the first to discover the real character of the Sloth.

It was he who found out that the previous ideas as to the
Sloth's mode of life were utterly erroneous, and that, instead of
being a sort of bungle, the Sloth was as perfect in its way, and
as well fitted for its mode of life, as the lion or tiger. He dis-

covered that the animal always hung from the branches, as shown in the illustration. In fact, as Sydney Smith remarked in his witty review of "Waterton's Wanderings," the Sloth

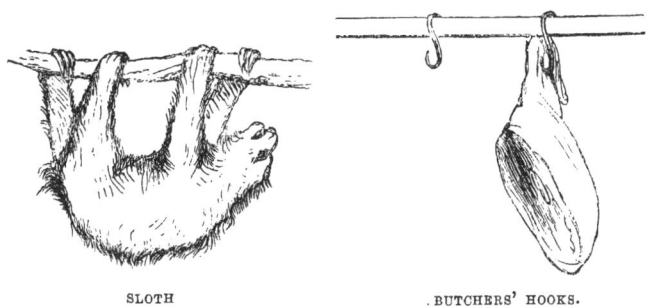

SLOTH   . BUTCHERS' HOOKS.

passes his whole life in suspense, "like a young clergyman distantly related to a bishop."

THERE are many other creatures which afford similar examples, though perhaps none are so striking as the Sloth.

For instance, there are the whole tribe of Bats, which, by means of the curved claws attached to their hind-feet, can hang themselves head downwards in the open air, and even swing in wind, without the least fear of falling.

USEFUL ARTS.

CHAPTER VIII.

DOMESTIC COMFORT.

How to make Home comfortable.—The Bed in its various Forms.—The Feather Bed of Man.—The Eider-duck and her Plumage.—The Rabbit and her Down.—The Long-tailed Titmouse and her wonderful Nest.—The Hammock of civilised Man and Savage.—The Sailor's Canvas Hammock.—The String Hammock of tropical America.—Nest of the Pensile Oriole.—Silken Hammock of the Tiger-moth and other Insects.—The Mat Bed.—Cocoa-nut Matting.—The Robber-crab and its Bed.—Strength and Uses of the Cocoa-nut Fibre.—The Surgeon's "Cradle" and the Pupa of Tabanus.—The Art of Sewing and the Tailor-bird.—Principle of the Umbrella and its Original Use.—Natural Umbrella on the Rosemary.—Servants and Slaves, and the Distinction between them.—The Use of Slaves in hot Countries.—Slavery in the Insect World.—The Ants and their Slaves.—Ornamental Gardening and Pleasure-grounds.—The Hanging Gardens of Babylon.—The Bower-birds and their Pleasure-grounds.

WE now come to a different branch of the same subject, namely, the means by which our dwellings are rendered comfortable.

After having procured a dwelling which can withstand the elements, we next look for a bed on which to repose, and which will ease the limbs and brain, wearied by the toils of the day.

Allusion has already been made to the ordinary feather bed and its multitudinous natural springs. We now have to see how the various kinds of beds are anticipated in Nature, and will begin with the feather bed.

As to our own beds, nothing need be said about objects so familiar, although, in order to preserve the parallelism, it is necessary to introduce an illustration on the right hand of the page.

On the left hand are shown two examples of natural feather

beds, selected from many others on account of the exact parallels which they afford.

We all know the wonderful warmth and lightness of the Eider-down mattress or quilt, though there are comparatively few who know how the Eider-down is procured.

In common with many other creatures, the Eider-duck forms a bed for her young by plucking the down from her own body. Rabbits do exactly the same thing, as all boys know who have kept them, the only difference being that fur is substituted for feathers. So do many insects, stripping themselves of their

LONG-TAILED TITMOUSE.　　　　　　　　FEATHER BED.
EIDER-DUCK.

own downy covering, and employing it for the comfort of their offspring.

The lower figure on the left hand represents the Eider-duck in the act of plucking the far-famed down from her breast in order to make a soft and warm couch for her young, and the amount of feathers which she will devote to this purpose is simply astonishing. Their weight is insignificant, but their bulk is wonderful.

Above the Eider-duck is shown the nest of the common Long-tailed Titmouse. It is the most perfect nest that is constructed by any British bird. Its shape exactly resembles that of an egg, and it has but one small aperture, as is shown in the illustration.

The Titmouse lays a vast number of eggs, and almost fills

the nest with soft downy feathers, on which they can rest.
If the finger be introduced into the nest through the aperture,
the tiny eggs can be felt reposing in their natural feather-bed.
In this case, however, the bird does not denude herself of
feathers, but has a way of picking them up wherever she can
find them.

Now we will take another form of bed, namely, the Hammock,
which is used in many parts of the world.

Putting aside the well-known hammock as used on board
our ships, we will take the same kind of bed as used among
the natives of tropical America.

In that wonderful part of the world, where water and
vegetation reign supreme, an aërial couch of some kind is
absolutely needful, and is supplied by the singularly ingenious
hammocks which are constructed by the natives. They are
made of a fine, but marvellously strong fibre, procured from
the aloe plant by the simple process of soaking the long leaves
in water, and dashing them against a stone. The soft green
parts are eaten away, and the tough fibres remain in all their
strength.

From these fibres are woven the strings of which the
Hammocks are made. I possess four of the Hammocks, all
made on different lines, but all based on the same principle.
In some the strings are laid parallel to each other, and connected by transverse strings at regular intervals, but in the
best specimens they are interlaced diagonally into a sort of
loose network without knots, so that it yields in every direction
to the outlines of the body.

It is one of the most comfortable couches ever invented,
especially when it is of considerable size. I have one specimen
which, even in its curved state, extends completely across a
tolerably sized room. I never use it because it is so comfortable
that the temptation to lie in it is almost too strong to be
resisted.

As to Hammocks in Nature, they are almost too many to
be computed.

So we will first take the nest of the Pensile Oriole, which is
shown in the illustration, and which is an admirable example

of the Hammock, being woven from long vegetable fibres intertwisted very much like the strings of the South American Hammock. And as if to increase the resemblance, the bird, whenever it can do so, will carry off hanks of cotton, linen, thread, or pieces of string, and weave them into its nest.

I have one of these nests, and, directly I saw it, was struck with its exact similitude to the Hammock of human manufacture.

There are many other birds in various parts of the world

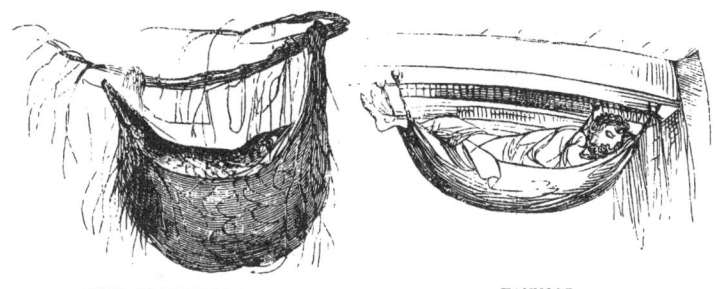

NEST OF PENSILE ORIOLE.   HAMMOCK.

especially in Australia, which make their nests on exactly the same principle, though in slightly varied forms.

Also, in the insect world, there are innumerable examples of the natural Hammock, the most common of which is that made by the caterpillars of the Tiger-moth, and in which it slings itself while undergoing its changes from the chrysalis to the perfect state.

It is made of silken threads, interwoven so slightly that the chrysalis can be seen through them, and so exactly like the Hammock of the South American Indian that if a drawing were made and enlarged, one might easily be taken for the other.

Now we come to the Mat Bed, which is so much used in the warmer parts of the world, where the earth is dry, and the air so warm that nothing is required but the slightest possible protection from the soil.

In inland places, such as Southern Africa, the bed is made of long grass-stems laid side by side, and sewn together with a

sort of twine.   One of these beds in my collection is some three
feet wide by seven feet long, and can be rolled up into a cylinder so compact and light that even a child could carry it.

Of course, when the Kafirs are on a journey, the women have
to carry the beds, together with the heavy wooden pillows and
other necessaries, the men carrying nothing but their weapons.
I have a pair of figures made by a native artist, representing a

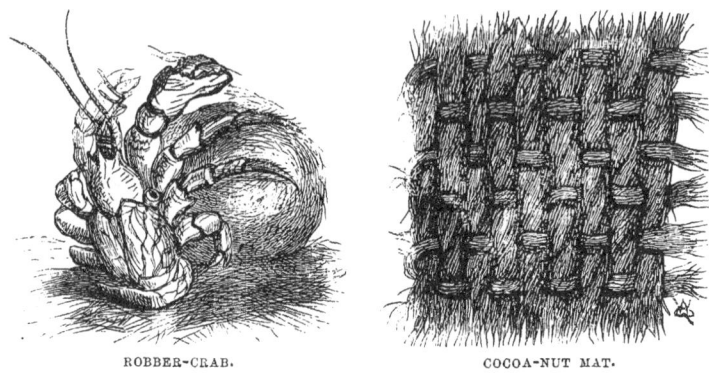

ROBBER-CRAB.   COCOA-NUT MAT.

Kafir man and woman on a journey, the woman staggering
under her heavy burdens, the bed being included, and the man
stepping lightly along, with nothing but his spears and knob-
kerries.

On the sea-coasts, however, where the cocoa-nut palm
grows, the fibre of the husk is the principal material for bed-
ding.   These fibres lie so parallel to each other on the surface of
the cocoa-nut, that they are easily stripped off, fastened together,
and formed into mats of any shape or thickness.   One of these
mats is shown on the right hand of the illustration, and the
reader will see how simple is its manufacture.

Owing to the ease with which it is made into a fabric, the
cocoa-nut fibre was in great use as armour before the bullet set
all armour at defiance.   It will be remembered that when Captain Cook was murdered, he committed the mistake of firing a
charge of small shot instead of a bullet, and the fact that
the cocoa-nut mat carried by the man at whom he fired
resisted the shot, encouraged the natives to attack and murder
him.

EVEN the cocoa-nut mat has its precursor in Nature.

There is a certain Crab inhabiting the cocoa-nut bearing parts of the world, which not only makes itself a bed from the fibre, but supplies it to mankind.

This wonderful Crab has the power of ascending the cocoa-nut palms, which is beyond the power of any man except a trained gymnast. It picks out the ripest fruits, and with its powerful claws tears off the fibre before breaking the shell and devouring the kernel, as is shown in the left-hand figure of the illustration.

After eating the kernel, which is at that time a soft, creamy substance, quite unlike the hard, indigestible material which we in England know by the name of cocoa-nut, the Crab carries off the external fibres into its den, and there makes its bed of them. So great, indeed, is the amount of cocoa-nut fibre thus collected that the natives are accustomed to save themselves the trouble of climbing the trees, and merely search for the holes in which these Crabs have made their nests, knowing the amount of ready-gathered cocoa-nut fibre that is always to be found in them.

ANOTHER modification of the bed needs a short notice, especially as I have practical and sad experience on the subject.

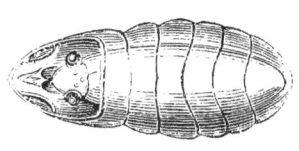

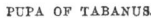
PUPA OF TABANUS.   SURGICAL CRADLE.

It is technically named the "cradle," and is used to keep the bedclothes from pressing on a damaged limb.

When a mere lad I contrived, at cricket, to dislocate the right ankle, and break the bone. An ignorant surgeon refused my request for a cradle, and absolutely tied the cover of a book to the sole of the foot. Of course this appliance was worse than useless. It acted as a lever, allowing the clothes to turn the foot round, and to the present day the right foot has never recovered its faculties. Had the simple "cradle" been used—

i.e. a few sticks bent into an arch-like shape, and tied together, so as to keep the clothes from even touching the foot—all would have been right.

On the right hand of the illustration is shown the surgical cradle, as a defence to a damaged leg. On the left is shown the curious natural cradle of the Gad-fly while undergoing its change into the perfect state. It is quite hard and rounded, being formed from the skin of the larva, and allows the pupa to lie within it, protected from any ordinary pressure.

ANOTHER point now comes before us.

We cannot well have our bedclothes—indeed, any kind of clothes—without the use of needles and thread. The simplest form of sewing is that which is adopted in many parts of the world, namely, of boring holes and pushing a thread through them, no eye being required in the needle. In this way the Kafirs of Southern Africa and the Esquimaux of the Polar regions make their beautiful garments of skins. I have for many years had in constant use two South African cloaks, or karosses, and one made by the natives of Vancouver's Island, and they are now as good as they were when they were first given to me. Naturally, such a mode of sewing consumes much time, but, as time is not of the least value to these native furriers, no harm is done, and the junctions of the different skins is absolutely perfect. Even where holes have been made in the skin, the native furrier has supplied their places with circular pieces so neatly inserted, that on the outside not a trace of the junction is visible, and even the very set of the hairs is preserved.

Our very modern needles, with their eyes which carry the thread, are but a modification of the original plan of boring holes, and pushing the thread through them.

NATURE has a singular parallel in the case of the Tailor-bird, which sews leaves together by their edges, and makes its nest inside them. It acts exactly like one of our own shoemakers, using its slender and sharply pointed beak in lieu of the awl, and employing a slight but strong vegetable fibre in place of the "waxed end" of the shoemaker, or the sinew-thread of the Kafir.

In the illustration an ordinary needle and thread are seen on the right-hand side, and on the left are two nests of the Tailor-bird, taken from specimens in the British Museum.

The mode of sewing is strangely like that which is employed by the uncivilised furriers who have been described, and much

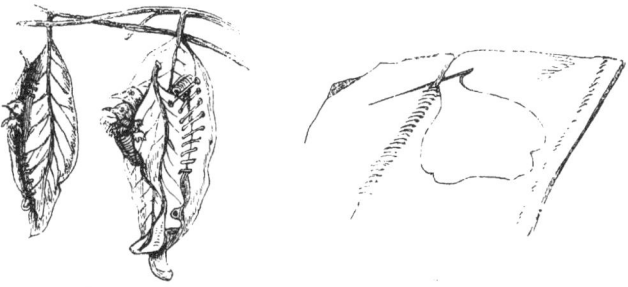

TAILOR-BIRDS AND NESTS.　　　　SEWING CLOTH.

superior to that which is seen in many other parts of the world. For example, I have a West African quiver made of hide sewn together with stitches infinitely more clumsy than those of the Tailor-bird.

The reader will also remark that I might have placed this singular nest in the category of beds, on account of the soft and warm lining on which the young repose. I have, however, thought that it more properly belongs to the present division of the subject.

SOMETIMES we require a temporary as well as a permanent shelter from the elements, and procure it by means of the Umbrella.

In many countries, especially those where the climate is hot, the Umbrella is almost exclusively used, as, indeed, its name denotes, to preserve its owner from the direct sunbeams, and is, in fact, the "parasol" of our European ladies. It also is a mark of dignity, the amount and quality of its decorations indicating rank, even though the man who sits under its shade is clothed in a modest cotton cloth wrapped round his waist.

For the purpose of shielding the bearer from the sun the Umbrella was first introduced, and the introducer incurred the obloquy usual in such cases. Now, however, the Umbrella has by common consent become a defence against rain and snow, the male sex leaving the parasol to the gentler half of creation, and submitting themselves to the chance of a sunstroke.

WE all know the ingenious Umbrellas of Africa, China, Japan, Siam, &c.; but there are few persons who know that a common

HAIRS OF ROSEMARY.   UMBRELLA.

magnifying-glass will disclose thousands of beautifully perfect umbrellas on the leaf of the Rosemary.

Pinch the Rosemary-leaf between the fingers, and a strong and peculiar perfume is evolved, just as when the peel of the orange is squeezed. The reason is the same in both cases, namely, the presence of multitudes of spherical vessels which contain their essential oil, secreted by the plant.

In the orange they are sunk below the surface of the skin, and are protected by it; but in the Rosemary they stand on slight footstalks, as shown in the illustration.

Being very delicate, and liable to be broken at the least touch, they are protected by a series of curiously formed hairs, which extend over them exactly as would an umbrella, and defend them from the elements.

The surface of a Rosemary-leaf affords a singularly beautiful sight, even with a common magnifying-glass, the tiny perfume-globes gleaming like little pearls in the broken lights that shine through the umbrella-like hairs.

Now we come to another part of domestic life, namely, Servants.

There is a diversity of ideas on this subject, as we know by the various discussions respecting "lady-helps" and "gentlemen-helps," which bid fair to initiate a revolution in domestic life. Servants are sometimes called the greatest plagues in life, but it is difficult to see what could be done without them.

Then there is the complaint that servants are not what they used to be—the faithful retainers of the household, and considering themselves members of it. Perhaps not, but I have had experience of several faithful retainers, and invariably found them to be unmitigated tyrants, assuming power, repudiating responsibility, and being practically the master or mistress of the household.

Then we come to the great question of slavery in its various bearings.

Putting aside the now acknowledged diversity of races, and the well-known fact that the negro in a state of slavery to a European is infinitely better off than he would have been in his own country, where there is no law but that of might, we must entertain the question of enforced servitude, *i.e.* where the servants have no choice either in entering or leaving their situations.

It is, of course, opposed, and rightly, to our modern English ideas that a slave, under such a name, should exist on British ground. Yet there are thousands of Englishmen who are more wholly enslaved than was any negro in the worst times of slavery. The chains may not be of visible iron, nor the whips of tangible thongs, but they are, perhaps, all the more galling and biting.

Some of my readers may be aware that slavery exists in the insect world, and probably existed long before man came on earth.

There are many species of Ants which are absolutely incapable of managing their own nests or rearing their own young, and which, in consequence, impress into their service the workers of other species of Ant, and hand over to them the entire labour of the establishment. They can fight, and they can establish fresh colonies, but they cannot build nests, nor nurse their young, and so they impress into their service those Ants whose instinct teaches them to do both.

Periodically the master Ants, if we may so call them, set off on a slave-hunting expedition. They find out the nest of the special Ant whose aid they need, penetrate into it, and bear off the pupæ, or "ants' eggs," as they are popularly called. These are carried to their new home, and are speedily hatched. They know no other home, and, led by instinct, set to work as industriously as if they had never been removed.

Those who have watched their habits are unanimous in declaring that they seem perfectly happy and contented. No

SLAVE-CAPTURING ANTS.   AFRICAN SLAVE-GANG.

compulsion is used towards them, and they work because told to do so by their own instinct. Work they must, and it does not in the least matter to them for whom the work is done.

ANOTHER branch of this subject is shown in the accompanying illustration, namely, the pleasure garden or playground.

This is, as we all know, a token of high civilisation, and even in the ancient times the hanging gardens of Babylon were reckoned as the greatest wonders of that great city, the then mistress of the world.

No savage ever dreamed of such a thing as a pleasure garden, nor could appreciate it if he saw it. Yet there are birds which far surpass the savage in this respect, and which build recreation grounds for the sole purpose of amusement.

These are the well-known Bower-birds of Australia, which I sincerely hope may not be extirpated by the white man, as has been the case with so many creatures, including the aborigines of Tasmania themselves.

The Bower-birds, which are distantly related to our thrush

and blackbird, but are about as large as jackdaws, have a curious habit of building arched bowers quite independent of their nests.

The shape of one of these bowers is shown in the accompanying illustration.

The bird first weaves a sort of platform of flexible sticks, and then fastens into them a number of other sticks, so set that they form a sort of arched gallery. Through this gallery the birds love to run, and they invariably decorate the ends with anything pretty that they can pick up, such as feathers, coloured

PLAYGROUND OF BOWER-BIRD.   GARDEN BOWER.

stones, shells, ornaments, and the like. So well is this proclivity known, that whenever any one who is living in the Bush loses any small piece of property, such as a pencil-case or watch-key, or even a tobacco-pipe, he always goes to the Bower-bird's pleasure garden, and mostly discovers the lost property.

At the Zoological Gardens these Bower-birds have long lived, and it is a most interesting sight to watch them weaving their platforms, raising the bowers over them, and then keep running in at one end and out at the other, like children at play, and with their burnished plumage gleaming in the sunbeams.

The right-hand figure simply depicts a modern pleasure garden, and needs no description.

USEFUL ARTS.

CHAPTER IX.

ARTIFICIAL WARMTH.—RING AND STAPLE.—THE FAN.

Various Modes of warming Houses.—The Fire of the American Indian and the Kafir. — The Oil-lamp of the Esquimaux. — The open Fireplace and Chimney Stoves.—The laminated Stove and its Powers. — Gills of the Lobster, Crab, and various Fishes.—Mode in which the Gills act.—Why Fishes lie with their Heads against the Stream.—Drowning a Fish.—The Ring and Staple, and their various Uses. — Head-bones of the Fishing-frog or Angler-fish. — The Fan and its Modifications. — Japanese and Chinese Fans. — The Feather Fan. — The Palm-leaf. — Indian Fans. — The Hive Bee and its Wings.—Fans of the Essequibo and South Sea Islanders.—The Fan Fire-guard.—Antennæ of the Cockchafer.—Burial.—Various Modes of disposing of the Dead. — Ordinary Habits of dying Animals.—Dead Insects.—The Funeral-ant and its wonderful Habits.

Artificial Warmth.

PASSING from the direct to the indirect comforts of a household, we will take Artificial Warmth.

The savage, as a matter of necessity, makes a fire in the middle of his hut, and lets the smoke have its own way. Sometimes, as is the case with the North American Indians, the top of the conical hut is open, and the whole edifice is a single chimney of large dimensions, something like the "chimney-corner" of past days, which only survives in such places as the New Forest.

Then there are the various Kafir tribes of Southern Africa. They have no aperture in their huts except the tiny doorway, which can only be entered on hands and knees. But they must have their fire. No argument can persuade them that they had better make their fire and cook their food outside the hut. So the wood-smoke fills the hut, coats it with a lining of soot, and gets out as it can through the sticks and withes of which the simple edifice is built.

As a contrast, we have the oil-lamp of Esquimaux-land, where there is no provision for ventilation, where the snow-houses are tightly closed and crammed with inhabitants, and where no one seems to need fresh air.

The next step in civilisation is to construct a tube for the purpose of carrying off the smoke, such as we know by the name of chimney or flue, and to place the fire within it. We English people have an ingrained love for the open fireplace, and though it really is an expensive arrangement, it is worth the cost. Granting that it carries much of the heat into the chimney instead of throwing it into the room, it has at least the advantage of acting as a ventilator, of ejecting air which has been rendered poisonous by respiration, and drawing a fresh supply from the outer atmosphere.

In some parts of the world, especially in Germany and the United States, the place of the open fire is taken by closed stoves, without any ventilation whatever, much to the discomfiture of ordinary Englishmen. Still, there are buildings, such as public halls and places of worship, in which open fireplaces are wholly impracticable, and where it is, therefore, necessary to make use of the stove.

It need hardly be said that in such cases the chief object is to procure the greatest amount of heat with the least expendi-

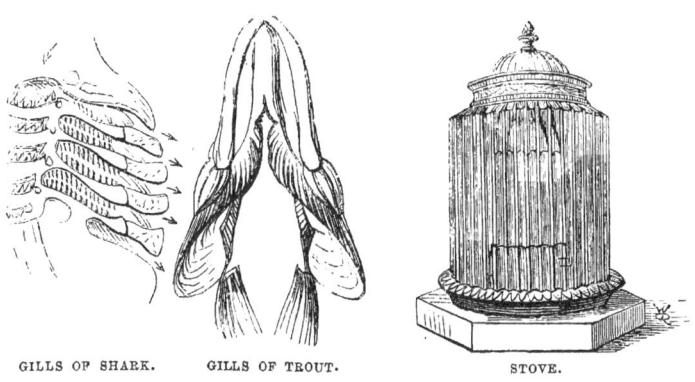

GILLS OF SHARK.   GILLS OF TROUT.   STOVE.

ture of fuel, and that object seems to be best attained by the Laminated Stove shown on the right hand of the illustration.

In this stove, the outer surface, instead of being plain, is divided into a number of perpendicular plates, which are heated

by the contained fire, and expose a very large surface of hot
metal to the air. Thus the heat, instead of being wasted by
being drawn through the flue or chimney, is thrown into the
room, and keeps up a perpetual supply of warm air.

THAT the invention of this stove is an ingenious one nobody
can deny. · But Nature has been long in advance of Art in the
way of exposing as large a surface as possible with the least
expenditure of space.

Very familiar examples of this structure may be found in the
many creatures which inhabit the waters and breathe by means
of gills, which extract the oxygen of the water.

Take, for example, a Lobster or a Crab, open it, and look at
the white, pointed, uneatable objects which are popularly called
"ladies' fingers." These are the gills, or breathing apparatus,
and their structure is really wonderful. They are composed
of innumerable laminæ, or very thin plates, covered with an
exceedingly fine membrane, and placed closely side by side,
but with sufficient distance between them to allow the water
to percolate the whole structure.

With the aid of an ordinary pocket lens the observer may
make out a most wonderful system of blood-vessels, which permeate every one of the myriad laminæ, and which extract the
life-giving oxygen from the water as it passes between them.

Then, to pass to animals of a higher order, take the gills
of fishes. Any fish will do, provided that it be fresh, and, if
it can be examined immediately after death, so much the better.
Taking things reciprocally, the gills of the fish and the laminæ
of the stove, are identical in principle, namely, the exposure of
much surface with little loss of space.

If possible, the observer should inject the blood-vessels of
the gills with the conventional crimson and blue wax, showing
the currents of the arterial and venous blood. Each lamina
forms a most wondrous object, and may be gazed upon for weeks
with increasing admiration.

Every one who has watched the habits of fishes must have
noticed that in running waters they always have their heads
against the stream, and do not greatly care about shifting their
positions.

In still waters, especially such as those of the ordinary glass aquaria, the fish are perpetually on the move, whereas in such a river as the Dove of Derbyshire, and even the Darenth of Kent, large trout may be seen almost motionless, but invariably with their heads directed up the stream.

The reason is evident enough. As long as the fish lies with its head up the stream the water flows through its gills, and enables it to breathe. Were the passage of the water stopped, the fish would be drowned. Consequently, all good anglers, when they hook a fish which is worth taking, keep its head down the stream, prevent the water from washing over its gills, and consequently render it so weak by deprivation of oxygen, that it becomes an easy prey, and is rendered subservient to a line of a single hair. Let the fish breathe, and a single struggle would smash a line of treble the strength. But keep it from breathing by directing its head down the stream, and it rapidly loses all strength, and can be directed into the landing-net, or brought within the scope of the gaff, without a chance of escape.

I NEED hardly remark that on the right-hand side of the illustration is shown a Laminated Stove, and that on the left are drawings of the gills of the Shark tribe and the common Trout. If the reader would really like to look into the subject for himself, I should suggest the purchase of a cod's head and shoulders and a lobster. The breathing apparatus can be removed from each for examination, and the remainder will serve as a first course for dinner.

RING AND STAPLE.

HUMBLE, and apparently insignificant, as the principle of the Ring and Staple may be, we owe no small amount of our domestic comfort to it. It meets us in all kinds of ways, in the hinges of our boxes, in the padlocks of our doors, in the inn-side fastenings for our horses, in the seaside fastenings for ships' cables, and in a thousand other ways too many to enumerate.

ON the right-hand side of the next illustration is shown the Ring and Staple as used for the purpose of mooring ships and boats, it being absolutely necessary that the machinery, simple as it is, must be capable of working in any direction, and with some latitude as to the extent.

On the left hand are shown two of the wonderful bones which are found in the head of the Fishing-frog or Angler-fish

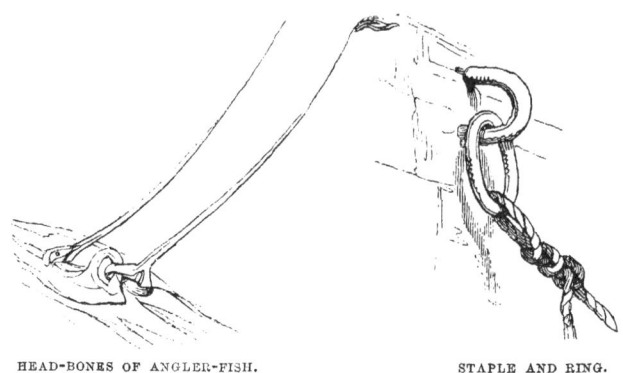

HEAD-BONES OF ANGLER-FISH.   STAPLE AND RING.

(*Lophius*), and which serve as decoys, by means of which the smaller fish are entrapped into the vast jaws of the Angler-fish.

It is clearly necessary that these singular appendages should be capable of movement in every direction, and this object is attained by the structure which is here shown, and which is almost equal to the ball-and-socket joint for its freedom of movement. It will even allow of partial rotation, so as to cause the little strip of skin at its end to assume the aspect of a living worm, and entice the smaller fish into the jaws of the dread trap that lies open before them.

A figure of this fish may be seen on page 92.

THE FAN.

EXCEPT in permanently cold countries, a Fan of some kind seems to be an absolute necessity. Sometimes, as in the greater part of Europe, it is used only by the softer sex. The harder sex would often be only too glad to use it if they dared, and the same observation is equally true with regard to the parasol.

But, in such lands as Japan and China, the Fan is an absolute necessity of existence. Men, women, and children alike carry their Fan, and almost perpetually use it. I remember, when the troupe of Japanese acrobats were in England, that one of them exhibited the national use of the Fan in an excessively ludicrous manner.

One of his comrades ascended to the roof of a lofty building, hung by his legs to one of the rafters, and held in his hands a bamboo pole which was twenty feet long. Another Japanese also ascended, climbed over his comrade, and settled on the bamboo pole, to which he clung only by the clasp of his bare feet. Suddenly he slipped down the pole, stopped himself when within a few inches of the end, squatted there with perfect unconcern, though at least forty feet from the ground, took his fan from the back of his neck, and fanned himself while gravely surveying the startled audience.

Perhaps some of my readers may remember Chang, the Chinese giant, who, by the way, in private life was a polished gentleman. He was never without his fan, always keeping it

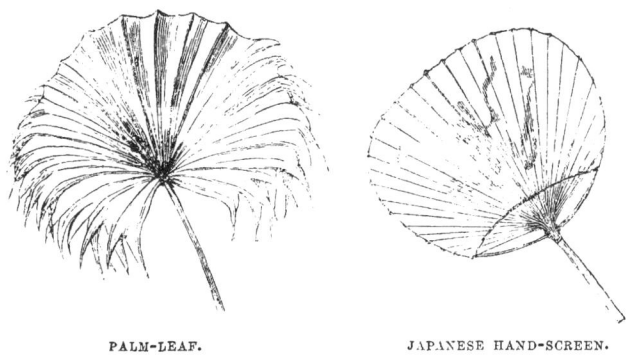

PALM-LEAF.　　　　JAPANESE HAND-SCREEN.

fluttering gently with an ease only to be acquired by a life-long practice, and I really think that if he had been deprived of it he would have been seriously ill. How he slept without it is a wonder, for in his own house the fan was incessantly in motion, and was worked with apparent unconsciousnesson his part.

I have often wished that in our country the ladies would manage their fans in the same quiet way when they are in a church or a concert-room, for the perpetual rattle of the joints is enough to distract any preacher or conductor, and very often does so.

As to the shape of the Fan, it varies greatly according to the country, but it may almost invariably be traced to some familiar object.

There is, for example, the common Japanese Fan or Screen, which is avowedly made on the model of the Palm-leaf, the ribs of the leaf being represented by split portions of a bamboo stem. The right-hand figure in the preceding illustration is taken from one of the common sixpenny Japanese fans that may be seen in many shop-windows.

There are exactly sixty ribs in the fan, all produced by splitting the bamboo into strips, kept in their place by a slight rod of the same material, and covered with two pieces of thin printed paper. Seeing that the original cost cannot be more than a penny, it is wonderful how such articles can be produced, and give a living to the makers.

The reader will observe that the shape of the Japanese Fan is almost exactly that of the Palm-leaf, with the exception of the jagged edges, and a better pattern could not be found. Then there are many Indian Fans framed on the same model, but which revolve on their handles, and are swung slowly round and round by the servants before the guests, and thus become miniature punkahs.

Here, again, we may find a parallel in Nature. The common hive bee ventilates its dwelling by using its wings in lieu of fans. When the hive is really in want of fresh air, the bees set to work, and wave their wings backwards and forwards for a considerable time, so that they necessarily expel the foul air from the interior of the hive, and create a partial vacuum, which can only be filled by fresh air from without.

Fans of very similar shape are in use among the South Sea Islanders and the inhabitants of the Essequibo district. They are often used as bellows when a fire has to be raised, but their primary object is to be employed as fans.

NEXT we come to those fans which are made of flattened sticks, which move on a pivot. This is, indeed, the ordinary form of the fan at the present day, the sticks being sometimes wide enough to constitute the entire fan, but mostly being connected with a sort of lining made with silk, paper, or feathers. Such fans as these can be moved on their pivots, so as to occupy a comparatively small space; and the same can be said of the modern fender-guards, which can be folded up when the room is unoccupied, and which form an effectual protection against the danger of ladies' dresses coming in contact with the fire.

Examples of such a screen, and two fans, are given on the right hand of the accompanying illustration.

On the left hand is shown one of the natural objects from which the fans, &c., might well have derived their origin. It is one of the antennæ—or horns, as they are popularly called

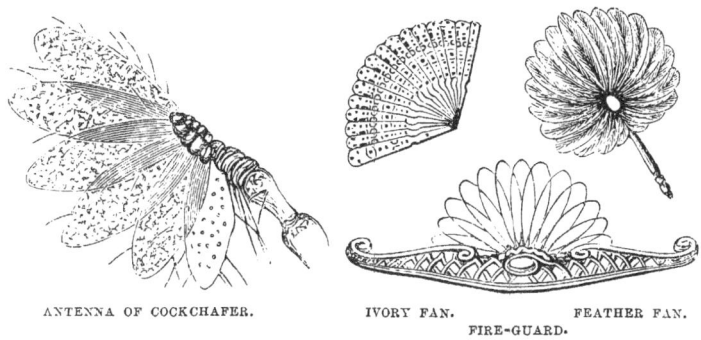

ANTENNA OF COCKCHAFER.   IVORY FAN.   FEATHER FAN.
FIRE-GUARD.

—of the common Cockchafer. The end of this antenna is composed of a number of flat plates, which work on a pivot exactly like the sticks of a fan, and, like those sticks, can be folded into a wonderfully small compass, or opened out into a fan-like shape.

BURIAL.

LAST scene of all.

I do not think that it matters very much to one who has " shuffled off this mortal coil " what becomes of the coil in which he had been imprisoned. Whether the abandoned body be buried in the earth, or sunk in the sea, or devoured by wild beasts, or consumed by fire, signifies nothing to him, though it may signify much to his surviving friends.

As a rule, the animals, of whatever kind they may be, contrive to dispose of their mortal remains in some mysterious manner, so that not a vestige of them is to be found. Take, for example, the domestic cat, and see how few bodies are found of cats which have died natural deaths.

For instance, there was my own cat "Pret," who lost his life from the bites of rats. He was blind, and so lamed that he could scarcely crawl. Yet, on the day of his death, he three times escaped from his comfortable bed in front of the fire, dragged himself through a hedge, down a steep bank, across a road, up another bank, through a crevice in a park fence, and curled himself up to die under a blackberry-bush.

Perhaps it was mistaken kindness on my part, and I should have acted better if I had left him to die in peace. But, though I carried him back three times, and though he was

BURYING-ANTS.   SAVAGE FUNERAL.

quite unable to see, he contrived to slip out of the house, and to find the same spot for his last resting-place on this earth.

I have heard that some cats have been known to bury their young, and Dr. J. Brown tells a most touching story of a dog that committed her dead puppy to the river.

But as to Insects, until a few years ago, no one ever dreamed that the principle of burial could be found among them. What millions of insects die in every year, and how seldom is a dead insect found! Flies, gnats, and the smaller insects might escape observation, but the large moths, butterflies, beetles, dragon-flies, &c., are scarcely ever found dead.

In my own neighbourhood, for example, the Stag-beetle, nearly the largest and most conspicuous of British insects, swarms to an almost unpleasant degree, especially in the summer evenings.

Yet I have never found a dead Stag-beetle that had not been killed by violence. What becomes of the bodies of the countless millions of creatures that annually pass into their other world is a problem which at present no one seems to be able to solve.

STILL, there are instances where even insects are known to bury their dead, and I scarcely need say that they are to be found among the Ants.

The story is a very curious one, and is narrated at length in the *Journal of the Linnæan Society*, vol. v. p. 217.

It happened that a lady found that her little boy was being stung by ants, and she at once killed them and threw their dead bodies away. After some time a number of ants came out of their nest, formed a procession as regularly organized as that of any undertaker's funeral, dug graves for each dead ant, laid the body in it, and covered it up again with earth.

They carried their organization to such an extent that they even had relays of bearers. But the strangest part of the story is that several worker ants would not assist in the funereal ceremonies. The soldiers at once set on them, killed them, and tumbled them all promiscuously into a common grave.

Such scenes were repeatedly witnessed by the lady, a Mrs. Hutton, who wrote the account while she was living in New South Wales.

USEFUL ARTS.

CHAPTER X.

WATER, AND MEANS OF PROCURING IT.

The Necessity of Water to Man.—Composition of the Human Body.—Natural and Artificial Distillation.—The Traveller's Tree.—Pitcher-plants and Monkey-pots.—Stomach of the Camel, and its Analogy to the Honey-comb.—Dew-drops.—Use of the Still at Sea.—Perspiration and its cooling Properties.—The Turkish Bath.—Perfume and Ether Spray.—Condenser of the Low-pressure Steam-engine.—The Dry and Wet Bulb Thermometer.—Ice produced in a red-hot Vessel.—Power of Water.—How Fountains are made.—Modern System of Hydrants. — Hydraulic Mining.—The Victoria and Niagara Falls.—Artesian Wells.—The Norton Tube, &c., in Abyssinia.—The Water-ram and Spout-hole.

IT has often been remarked that man can live a comparatively long time without solid food, providing that he can
only obtain water, of which the chief bulk of the human body is
made. Dying by thirst is a horribly painful death, but,
according to Mr. Mills, the ill-fated Australian traveller,
"starvation on nardoo (an innutritious plant) is by no means
unpleasant, but from the weakness one feels, and the utter
inability to move one's self."

Those who have been shipwrecked, and unable to obtain
fresh water, have always found that the tortures of thirst were
infinitely harder to endure than those of hunger ; and the
reader will probably remember that those who perished in the
Black Hole of Calcutta owed their deaths chiefly to thirst,
their bodies being exhausted of moisture by the heat of the
room, and no fresh supply attainable.

Civilisation especially shows itself in the way in which
water is brought within the reach of every one, even in the
most crowded of cities. The reader may probably call to
mind the wonderful aqueducts of ancient Rome, the gigantic

remains of which still exist. Then, as to our own country, we are all practically acquainted with some water company, by which the water, more or less purified, is brought into our houses, and can be obtained by the mere turning of a tap.

Yet all this ingenuity is but a following of natural proto-types, as will presently be seen; and even the familiar Water-tank, as shown at the right hand of the illustration, has been anticipated by Nature.

On the left hand of the illustration there are three examples of natural water-tanks, two belonging to the vegetable, and one to the animal kingdom.

That on the extreme left, with a number of radiations, repre-sents a portion of a Madagascar palm, popularly called the

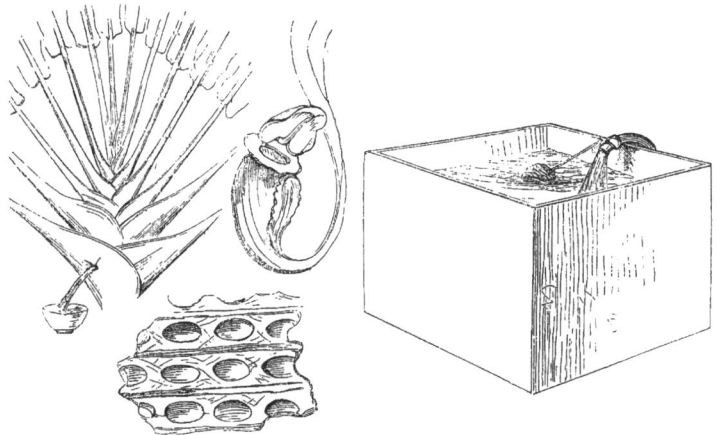

TRAVELLER'S TREE.   STOMACH OF CAMEL.   PITCHER-PLANT.   CISTERN.

Traveller's Tree. Having very large leaves, arranged in the manner there shown, the Traveller's Tree condenses the nightly dews, and allows them to trickle down into the hollows of the leaf-stems.

There the water remains, out of the reach of sunbeams or wind, and if a traveller happens to be thirsty, all he has to do is to pierce the base of one of these gigantic leaves, and out rushes a stream of the purest water, as is shown in the illus-tration.

NEXT to the Traveller's Tree is shown one of those extraordinary vegetables called Pitcher-plants, from the strange conformation of the leaves. They inhabit Borneo, Siam, and other hot countries. In these remarkable plants some of the leaves are developed into suitable pitchers, with hinged lids, exactly like our hot-water jugs. They serve, however, a different office, and contain cold water which the plant has distilled from the dew.

As the monkeys are in the habit of resorting to these plants when thirsty, they are sometimes called Monkey-pots. There is an admirable account of the Pitcher-plants and their development in the *Transactions of the Linnæan Society*, vol. xxii. part iv. The scientific name of those plants is Nepenthes.

BELOW the vegetable comes a rather celebrated animal cistern, namely, a portion of one of the stomachs of a Camel.

It exactly corresponds with that part of an ox which butchers call "honey-comb tripe," and consists of a multitude of cells, which can be closed or opened at will. When the camel takes in its provision of water, it can treat this portion of the stomach much as the hive bee treats the honey-bag, and fill its cells with water.

By degrees, when it finds the necessity for moisture, it can squeeze the water out of these receptacles into the digestive portion of the interior, and so can sustain life for a wonderfully long time under conditions which would kill any other animal. I may remark, by the way, that the amount which a camel can drink, and the length of time through which it can endure its desert life, have been much exaggerated. There is another point to be considered, namely, the curious resemblance between these cells and the honey-comb of the hive bee. Every one knows that honey, no matter how tightly closed, will crystallize and lose its best qualities if kept in jars, whereas if it be allowed to remain in the waxen comb, where it is divided into very small portions, it will remain good for years.

It is just the same with the cells of the camel's stomach, they being able to preserve water in a pure state by distributing it among a number of small cells, which can be opened or closed at will.

Then we come to the various means of obtaining water.

Reference has already been made to the Filter, by which foul water can be made pure for human consumption, and we will therefore pass to another mode of obtaining pure water, namely, the Still.

In former days, if there were a failure of the supply of fresh water on board ship, the whole of the occupants must necessarily perish. Now, however, no such danger exists, as every well-furnished ship carries at least one Still, by means of which the sea-water can be made to abandon its salt, and to give out nothing but pure water fit for drinking.

Even in cases where no regular Still has been on board, an extemporised Still has been made from a kettle, a gun barrel, or piece of lead piping, or anything of a similar nature.

The principle of the Still is simple enough, and is shown by the diagram, rather than drawing, on the right hand of the illustration. There is a vessel in which liquid is boiled. From the upper part of it rises a tube through which the

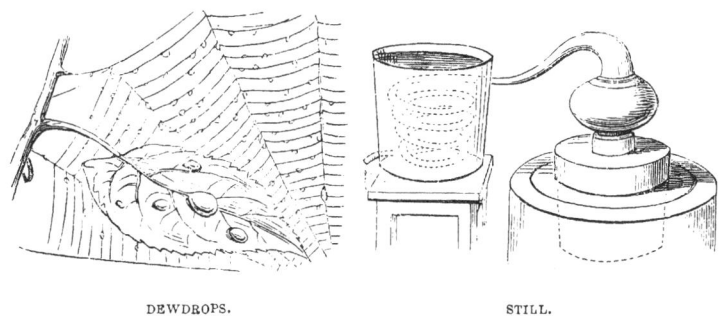

DEWDROPS.   STILL.

steam must pass as it is generated. The tube in question is generally of considerable length, and is coiled inside a vessel filled with cold water, rendered colder by ice, if possible.

As the steam passes through the cold tube condensation takes place, and it becomes liquid again, but deprived of its heavier particles, so that if sea-water be placed in the still, the salt is left in the vessel, and nothing but pure water passes through the tube. In dissecting-rooms a small still is almost invariably kept. Many preparations are of such a nature that the spirit in which they are placed becomes discoloured, and has to be repeatedly changed. Now, even methylated spirit is an expensive article, and therefore, instead of being thrown away, the discoloured spirit is placed in the still, and reproduced in a clean and transparent state.

Nature affords innumerable examples of distillation, the chief of which are the Dewdrops which have already been mentioned. During the daytime the air is full of moisture drawn by the sunbeams from ocean. We cannot see it, but it is there, and when the chill of night cools the various trees, herbage, and other such objects, the aërial moisture is condensed upon them, which is then known by the name of Dew.

On the left hand of the illustration are shown the tiny Dewdrops as hanging on the slight threads of a spider's web, and collected in larger drops upon a leaf.

THERE are many other familiar examples of the principle of condensation, the commonest of which is the so-called steam as it pours from the spout of a kettle. In point of fact, it is not steam at all, but only water condensed into very small drops. At the orifice of the kettle it is quite invisible, but when it passes into the air, and is condensed, the tiny globules become visible. The same fact may be noticed in the Napier's Coffee Machine, which has already been mentioned. When the water is boiling in the glass globe no steam is visible, though the upper portion of the globe is entirely filled by it. But, no sooner is the cork removed, and the steam allowed to escape, than it at once becomes visible as a white cloud, being, indeed, a miniature copy of the rain-clouds that float above us.

THEN there is that mostly invisible passage of liquid through the multitudinous pores of the body, which is generally known as perspiration. It is invisible in warm weather, but on a cold day is as visible as a rain cloud.

The Turkish Bath affords a good example of this fact. Sometimes the hottest room attains a temperature of 250° or more, water boiling at 212°. When a bather goes into that room, he appears to have a perfectly dry skin, the moisture being in the form of invisible steam, and swept off as soon as it is generated.

But, if he passes at once into the cold room, he is so enveloped in vapour that for a few moments he is wrapped in it as in a cloud, and can scarcely be seen, the vapour having been condensed by the cold air.

A very familiar instance of this sudden condensation may be seen in the streets of London on any winter day. There may be a couple of omnibus horses, nearly at the end of their day's work, and quite tired out. Suddenly they are pulled up by the

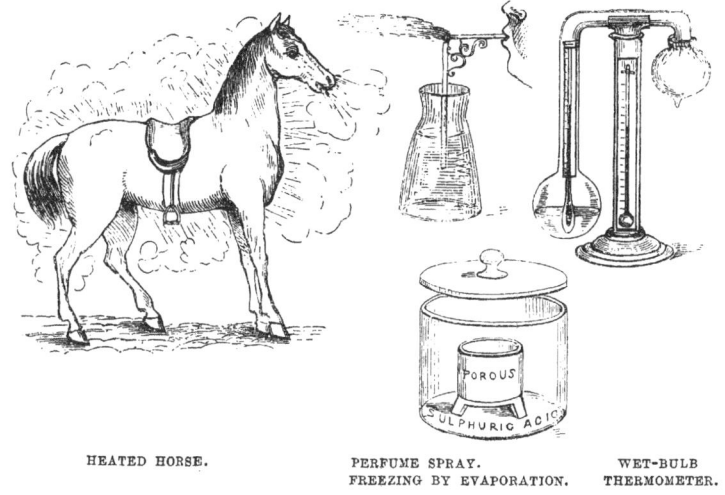

HEATED HORSE.

PERFUME SPRAY.
FREEZING BY EVAPORATION.

WET-BULB
THERMOMETER.

driver, and as suddenly disappear for a moment or two, being concealed in a cloud of moisture proceeding from their bodies. Of course in a hot day there is more of the moisture, but the warmth of the atmosphere prevents it from condensation, and so it is not visible.

One valuable property of the system of evaporation and condensation is its cooling power. Thus it is that a person who is ill with fever tosses about with a burning skin until the pores of the body act, and allow the normal moisture to pass through them. Then the body cools by evaporation, and the patient begins to amend.

So it is that the bather can endure in the Turkish bath a heat so great that a glass of water, if held in the hand, would speedily boil, and a piece of meat be cooked in about the same period. But, if the air were not dry enough to carry off the perspiration, the bather would be scalded to death.

A most valuable adaptation of the principle is shown in the

little glass machine for dispersing perfumes in the form of spray. In cases of headache it is almost invaluable, the spray cooling the heated forehead like magic, and at the same time filling the room with the grateful perfume.

It has even a greater claim to human gratitude, as I can personally testify. I have the strongest objection to a surgeon's knife, especially when I know, from sad experience, that he is going to make very free use of it. But, on the last occasion, I cared nothing for it, owing to the happy invention called Ether Spray.

The effects were remarkable. First, a delicious cooling of a spot raging with internal fires. Then it was rather colder than I liked. Then it was much colder than I liked. Then it became almost too cold to bear, reminding me of my child-hood's feet on the outside of the Birmingham coach in the depth of winter.

Suddenly all sensation ceased, and the skin became white as parchment. Out came the surgeon's bistoury, and I looked at him with as calm composure as if he had been whittling a deal plank. There was absolutely no feeling whatever, the local nerves having been temporarily frozen, so great is the power of evaporation. If it ever be my lot again to endure cold steel, I shall have the ether spray.

On the extreme right of the illustration is seen the " Wet-bulb " Thermometer, which carries out the same principle, the thermometer being double, and one bulb being covered with a wet envelope, while the other is dry.

Below is one of the many inventions for making artificial ice, all of them depending on the cooling power of evaporation. Perhaps some of my readers may have seen molten iron poured over the human hand without doing the least harm, or mercury frozen in a red, or rather a white, hot vessel. Both these phenomena are due to the cooling power of evaporation, which is made to act with extreme rapidity, and so absorbs the heat until even mercury is rendered solid, and can be cast in a mould like a leaden bullet.

In the accompanying illustration we have an example of the Condensating principle as applied to the steam-engine, and popularly known as the "Low-pressure Engine." In this case force is reconverted, so to speak, and, if a cubic inch of water has been converted by heat into a cubic foot of steam, creating a pressure in one direction, it can be reconverted by cold, and so produce a pressure in another direction.

It is owing to this fact that some parts of the world are always hot and always wet, Guiana being a striking example.

The wind blows over the ocean, absorbing moisture as a sponge does water. As it passes from the sea over the land, it is met by secondary mountain ranges, too low to arrest its pro-

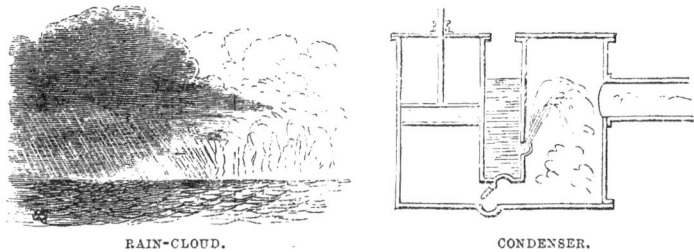

RAIN-CLOUD.   CONDENSER.

gress altogether, and high enough to have their summits clothed in eternal snows. As soon, therefore, as the warm, water-laden winds pass over these mountains, the moisture is condensed by their frozen tips, and down rushes the rain in torrents.

Even in our own temperate land we can often trace the cause of a heavy rain to the presence of a lofty hill, or even an exceptionally tall spire. The moist climate of Oxford has been attributed by scientific men quite as much to its spires and towers as to its low-lying situation.

Now we come to the various modes of extracting the water which is laid up within the earth, and which only slowly ascends to the surface when drawn up by the heat of the sun.

Water is everywhere, but the depths at which it is found are vastly different. For example, at one house in which I lived it was not possible to dig for three feet without coming to water. In another, no water was found within some two hundred feet, and, as I several times relieved the old gardener of the task of drawing the water for the day's consumption, I have reason to remember the depth.

The pail, rope, and winch which were in use at that time —and may be still, to the sorrow of the gardener—are but a sort of semi-savage way of procuring water from the depths of the earth. It is a well-known fact that under certain conditions water always finds its own level, *minus* the friction of the

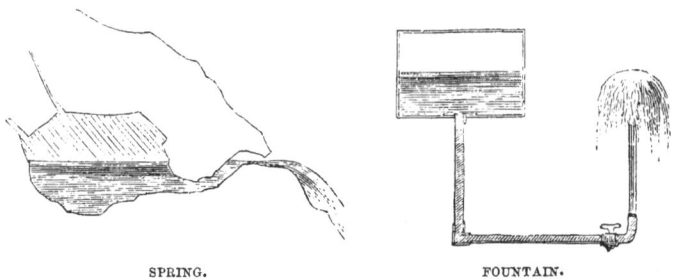

SPRING.   FOUNTAIN.

channel through which it passes. On this principle all fountains are made. Those, for example, at the Crystal Palace, which fling their waters to such a height, are fed from tanks on the summit of the two great water towers. And, were it not for the friction of the water in the tubes, and that of the air, the fountains would rise as high as the tanks from which they are fed.

Such is the case with springs, especially with those of an intermittent character, in which latter instance the rushing of the water is exactly coincident with the filling of the hidden tank which supplies it.

The modern Hydrant system, which bids fair to supersede the cumbrous machinery of fire-engines, even when worked by steam, is based on the same principle. The water-tanks are placed at such a height that, when a hose is attached, and the tap turned, the water can be thrown over the roof of the highest building. Such hydrants have been attached to Canterbury Cathedral since the fire which so nearly consumed that magnificent and venerable building.

A VERY remarkable use has been made of this power of water in mining operations. Most of my readers know that in gold mines the metal is chiefly found scattered among quartz, one of the hardest of the minerals. The usual plan has been to dig out the quartz, pound it to powder with specially devised machines called "stamps," to pass the powder through mercury, which amalgamated with the gold, and gave it up again on being heated to a certain temperature.

Now a different mode of mining is brought into operation, the pickaxe, spade, and stamps, with all their expensive machinery, being abandoned, and water made to do the duty of all three, some ingenious individual having noticed the effect which water has on the hardest rock.

Such, for example, is the case with those wonderful Victoria Falls of Africa, where the rushing water has cut its sinuous channel through so many hundreds of yards of rock. Such, also, is the case with the more celebrated, but not so wonderful, Falls of Niagara, which have been gradually working their way backwards, having worn away the rocks over which they fall, and which are shown to be many miles away from the spot where the river first discharged itself over the cliff.

In fact, it is well known that the Falls are receding at a definite rate annually, and that the rate has been calculated

HYDRAULIC MINING.

WATER-FALL.

with scientific accuracy. The cliffs of our own coasts—say of Margate or Ramsgate—crumble away with equally calculable speed.

In the hydraulic mining system large tanks are erected, at least two hundred feet above the level of the mine. From these tanks proceed pipes, terminated by hose, just like those of our ordinary fire-engines. The miners, instead of using pickaxe or crowbar, simply direct the streams of water against the solid rock. Their effect is tremendous. They tear it to powder, and carry it down the wooden troughs called

" flumes," in which the mercury is so arranged that not a single atom of quartz rock can pass without having its gold extracted.

The following graphic account of Hydraulic Mining at Nevada is taken from Mr. J. K. Lord's "Naturalist in British Columbia :"—

"Near Nevada are the famed Hydraulic washings. The gold is disseminated through terraces of shingle conglomerates, often three hundred feet in thickness. These terraces are actually washed entirely off the face of the country by propelling jets of water against them, forced by pressure through a nozzle.

"To accomplish this, the water is brought in canals, tunnels, and wooden aqueducts, often forty miles away from the 'draft.' This supply of water the miners rent.

"As we near the washing spot, in every direction immense hose, made of galvanized iron, and canvas tubes six feet round, coil in all directions over the ground like gigantic serpents, converging towards a gap, where they disappear.

"On reaching this gap, I look down into a basin or dry lake, three hundred feet below me. The hose hangs down this cliff of shingle, and following its course by a zigzag path, I reach a plateau of rock, from which the shingle has already been washed.

"A man stands at the end of each hose, that has for its head a brass nozzle. With the force of cannon-shot, water issues in a large jet from this tube, and propelled against the shingle, guided by the men, washes it away as easily as we could sweep a molehill from off the grass.

"The stream of water, bearing with it the materials washed from out the cliff, runs through wooden troughs called 'flumes,' floored with granite. These 'flumes' extend six miles. Men are stationed at regular distances to fork out the heavy stones.

"Throughout its entire length, transverse strips of wood dam back a tiny pond of mercury. These are called *ruffles*— gold-traps, in other words, that seize on the fine dust-gold distributed through the shingle. The flumes are cleaned about once a month, and the gold extracted from the mercury.

"I try with a powerful lens to detect gold amidst the material they are washing, but not a trace is discoverable, and yet it pays an immense profit to the gold-washers."

THERE are two more modes of extracting water, which will be but cursorily mentioned.

The reader will remember that water finds its own level, and that the terrific power of hydraulic mining is owing to the fact that the water expends its force against the solid rock instead of ascending into the air.

It is now found that, even without artificial assistance, water has a habit of finding its own level, and that, if it be allowed its own course, it will contrive to find its way nearly to the highest point whence it derived its origin. On this principle

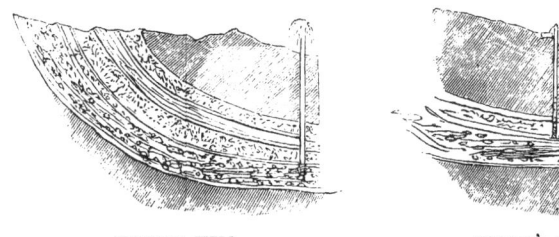

ARTESIAN WELL.   NORTON'S TUBE.

are based the Artesian Wells, which, when they "strike water," spurt it up in a torrent, as is the case with the now celebrated Norton Tubes, which are screwed down into the earth like hollow gimlets, and which always contrive to extract the water hidden beneath the surface of the earth.

The success of our army in Abyssinia was greatly owing to these Norton Tubes, which, being of small diameter and of peculiar make, could be screwed into the ground when the troops made a halt, unscrewed when they left the spot, and used again for the next halt.

Similarly, the French used the Artesian-well system with wonderful success in Northern Africa. Water is the chief necessity of life in that part of the world, and a nation who could cause pure cold water to spring out of the hot and thirsty sands was naturally looked upon as something more than human.

Yet the principle was exactly the same in both cases. Water

is always latent somewhere beneath the surface of the earth, and, if a tube can be driven deep enough, the water will come up it.

The accompanying illustration shows the Artesian Well and Norton's Tube, and their similitude in principle, the tube penetrating through various layers of soil, until it reaches the water which it seeks.

THEN there is another way by which water can be made to force itself to a considerable height. Not being much of a mathematician, I do not recollect the exact proportional height to which a stream of water may raise itself, but if any one can secure a fall of some eight or ten feet, he can furnish his house with water by means of the "Ram," a chart of which is shown in the illustration.

The principle of the Ram is, that the water is allowed to flow down a tube, when it meets with a valve. This valve is

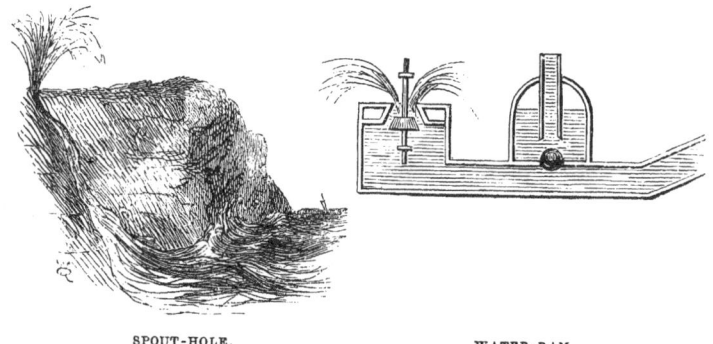

SPOUT-HOLE.   WATER-RAM.

suddenly closed by the pressure, and the water is forced onwards by the shock. Much water escapes at each blow of the valve, but that does not signify.

The force of water thus suddenly stopped is hardly appreciated. Even in ordinary houses the sudden turning of a water-tap has been known to burst the pipe and deluge the house with water.

IN Nature a similar effect is produced, called popularly the "Spout-hole."

It is a hole or tunnel on the seashore, passing upwards from the level of the sea to the summit of the cliff.

When the waves are urged against the tunnel by the wind, the water is dashed into it. Being partially checked by the friction, which acts exactly like the water that is checked by the Ram, the wave hurls itself up the channel, and flies out in showers of spray, high above the level of the original wave which caused it.

In the illustration are shown the Water-ram with its globular valve, and the safety or escape valve of the waste water. On the left is shown one of the natural Spout-holes, with the water dashing through its tunnel into a mass of spray.

USEFUL ARTS.

CHAPTER XI.

AËROSTATICS.—WEIGHT OF AIR.—EXPANSION BY HEAT.

Ascent and Descent.—The Balloon and the Parachute.—Description of the
Balloon.—The Montgolfier Balloon.—Causes of its Abandonment.—The Gas
Balloon.—Hydrogen Gas and its Manufacture.—The Gossamer Spider.—
Reasons of its Ascent and Descent.—Many Species of Gossamers.—Descrip-
tion of the Parachute.—Its Mode of Action.—A Balloon converted into a
Parachute.—Toy Parachutes.—Natural Parachutes.—The Dandelion Seed
and its Structure.—The Flying Squirrel.—The Flying Monkey.—Flying
Mice and Flying Opossums.—The Flying Dragon and its Pseudo-wings.—
The Flying Frog.—Weight of Air.—Pressure per Square Inch.—The Air
Ocean and its Storms.—Principle of Air-currents.—The Sun, the Earth, and
the Air.—Ventilation of Mines.—Choke-damp and Fire-damp.—The Air-
shafts.—Chimneys of Factories.—The Steam-blast.—The Barometer, and
Mode of its Construction.—Water and Mercury.—Sucking Eggs and Sugar-
cane.—Expansion of Water and Metals by Heat.—The Thermometer.—
Wheel-making.

AËROSTATICS.

WE will begin this chapter with the only two modes at
present known by which man can ascend from the earth
or descend to it with safety, namely, the Balloon and the
Parachute, the latter being generally attached to the former,
and detachable at pleasure.

The Balloon is, in fact, as its name imports, a large, hollow,
air-tight ball, filled with some substance lighter than ordinary
air. The original Balloons by Montgolfier were filled with
heated air exactly like our toy fire-balloons. Just as the
supply of hot air is kept up in them by a sponge dipped in
lighted spirits of wine, so in Montgolfier's balloons the same
object was attained by straw which was kept continually
burning in a grate.

There were, however, two disadvantages about this plan. The
first was the great danger of fire, which on one occasion did

ignite a balloon when at a great height. The second was the perpetual labour required in keeping the fire alight. Straw burns very rapidly, and so the aëronaut had no opportunity of making those meteorologic observations in which consist almost the entire value of the balloon.

Then it was thought that hydrogen gas, being about fourteen times lighter than ordinary air, would answer the purpose, and such has proved to be the case. Formerly the gas was

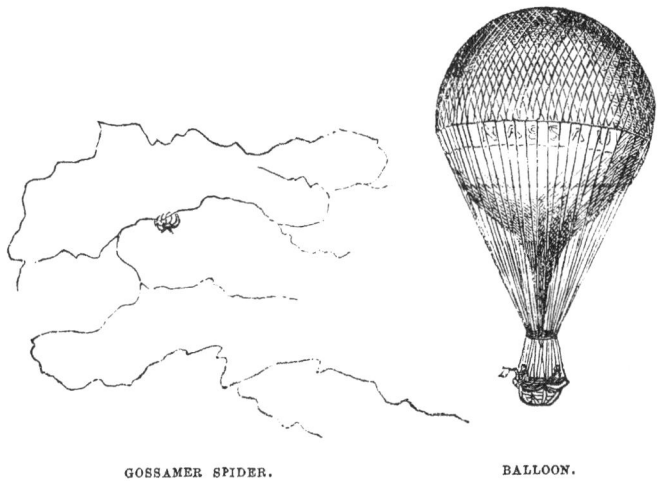

GOSSAMER SPIDER.   BALLOON.

made at great expense from sulphuric acid and zinc, but it is now found that the common coal-gas is quite as efficient, very much cheaper, and fills the balloon much more rapidly.

THE same principle, though not the same form, is found in Nature.

There are certain tiny spiders called Gossamers, which have a curious power of floating in the air. They have been seen on the tops of lofty spires, and they are sometimes so numerous that the air is full of their floating webs, and the ground is white with those that have descended.

Their mode of ascent is this. They climb to the top of some elevated object, if it be only a grass-blade. They then pour out a long, slender, thread-like web, which shortly begins to tend upwards. As soon as the Spider feels the pull, it crawls upon the web, and sails away into the air. The duration and height of the ascent depend much on the wind and character of the atmosphere.

The web ascends because it is for the time lighter than the atmosphere. But, as it gradually becomes laden with the moisture that more or less fills the air, it becomes heavier than the atmosphere, and gently sinks to the ground.

What may be the object of these aërial voyages no one knows. They may be for the purpose of capturing minute insects, or they may be for mere amusement. But in either case they are highly instructive, as showing the principle on which the balloon was framed.

The little Gossamer Spider is shown on the left hand of the illustration, clinging to its floating web. I believe that the Gossamer is not a single species of Spider, but that there are many species which deserve the name, being able to float in the air when they are small, but losing that capacity as they increase in size and weight.

Now we come to another branch of the same subject, namely, the safe descent from a great height by means of the Parachute.

On the right hand of the illustration is the ordinary Parachute as it appears when open and closed, in either case having somewhat the appearance of a large umbrella. It is hung to the balloon in its closed state, and when detached it falls rapidly for a yard or two with startling rapidity. The pressure of the air thus forces the ribs open, and gives sufficient assistance to the atmosphere to insure a gentle fall.

On one memorable occasion, when the late Albert Smith was in the car of a balloon upwards of a mile from the ground, the balloon burst. Fortunately it burst so completely, that the silk was driven into the closely meshed netting, and formed an extemporised parachute, which took the voyagers to the earth with safety, except some rather severe bruises.

Children often amuse themselves with miniature parachutes. They take a square piece of thin paper, tie threads to the four corners, and then bring the ends together, a cork taking the place of the car. They then launch it from a high window, and should there be a favourable breeze, it is wonderful how far it will be carried before it comes to the ground.

Once, when a boy of eleven, and consequently thoughtless,

I set a chimney on fire by one of these Parachutes. I wished to see whether it would go up the chimney, and come out at the top. Unfortunately it was caught by a flame as it was launched, flew up in full blaze, and, as the chimney needed sweeping, the result was inevitable.

In the centre of the illustrations, and at the top, are two examples of a well-known natural Parachute called the

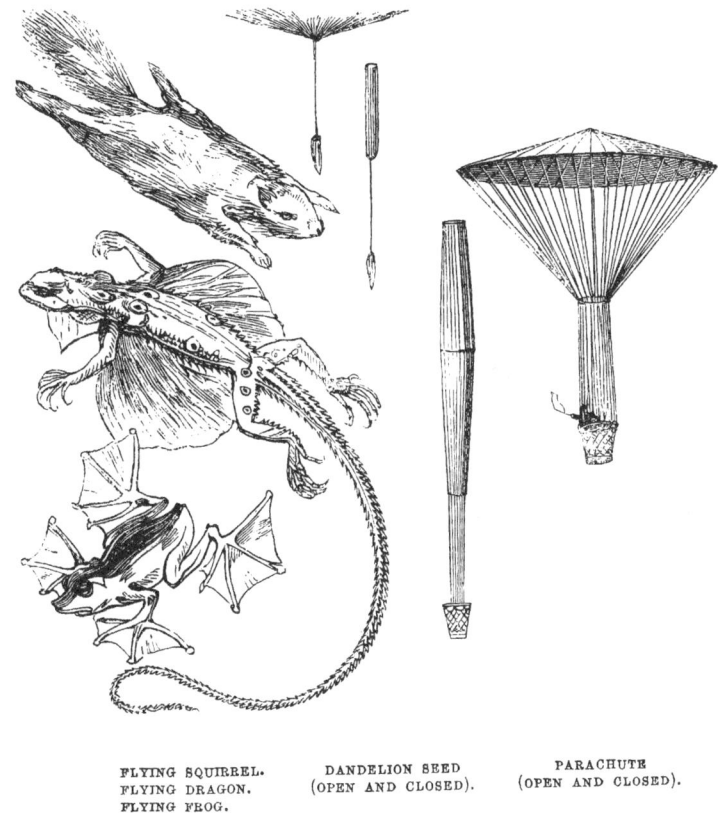

FLYING SQUIRREL.   DANDELION SEED   PARACHUTE
FLYING DRAGON.   (OPEN AND CLOSED).   (OPEN AND CLOSED).
FLYING FROG.

Dandelion seed. The resemblance to the real Parachute is wonderful, the actual seed occupying the place of the car, and fulfilling the same office, i.e. keeping the seed upright until it reaches the ground.

When the tuft is closed, as is the case before the pretty ball of seeds bursts from the green envelope in which they had been confined during the process of development, its form bears the same startling resemblance to the Parachute.

PASSING from the vegetable world, there will be seen three examples of Natural Parachutes. Several others will be mentioned, but we have no space for description or figure. It will be seen, however, that the one principle which characterizes them all is the exposure to the air of a flattened and large surface, in proportion to the size of the object.

Before beginning the description, however, I must mention that nearly all animal parachutes can to a certain extent guide their course, while neither the balloon, the gossamer, the parachute, nor the various winged seeds have the least power of guidance, but must follow every current of air in which they may happen to float.

THE upper figure represents a Flying Squirrel.

There are many species of Flying Squirrel, but they all agree in one point. The skin of their sides is modified into a very thin fold, which extends as far as the feet.

It is very elastic, so that when it is not in use it falls into folds or wrinkles, and is hardly perceptible. But should the Squirrel wish to pass from one tree to another, without coming to the ground, it spreads its legs as widely as possible, so as to stretch the membrane into a wide, flat surface. It then boldly springs into the air, and sweeps upon its mark with a sort of skimming movement. Except that it does not revolve, it passes through the air much after the fashion of an oyster-shell when thrown horizontally.

Many mammalia are constructed after a similar fashion, such as the Colugo, or Flying Monkey, the Flying Mice, and the Flying Phalangists, or "Opossums," as they are popularly called.

IN the centre is the Flying Dragon, or small lizard, which very probably gave rise to the fabled Dragons in which our ancestors so devoutly believed. Indeed, on looking back at the old illustrated works on Natural History, there can be but little doubt on the subject.

In this creature, the ribs, instead of the legs, carry the

flat and elastic membranes. When simply crawling on the branches, after the manner of tree-lizards, the ribs lie flat against the sides, and the membranes collapse, so that the shape of the body is little different from that of any crawling lizard.

But the ribs are movable at will, and, when the creature wishes to pass from one tree to another, it extends the ribs, stretches the membranes, and launches itself into the air, exactly as has been narrated of the Flying Squirrel.

THE lowest figure represents a most extraordinary animal, called the Flying Frog. Only one specimen is believed to be known, and that was discovered in Borneo by Mr. Wallace.

Here we have an analogy with the bats of the present day and the pterodactyles of the past, namely, the elongation of the toes, and the stretching of a web between them. In the two latter animals, however, only the toes of the two fore-legs are elongated, whereas, with the Flying Frog, the elongation is found in both pairs of limbs. The ends of the toes are furnished with adhesive pads, like those of the tree-frogs, to which it is probably related.

By means of the four membranes, the creature s able to sweep through the air for some distance, and, indeed, this power was the reason why it was caught. It was seen to skim from one tree to another, and was immediately secured. Had it remained sticking on the tree, it would probably have escaped observation.

WEIGHT OF AIR.

WE have already noticed that hydrogen gas is fourteen times lighter than air, and infer necessarily that the weight of the atmosphere must be very considerable if so heavy an object as a balloon, with its car, instruments, sand-bags, and passengers, can rise and float in it.

We are not conscious of its weight, because it permeates us, and the pressure is neutralised. But, in fact, we live at the bottom of a vast ocean which we call the atmosphere; and as, on an average, there is a pressure of fifteen pounds on every square inch of surface, we have to sustain an almost incredible weight.   Let, for example, any one measure the surface of his own hand, reduce it to square inches, add together fifteen pounds for every square inch, and he will then appreciate the weight of the atmospheric ocean in which we live.   On an average, every human being endures a pressure of some ninety thousand pounds.

This ocean is in perpetual movement, sometimes violently, which we call storm ; sometimes gently, which we call breeze ; and sometimes very gently, which we call calm.   There are air-spouts as well as water-spouts ; and, in fact, the water-spout is nothing but a continuance of the air-spout, as is shown by the moving sand-columns of the desert.   Whatever may be the character of the winds, as we call this movement, the air is never for a moment still ; and, indeed, were it to be still for any time, the whole human race would perish.

How winds are caused we shall see by the aid of the diagram on the left-hand side of the illustration.

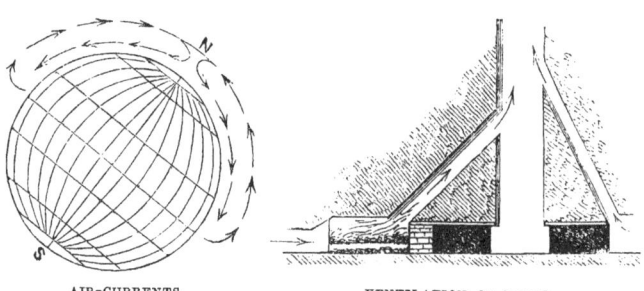

AIR-CURRENTS.   VENTILATION OF MINES.

The original cause is the sun.   His rays fall upon the earth, heating it, and so by radiation heating the air.   Now, as has been remarked, heated air will cause a heavy balloon to float through ordinary air, and to carry up a considerable amount of dead weight besides ; consequently the heated air must ascend, while cool and heavier air rushes in to take its place, and thus the currents are produced.   Were the earth set straight upright, the currents would invariably run in one direction ; but, as it is tilted on one side, the needful variety is obtained, and we find the winds blowing from all parts of the compass.

The principle, therefore, of all winds is, that heat expands, and therefore becomes lighter than air at an ordinary temperature.

WERE it not that man has taken advantage of this principle, there could not be a deep mine in England. In any deep excavation, even though it be a well, foul air, mostly composed of carbonic acid gas, always collects, and, being much heavier than atmospheric air, lies at the bottom of the pit as surely as hydrogen would rise out of it. To breathe this air is as certain and as sudden death as to take prussic acid, and no mine can be worked as long as " choke-damp " is in it.

In coal mines there is an additional source of danger, namely, the coal gas, which is nearly identical with our coal gas of the streets, and takes fire when brought into contact with flame. To rid the mines of these gases, a simple, ingenious, and effectual remedy is used. A ventilating shaft is made, which reaches from the bottom to the mouth of the pit. At the bottom, diagonal shafts are made, entering the main shaft, as shown on the right hand of the illustration. One of these is connected with a furnace, and the other, or others, open into the mine.

The heat of the furnace rarefies the air in the shaft, causing it to rush upwards with great violence, and so, by creating a partial vacuum, to force the air in the shaft to follow it. The loss of air thus caused is supplied by fresh air from above, which, by the law already described, is obliged to take the place of that which was driven out. Thus a complete circulation of air is kept up, and a well-managed mine has a fresher atmosphere than many houses in which the windows are mostly kept shut, and the only ventilation is accomplished by occasionally open doors.

The " draught " of our domestic chimneys is owing to this principle, and the reason why factory chimneys are built of such enormous height is, that the column of heated air may be increased, and consequently that the draught may be stronger, and the heat of the furnace made fiercer.

The "Steam-blast," by which the escape steam of engines is sent into the chimney, is another example of this principle, the steam taking the place of the hot air.

Further examples of the weight of the atmosphere are given

in the illustration.   That on the right represents the common
Wheel Barometer, which marks the weight of the air by a
hand moving in front of a dial.   If the hand move towards the
right, the weight of the air is increasing; if to the left, it is
decreasing.

There are certain words, such as Wet, Change, Fair, Dry,
&c., on the face of the dial, but they are only conventional, the
real test of the weather being the direction in which the hand
moves.   For example, if with a west wind the hand moves
from Dry towards Fair, rain may be expected; whereas, if it
should move from Wet to Change with an east wind, we may
reasonably think that fine weather is coming.

The whole cause of this revolution of the hand may be found
in the weight of the atmosphere.

It is found that a column of water thirty feet high, or a
column of mercury thirty inches high, is exactly equal in
weight to a column of air of the same diameter, but some forty
odd miles high, so that the two columns precisely balance each
other.

Suppose, then, the water or mercury to be placed in tubes
closed at the top and open at the bottom, the water or mercury
will exactly balance the air, and will not escape from the tubes.

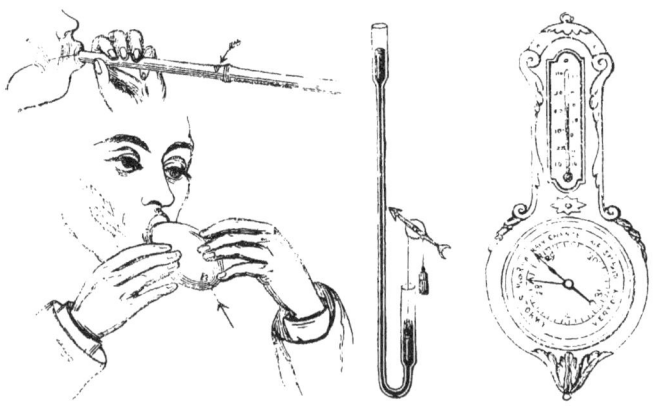

SUCKING SUGAR-CANE.   SUCKING AN EGG.   BAROMETERS.

It necessarily follows that if the air be heavier than usual, it
will force the liquid higher into the tubes, and, if it be lighter
than usual, will allow them to fall lower.   This is the principle
of the Barometer.

The mechanism of the hand and dial is shown in the diagram which occupies the centre of the illustration. For convenience, sake the mercury column is mostly employed, but several Water Barometers, some thirty feet in length, have been constructed.

On the left hand is seen a boy engaged in sucking an egg. The plan employed is simple enough. A tolerably large hole is made at one end, and a very small one at the other. The yolk having been broken up by a long needle, or similar implement, the larger hole is placed to the lips, and, suction being used, the contents pass into the mouth.

Were it not for the hole at the end opposite the mouth, it would be impossible to extract the contents, but the air rushes through the aperture, and so forces out the contents of the egg.

Above is a representation of the way in which Sugar-cane is sucked. The reader probably knows that the Sugar-cane, like the wheat-stem, has knots at certain intervals, which divide the cane into a number of separate parts.

There is quite an art in sucking the Sugar-cane. If a joint be cut off, and the lips applied to the end, not a drop of the sweet juice would be extracted. But if a notch be cut close to the joint, as shown in the illustration, the air can gain access, and then the juice flows easily enough.

It has already been mentioned that air expands when heated. The same rule holds good when applied to other objects, such

BOILING WATER.

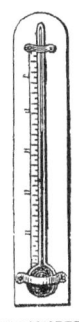

THERMOMETER.

as the various liquids, metals, &c. A very familiar example of this fact is the " boiling over " of water, when the vessel has been filled too much to allow for the expansion of the heated liquid.

Advantage has been taken of this principle in the formation of the Thermometer, a word which signifies " heat-measurer." Liquid of some kind is placed in an hermetically sealed tube, generally terminating with a bulb, and in proportion to the heat the liquid expands, and is forced up the tube.

Any liquid will answer to a certain extent, but, as water freezes at $32°$, it would be useless for measuring degrees of cold below the freezing point. Coloured spirits of wine are used ; but the very best liquid is mercury, which is a metal in a state of fusion.·

This expansion by heat is so powerful in iron, that it is utilised in several ways.

Take, for example, wheel-making. The iron tire is made rather smaller than the wheel, and is then placed in a fire until it is red-hot. It then expands so much that it can be easily slipped over the wheel as it lies on the ground. Cold water is then dashed on it, and the tire contracts with tremendous force, binding the parts of the wheel firmly together.

In all buildings where iron is much used, such as iron bridges, iron beams, &c., it is necessary to make allowance at both ends, so as to permit the iron to expand on a hot day and contract on a cool one. Buildings formed of stone and iron were once thought to be safe in case of fire. They are now known to be just the contrary, the stone flying with the heat, and the iron expanding.

USEFUL ARTS.

CHAPTER XII.

The Cassava Press and its Structure.—Mode of using it.—The Siamese Link.—An ingenious Robbery.—Muscles and their Mode of Action.—Human Arms and Steelyard.—Change of Direction.—The Human Hand and Wrist.—Story of a Carpenter.—The Pulley.—Reduction by Friction.—Past and present Engines.—Oiling Machines.—Treatment of the Sewing Machine.—Use of Paraffine.—Disuse of Machine hurtful.—Human Joints.—Synovia and its Value.—Disuse of Joints hurtful.—The Lazy-tongs and its Usefulness to Invalids.—Suggestions for Improvement.—Larva of the Dragon-fly and its Mask.—Curious Mode of seizing Prey.—Proboscis of the House-fly, and Mode of using it.—The Apple-parer.—Squirrel and Nut.—Structure of Teeth.—Rock-splitting.—Powers of Ice.—How the Pebble-ridge is formed.—Splitting Stones by Moisture.—The Diamond Drill.—Ovipositor of the Gad-fly.—Curious Similitude of Structure.

MEANS AND APPLIANCES.

IN this chapter we will take some miscellaneous appliances of force both in Art and Nature.

In the accompanying illustration is shown the Cassava Press of Southern America, a most effective and simple instrument for extracting the juices of the root. These juices are poisonous when raw, but, when properly boiled and cooked, they make an excellent sauce.

The press in question is an elastic tube made of flat strips of cane woven together exactly like the "Siamese Link," which will be presently described. The cassava root, after having been scraped until it resembles horseradish, is forced into the press until it can hold no more. The result is, that the tube is shortened and thickened, being widest in the middle.

It is then hung by its upper loop to the horizontal beam of a hut. A long pole is passed through the lower loop, the short end is placed under a projecting peg on the upright post of the house, and a heavy weight attached to the longer end. A powerful leverage is thus obtained, the tube is forcibly shortened, and the juice exudes through the apertures of the woven cane.

When it begins to run slowly, a woman seats herself at the end of the pole, so as to increase its weight. I must mention here that in the illustration the press is too near the middle of the pole. This is because the exigences of our page do not

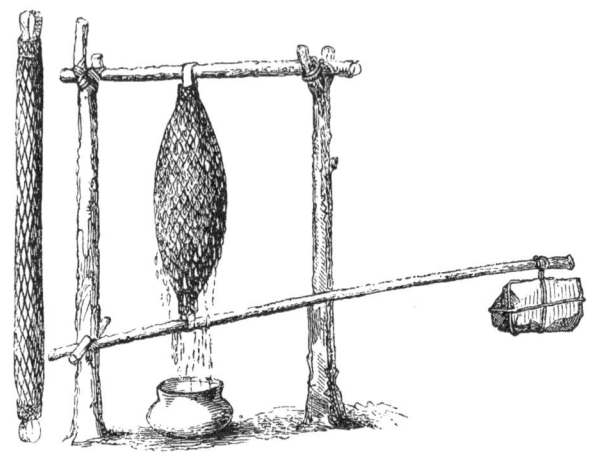

CASSAVA PRESS.

admit of the requisite length. But if the reader will kindly assume the end to which the stone is attached to be three or four times longer, he will have an idea of the great power which is exerted upon the cassava.

On the left hand of the illustration is the same cassava press as seen when empty, and both figures, as well as that of the pot for receiving the juice, are taken from specimens in my collection.

On the right hand of the following illustration is the Siamese Link, which caused such a sensation when it first came out.

A finger is inserted at each end, and, when the owner attempts to withdraw them, the Link contracts, and the harder the pull, the tighter is the hold. If the fourth instead of the first finger be employed, the hold of the Link is exceedingly strong.

The only mode of release is by pushing the fingers together, when the Link will relax. It should then be held by the remaining fingers of one hand, so that it shall not contract again, and the finger of the other hand comes out at once.

An ingenious robbery was once committed by means of the Siamese Link. A man of good address struck up an acquaintance with a jeweller. One day he produced a Siamese Link, and challenged him to get his fingers out when once they were in. So the jeweller was told to put his hands behind his back, and push his little fingers as far in as he could.

This he did, when the treacherous friend made a clean sweep of all the rings, brooches, ear-rings, and such jewellery as was within his reach, while the unfortunate jeweller was

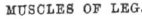
MUSCLES OF LEG.   SIAMESE LINK.

vainly tugging at the Link. This only occupied a few seconds for a practised hand, and the thief quietly opened the door, shut it, and was lost in the passing crowd before the jeweller could recover from his surprise.

On the left of the same illustration is a view of the muscles of the human leg, which, as the reader will see, are curiously like the distended cassava press. Although the mode of applying the force differs, the principle is the same.

In the latter case an external force is applied to the press, but in the latter an internal, or rather a central, force is

applied to the bones.   It is evident that if a similar process
were carried on with the cassava press, and the central
portion forcibly distended, the supports at either end would be
drawn powerfully towards each other.   Substitute the muscle
for the press, and the bones for the poles, and this is muscular
action.

HERE we have a diagram which speaks for itself, as far as
muscular action is concerned, but there is another point to
which we shall presently pass.

The muscle of the arm is seen running along the bone,
passing over the elbow, where it is held down by a tendinous
band, and, by its contraction, enabling the arm to be bent so as
to uphold a considerable weight.   The mechanical analogy

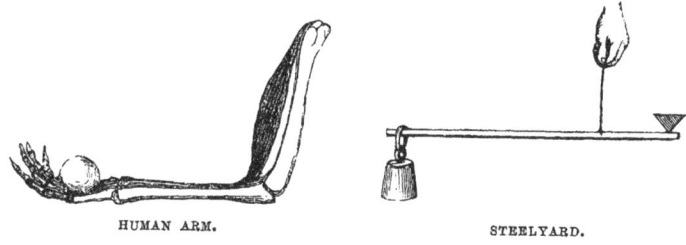

HUMAN ARM.   STEELYARD.

between this arrangement and the common Steelyard is too
evident to need any explanation except inspection of the
diagram.

THERE is, however, another point which is worthy of consideration.   The muscle does not proceed at once from the
shoulder to the wrist, but passes under the tendinous band
above mentioned, and so produces a change of direction when
the arm is bent.

There is a more complicated arrangement of a similar
character in the human hand, a diagram of which is given in
the left-hand figure of the accompanying illustration.

The fingers are, of course, moved by a set of tendons, and
the muscles, from which these tendons spring, are attached to
the fore-arm (I purposely omit the scientific titles, though
they would be much easier to write).   Any of my readers can
prove this for themselves.

Let him first grasp the upper arm firmly, and bend the limbs, and he will at once find that the swelling of the muscle shows the source of power.

Then let him do the same, but grasp the fore-arm, and he will find that the muscles are quiescent, showing that the former set of muscles belong to the entire arm, and not to the fingers, while the muscles of the lower arm have nothing to do with the bending of that limb.

Now let him grasp the fore-arm, and open and close the fingers, and he will feel a whole set of muscles rise, and swell and harden under his grasp. Next let him bend his hand inwards, and he will find that the fingers work perfectly well, though the direction of force is changed.

This is owing to a band of tendons passing across the wrist, under which the finger-tendons play. The course of the tendons is marked in the illustration by leaving them white.

The wondrous structure of the human hand and its multitudinous tendons can only be appreciated by actual dissection, but an idea of their variety and use may be obtained by watching the hands of a skilful pianoforte-player. This struck me forcibly the first time that I ever heard Thalberg play.

While on the subject of tendons, I once met a curious case. A journeyman carpenter missed a blow with his axe, and struck his left hand at the junction of the thumb and wrist. The important tendon was severed, and the inner muscles, having no counteracting force, dragged the thumb into the hollow of the hand.

To all appearance, the man could no longer earn a living as a carpenter. But he would not be discouraged, and while he was in hospital he borrowed a book, and studied the anatomy of the human hand. By means of this knowledge he constructed a sort of semi-glove, in which he introduced pieces of watch-spring, that supplied the place of the lost tendon.

Not content with this, he studied Euclid for the purposes of his trade, so as to get the most possible out of a piece of wood of given dimensions, and be able to go straight to his mark by a problem, instead of doing it slowly and clumsily with a two-foot rule and a pair of compasses. When I saw him last he was a master carpenter in a large and increasing business.

MAN has unconsciously imitated Nature in the invention of the Pulley, whereby the direction of force may be altered almost at will. In this case the cord takes the part of the working tendon, and the Pulley of the fixed tendinous cross-bar. There is much matter of interest in the tendons, but, as our space is fast waning, I must resist the temptation of describing them.

IN all machinery one of the chief objects of the machinist is to reduce friction as much as possible. He makes all the joints as smooth as tools can polish, and always introduces oil or some lubricating substance into the joints. Otherwise the

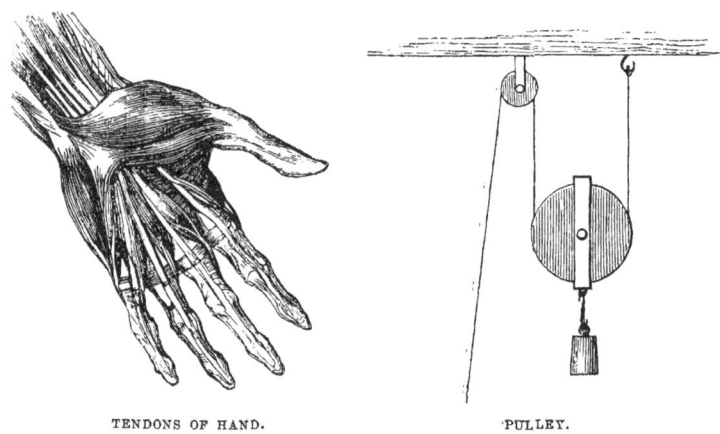

TENDONS OF HAND.   PULLEY.

engine rattles with a noise proportionate to its power, and wastes its force on the friction.

In my childish days a steam-engine of any kind used to rattle so loudly that conversation was almost impossible. Now they are made with such perfection, that the vast engines in use at the pumping stations of the metropolitan drainage are almost absolutely silent.

There is the enormous hall, filled with gigantic beams and rods, and cranks, and wheels. A single man turns a little handle, and the whole machinery starts into life. Beams rock, cranks and wheels revolve, rods slide up and down, and all in a silence which is nearly appalling in its manifestation of unassuming strength. Indeed, many a hand sewing machine makes far more noise than one of those giant engines, and all because in the latter friction is avoided as far as possible, every screw is well braced up, and every joint is kept well lubricated.

Here I may observe that few sewing machines get fair play. They rattle, they squeak, they become stiffer daily, they snap the thread, and then decline work altogether. And in almost every case this is done by neglect on the part of the owner, who does not lubricate every point of the machine which works upon another.

Ladies especially are very careless in this respect, and will mostly omit three or four of the oiling points. They might just as well omit them all, as a single unoiled point will disarrange the harmonious motion of the whole machine. I have often been called in as surgeon in such cases, and have almost invariably

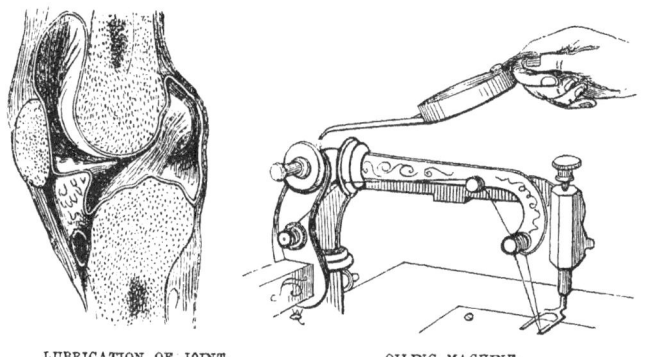

LUBRICATION OF JOINT.   OILING MACHINE.

been able to point to several spots which needed oil, and did not get it. Sometimes, out of false economy, an inferior oil is used, which speedily clogs and hardens, and stops all movement. In such a case the best remedy is to apply paraffine liberally, and use it for a quarter of an hour or so. It will soon dissolve the clogged oil, which may be worked out by turning the handle or crank of the machine.

Of course the best remedy is to take the machine to pieces, polish the joints, lubricate them, and put it together again. But this is a perilous process, and an amateur, if he tries it, will generally find himself with half-a-dozen pieces for which he can find no place. Paraffine will answer every purpose, and I have released many a stiffened machine by its use.

Then some people leave their machines untouched for days, or even weeks, and then wonder that they work stiffly. Every day the machine should be worked, if only for a few seconds, and then it will seldom stiffen. It is just the same with steamers. When they are in harbour, though the fires be out, and they are not meant to move for weeks, the engines are always turned round at least once daily.

BOTH these rules hold good in the animal kingdom.

To every joint there are attached certain glands that supply a kind of oily substance technically named "synovia," which acts exactly the same part as the oil or grease of machinery. If these glands do not do their duty, and the supply of synovia be defective, the joints become stiff, painful, and crackle when they are moved.

Then, exactly as the joints of a machine become stiff from non-usage, so do those of a human being. We will take, for example, the Indian Fakirs who vow that they will not move some limb from a definite posture. At first the exertion is trying and painful, but by degrees the disused joints lose their faculty of motion, and, even if their owner wished to move a limb, he could not do it.

The right-hand figure of the illustration represents the lubrication of an ordinary sewing machine, and the left-hand figure is a section of the human knee-joint, showing the gland which supplies the synovia.

PERHAPS some of my readers may think that such a subject as the "Lazy-tongs" is too trivial for a work which deals, however lightly, with science. But there may be some who know the inestimable benefit of Lazy-tongs under certain conditions.

There are many cases where a severe injury has occurred, or where rheumatism has fixed its tiger-claws in the joints, so that movement is all but impossible. There may be no one in the room to help the invalid, and even to stretch the arm over the table is as impossible as to jump over the house.

Then it is that the real value of the Lazy-tongs becomes manifested, and that it shows itself in the light of a supplementary limb. With a mere movement of the fingers it can be stretched across any table which is likely to be placed before an invalid, and seize the required object by the tongs at the further end.

The only drawback to its use is, that the instrument cannot be shortened without opening the tongs. But, if some plan could be devised whereby the tongs could retain their hold under those conditions, the instrument would be a perfect one.

EXACTLY such a Lazy-tongs we have in Nature, in the well-known "mask of the larva and pupa of the Dragon-fly." It is called a mask because, when closed, it covers the face.

It chiefly consists of two flat, horny plates, hinged in each other like a carpenter's two-foot rule, and being capable of extension to a considerable length. The end is widened, and

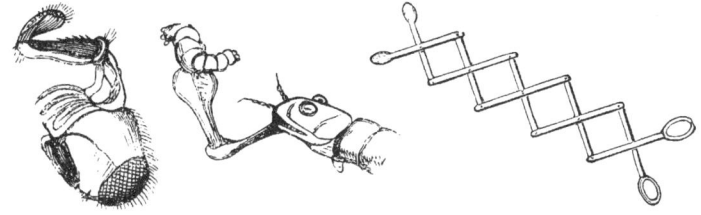

PROBOSCIS OF HOUSE-FLY.   MASK OF DRAGON-FLY LARVA.   LAZY-TONGS.

furnished with two jaws, which take the part of the tongs in the instrument above described.

This curious apparatus is used for the purpose of securing prey.

I have kept many of these creatures, and watched their mode of feeding. As has already been mentioned, they have two modes of progression, *i.e.* walking by means of legs like those of ordinary insects ; and driving themselves along by ejecting water from the tail, on the principle of the rocket. As far as I have seen, the latter mode is always used in taking prey. The Dragon-fly larva always lives at the bottom of the water, though it can force itself to the surface if needful. And, like the dreaded ground-shark, it seizes its prey from beneath.

Its favourite food is the larva of the whirlwig-beetle, a fat white grub, with a number of white, soft, feathery gills fringing its sides. In order to produce a current of air over these gills, the larva wriggles itself up to a height of several inches, and then sinks slowly down, with the white gills floating on either side.

Should a Dragon-fly larva be near, it sees the grub ascending, glides quietly under it without using its legs so as to cause alarm, waits for it to sink, darts out the mask, seizes it in the jaws, drags it to its mouth, and the grub is seen no more. So voracious are these larvæ, that, if only two are kept in the same vessel, one is sure to devour the other.

ANOTHER good example of the Lazy-tongs is the Proboscis of the common House-fly. We have all seen these insects alight near sugar, or any other tempting food, unfold the proboscis, pour a drop of liquid in the sugar, dissolve it, suck it up, and then shut up the proboscis as if by hinges.

ANOTHER labour-saving machine is the Apple-parer, a comparatively modern invention. The principle is, that a knife is

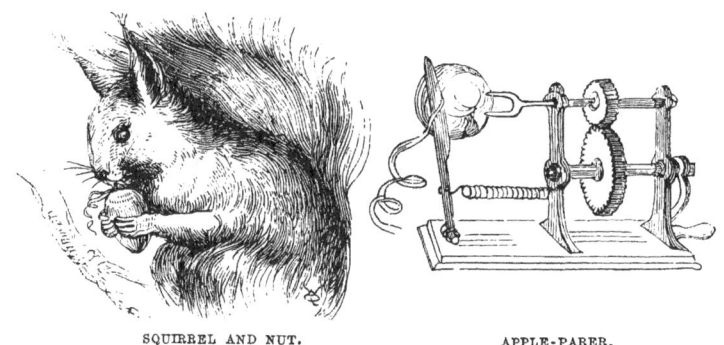

SQUIRREL AND NUT.   APPLE-PARER.

pressed lightly by a spring against a revolving apple, and set at such an angle that nothing but the outside peel can be removed. Where large numbers of apples have to be pared, as in making preserves or in hotels, this is a most useful invention.

When I first saw it at work, the operation seemed familiar to me, but I could not at first remember the parallel. At last it flashed across me that a Squirrel eating a nut was the natural parallel of the Paring Machine.

After splitting the shell and extracting the kernel, the

Squirrel takes the latter between its fore-paws, presses it against its upper incisor teeth, and makes it revolve rapidly. In a second or two the kernel is perfectly peeled, and is then eaten.

In this case the incisor teeth of the Squirrel take the part of the knife, the muscles of the leg that of the spring, and the sharp edges of the upper teeth that of the knife. The structure of the Rodent teeth has already been explained in page 233.

THE wonderful effects of water in breaking up the hardest rock have already been described. We will now proceed to another branch of the same subject.

Perhaps some of my readers may have wandered along our

FROST-CLEFT ROCK.   STONE-SPLITTING.

rocky coasts, and have seen how large masses of rock are continually detaching themselves, though they are so hard that a cold chisel is needed to make any impression upon them.

Then they fall into the sea, and are rolled backwards and forwards until they become smoothed and rounded, and are called pebbles, while the portion that is rubbed off them is called sand. The phenomenon is well shown in the wonderful Pebble Ridge of North Devon.

The real agent is ice.

We all know that, when water freezes, it expands considerably. This accounts for two phenomena.

First, as it expands, it becomes lighter than water, and consequently floats on the surface.

Next, there are few of us who have not seen water-bottles

cracked by the freezing of the water. The most common, and perhaps the most unpleasant, example of this propensity is the bursting of water-pipes in the winter, followed by a flooding of the house when the thaw comes.

This is caused by the expansion of the frozen water, which will burst not only a thin leaden tube, but a stout iron vessel. Care should therefore be taken, at the beginning of winter, to cover up all exposed portions of leaden pipes, and there will then be no danger. There was one pipe in my house that was always bursting, but after I covered it with two or three layers of carpet placed loosely over each other, so as to entangle the air and form a non-conductor, the pipe has never frozen, and the water supply has been uninterrupted by the severest frosts.

I am told that a still better plan exists, especially in places where the pipes cannot be thoroughly protected by external wrappings. Let six inches or so of the leaden pipe be removed, and its place supplied by a vulcanised india-rubber tube.

The ice *must* expand somewhere, and chooses the spot where least resistance is offered to it. Consequently, it expands in the india-rubber tube, but does not break it, and, when the thaw comes, there is no overflow of water.

MAN utilises this power of ice in stone-splitting. Instead of taking the trouble to cut the stone by manual labour, the workmen bore a series of holes, fill them with water, insert tightly a wooden plug to prevent the ice, when formed, from oozing out of the holes, and leave the rest for the frost to do.

A like effect is produced in the warm weather by substituting similar plugs, but quite dry, having been baked for hours in an oven, for the purpose of driving out every particle of moisture. These plugs are hammered into the holes as deeply as they will go, and there left. Even if there be no rain, the nightly dews make their way into the pores of the dry wood, and cause it to swell with such irresistible force that the stone is split with scarcely any manual labour on the part of the workmen.

YET another plan for cutting hard stones. Some of my readers may be aware that a singularly ingenious instrument has been invented for cutting holes in granite and other hard rocks. It is called the Diamond Drill, because its tip is armed with uncut diamonds.

It is necessary that the diamond should not be cut, as the natural edges are needed. A glazier's diamond, for example, is always set as it came out of the mine. The stories that are told about cutting out panes of glass with a diamond ring are all absurd. A diamond, when it has once passed through the hands of the jeweller, cannot cut glass. It can scratch glass, but not one whit better than a flake of ordinary flint.

It is found that the Diamond Drill works with wondrous rapidity, cutting away the stone with ease, and suffering

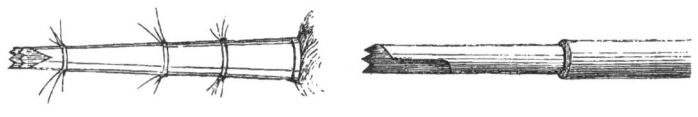

BORER OF ŒSTRUS.   DIAMOND-HEADED BORER.

scarcely any damage itself. The tube to the end of which the diamonds are fixed is generally made in telescopic fashion, so as to allow it to penetrate deeply into the rock, without the necessity of shifting the machine by which it is turned. I need hardly say that its rate of speed is very great indeed.

OUR old friend, the Gad-fly, again affords an example of a parallel.

The ovipositor is tubular, telescopic, and furnished at the top with five little hard, sharp, scaly knobs, which act the same part as the diamonds of the mining tool. Even the scoop-like shape of the tip, and the telescopic shaft, are almost identical in both instances.

USEFUL ARTS.

CHAPTER XIII.

TELESCOPIC TUBES.—DIRECT ACTION.—DISTRIBUTION OF WEIGHT.—TREE-CLIMBING.—THE WHEEL.

Telescopic Tubes, their Structure and Uses.—The Japanese Fishing-rod.—The Tripod Wheel-bearer and its Telescopic Structure.—The Rat-tailed Maggot. —Locomotion.—Direct Action.—The Rocket, the Water Tourniquet, and Electric Tourniquet.—Cuttle-fish.—The Flying Squids.—The Paper Nautilus.—Proceedings of newly-hatched Calamaries.—Larva of the Dragonfly.—Distribution of Weight.—The Snow-shoe, its Structure and Mode of using it.—The Skidor of Norway.—A formidable Rifle Corps.—The Mudpatten.—Foot of Duck tribe.—Foot of Jacana.—Locomotion of Watergnat.—Tree-climbing.—Mode of ascending Palm-trees.—The Value of a Hoop.—The "Girt Pupa" and Butterfly.—Principle of the Wheel.—The primitive Wooden Wheel.—Spoked Wheels.—Driving Wheel of the Bicycle. —Naturally spoked Wheel of the Chirodota.

MEANS AND APPLIANCES (*continued*).

WE will now treat rather more in detail the two subjects which were lightly touched upon at the end of the last chapter.

The reader will remember that the diamond-headed borer is made in telescope form, so as to be adjustable at pleasure. It was also remarked that the ovipositor of the Gad-fly was made in a similar fashion, so as to be withdrawn within the body of the insect when not needed, and protrusible to a considerable extent when the Gad-fly wishes to deposit her eggs.

As to our modern telescopes and opera-glasses, they are so familiar that there is little use in describing them, except to say that their framework consists of a number of tubes of gradually lessening diameter, the one sliding within the other, so that the instrument can be lengthened or shortened at will, so as to suit the focus of the observing eye.

A very ingenious adaptation of the telescopic principle is seen in the Japanese fishing-rod, which is now tolerably well known. Our own telescopic rods require to be withdrawn at the butt-end, and then fitted together in front. But the Japanese rods are so made that, after taking off the ferrule of the seeming walking-stick, a mere fling of the hand will send joint after joint flying out, and fixing themselves in regular succession. So admirably are these rods made, that even blowing into the butt-end will have the same effect.

ONE of the most perfect, if not the most perfect, example of the telescopic tube is to be found in the Tripod Wheel-bearer (*Actinurus*), one of the numerous aquatic Rotifers.

It is not usually so small as the generality of its class, being nearly one-twentieth of an inch in length, and visible to the

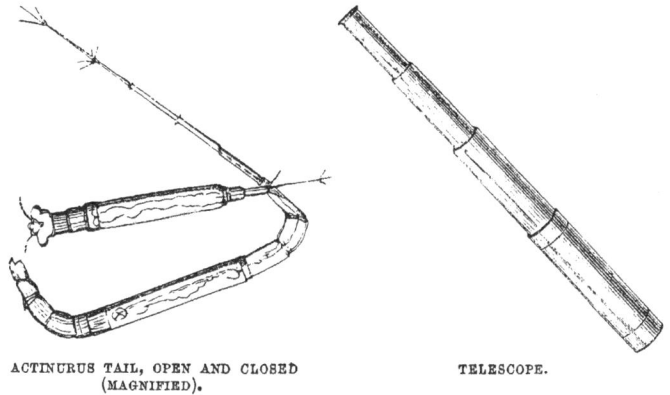

ACTINURUS TAIL, OPEN AND CLOSED (MAGNIFIED).

TELESCOPE.

unassisted eye, provided that the owner of the eye in question knows how to use it.

When placed under a miscroscope of moderate power, the Actinurus is seen to be built almost wholly upon the telescopic pattern. Only the centre of the body remains stationary, the two ends being framed on the principle of the telescopic tube, and capable of being enclosed within the central portion, just as is the case with the Japanese fishing-rod.

In the illustration the Actinurus is shown in two attitudes. In the upper figure it is represented as having the fore-part of the body entirely, and the tail part nearly, withdrawn within the central portion. The lower figure shows the same specimen with all its telescopic tubes drawn out to full length.

The creature is perpetually elongating and contracting its body by means of these tubes, so that a measurement of its length is not easy to obtain.

A full and interesting description of this curious Rotifer may be found in Gosse's "Evenings at the Microscope," p. 300. The long tails of the Rat-tailed Maggot, already described under the head of Diving, are good examples of the draw-tube as found in Nature.

LOCOMOTION.—DIRECT ACTION.

THE second point which has to be elucidated is that of progress by means of Direct Action.

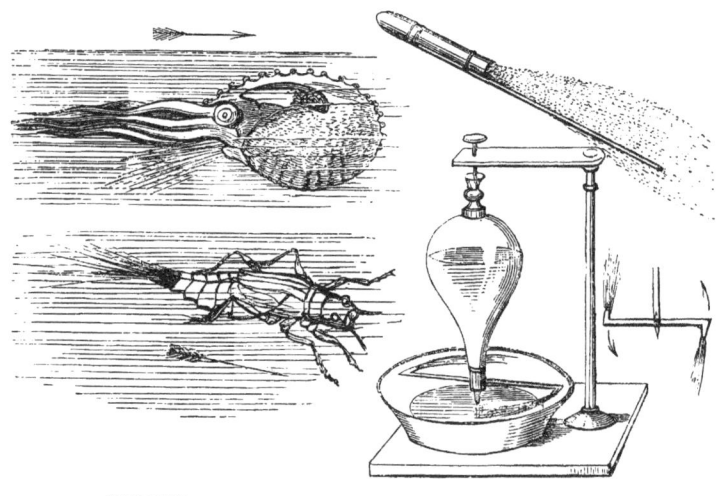

NAUTILUS.
LARVA OF DRAGON-FLY.   WATER TOURNIQUET.   ELECTRIC TOURNIQUET.

We have already seen how vessels can be propelled by sail, oar, paddle, or screw. We have now to consider a mode of progress which requires none of these things, but which works by means of Direct Action.

Such, for example, is the progress of a Rocket through the air.

The heated gases rush out with tremendous violence, and, by their pressure, urge the heavy rocket into the air with the rush, roar, and bang so familiar to all who have witnessed a good display of fireworks.

A rocket in the act of ascent is shown in the uppermost figure of the accompanying illustration.

Below it is shown the Water Turbine, the principle of which is evident from the sketch.

From each of the apertures a stream of water is forcibly directed, and, by its resistance, spins the vessel round and round. There are several shops in London in which this instrument may be seen at work.

Although in such positions it is necessarily a mere toy, it carries with it, in common with many other toys, the germs of valuable inventions. Indeed, there have been attempts to utilise the principle of Direct Action in the propulsion of vessels, but as yet the mechanical difficulties have proved practically insuperable, and, although a vessel has been thus propelled, the expense has been heavier than that of the paddle or screw, and the speed not nearly so great.

On the right hand of the illustration is another example of Direct Action, called the Electric Tourniquet.

In the two previously mentioned instruments the motive power is visible, but in this it is invisible except in the dark.

The principle is exactly the same as in the pocket or water tourniquet; but, instead of heated air or a stream of water, electricity is used. The instrument is attached to an electric machine, and fully charged. The electric fluid rushes out of the points, forces itself against the air, and so, by its recoil, drives the machine round and round upon its pivot.

WE will now take two examples of Direct Action as found in Nature.

Perhaps many of my readers have seen the Octopus, and admired the manner in which it glides through the water, trailing its long arms behind it. Whence the force comes is not easily seen, and the creature appears to move almost by volition. In reality, however, it employs Direct Action. It takes water into the body, and then it ejects it through a tube called the "siphon" with such force that the animal is propelled backwards through the water.

Some of the creatures belonging to the Cuttles, and popularly called Squids, can use such extraordinary powers that they can project themselves far out of the water. In consequence of this power, they are sometimes called Flying Squids, and, as they have been known to shoot themselves completely over the hull of a large ship, they well deserve the name.

The common Squid of our coasts, which furnishes the so-called Cuttle-bone, affords us a good example of Direct Action. I once hatched a number of young Squids from the grape-like eggs, and it was most curious to see how the little creatures shot about as soon as they escaped from the egg.

They also utilised the siphon in another way. Poising themselves just above the sand with which the bottom of the vessel was covered, they directed a stream of water upon it, and thus formed little cavities into which they settled like birds into their nests.

The figure represents the Paper Nautilus as it appears while passing through the water. Just at the base of the tentacles is seen the short siphon, from which it is pouring the stream of water which drives it along.

Below the Nautilus is seen the larva of the common Dragon-fly. We have, when treating of the Lazy-tongs, already described the mode in which the insect takes its prey, and our object could not be served by repetition. Suffice it to say that the insect is shown in the act of ejecting water, and so shooting itself along in preparation for seizing prey.

DISTRIBUTION OF WEIGHT.

BEING on the subject of locomotion, we will examine a few of the contrivances by which a man is enabled to pass in safety over soft substances into which he would otherwise sink.

The first and best-known of these is the Snow-shoe of Northern America. It is a framework of wood, shaped as shown in the upper figure on the right-hand side, and strengthened by two cross-bars. The interior of the " shoe " is filled in with hide thongs arranged much like those of a racket, and stretched as tightly. The front of the snow-shoe is slightly turned up, so as to avoid the danger of the point sticking in the snow, an event which, however, generally happens to a novice.

These instruments are of considerable size, a specimen in my collection measuring exactly five feet in length, by fifteen inches in width.

Supported on the snow-shoe, the hunter is enabled to glide unhurt over the deep snow in which he must have sunk without some such aid. He can thus hunt the bison, the wapiti, or any of the larger animals, being able to pass rapidly over the surface, while they are laboriously ploughing their way through the snow-drifts.

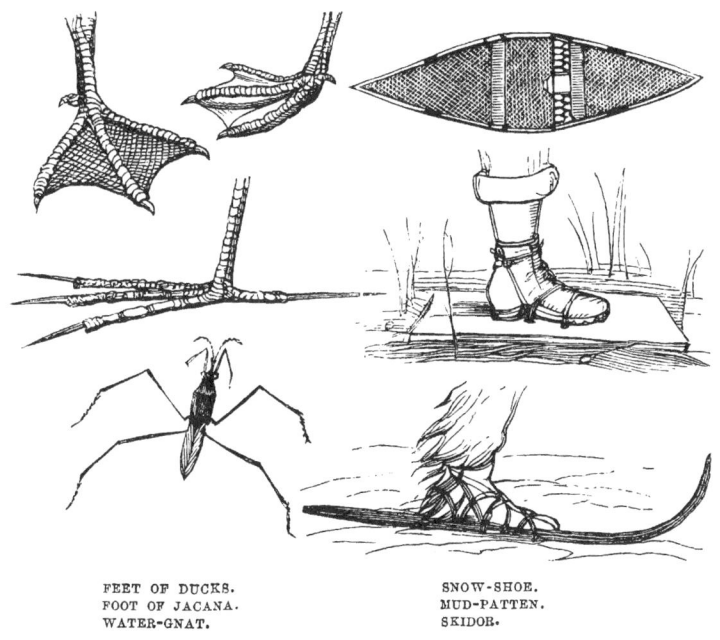

FEET OF DUCKS.   SNOW-SHOE.
FOOT OF JACANA.   MUD-PATTEN.
WATER-GNAT.   SKIDOR.

It occasionally happens that the snow falls before the shoes are ready. In this case the hunter is obliged to extemporise snow-shoes by cutting them out of thin boards.

Several years ago, when snow fell heavily and remained unmelted for many days, some Canadians, who were visiting England, made quite a sensation by donning their snow-shoes, and travelling over the snow-clad country. It was very pretty to see the easy way in which they could shoot down a hill, and to watch the peculiar gait which is needed by the snow-shoe.

AT the bottom of the illustration is shown a portion of a curious skate used in Norway, and called Skidor.

These remarkable implements achieve by means of length the task which the snow-shoe accomplishes by width. They are made of wood, and, though but a few inches in width, are ten feet or more in length. One is always a few feet shorter than the other, for the convenience of turning. Much practice is needed for the management of the Skidors, but, when they are fairly mastered, they enable their owner to travel at a wonderful pace.

The Norwegian hunter is quite as dependent on his Skidor as the North American on his Snow-shoe, and uses it for exactly the same purpose. A corps of these hunters has been organized for war, and very formidable they were, hanging on the skirts of the enemy, and giving him no rest, day or night. They never came within fifty yards of each other, so that even cannon were useless; and, as soon as they thought that they were endangered, they dispersed in all directions, only to reunite and swoop down again on the enemy at the first opportunity.

THE central figure represents the Mud-patten, which, as its name implies, plays the same part towards mud that the snow-shoe and skidor do to the snow. Like them, also, it is not easy to manage; and a novice is tolerably certain to drive the front of the patten into the mud, and so get an awkward and not aromatic fall.

This patten, which is merely a square piece of board attached to the foot, is in use on many of our coasts where the ebbing tide runs out to a great distance, leaving a vast expanse of soft mud. Like the skidor and the snow-shoe, it is mostly used by sportsmen, especially in the winter, when wild-duck shooting sets in.

Aided by the pattens, a sportsman can travel for miles over mud that would otherwise swallow him up, shoot his birds, and secure them when fallen. While engaged in winter shooting on the Medway, we have often lost birds because they fell beyond a deep mud-bank, and we had no means of crossing it.

ON the left hand of the illustration are some natural parallels of these artificial aids. The two upper figures represent two forms of webbed feet, and the analogy between them and the snow-shoe and mud-patten is too obvious to need explanation.

In the centre is the foot of the Jacana, an Asiatic bird. Its foot may well be taken as the analogue of the skidor, length taking the place of breadth, and enabling the weight to be distributed over a large surface.

This bird finds its food in rivers and lakes, and, by reason of its enormously long toes, can walk with safety over slight floating vegetation, which would give way at once under the tread of any bird except a Jacana. Very good representations of this bird are to be seen in Japanese works of art, especially those which are mounted as screens. Even the peculiar gait of the bird is given with marvellous truth.

The last figure represents the common Water-gnat (*Gerris*), which may be seen in almost any piece of fresh water, however small. Ponds that are open to the south, and sheltered from the north wind, are its favourite localities.

It is a carnivorous being, feeding almost wholly on insects that fall into the water. In order to capture them, it runs rapidly over the surface of the water, the long slender legs distributing its weight over a large surface, and so keeping it from sinking. Only the last two pairs of legs are employed for this purpose, the first pair being held in front of the body, and used for the purpose of capturing prey.

TREE-CLIMBING.

ANOTHER curious aid to locomotion is shown in the accompanying illustration.

In many parts of the world, where the cocoa-nut palm grows, the natives have invented a simple, but ingenious, plan for ascending the tall, curved stem. Such a thing as an upright palm-tree is unknown, and consequently the ascent of the branchless stem is not an easy task without artificial assistance.

When I treated of Warfare and the different modes of scaling walls, the climbing-spur was casually mentioned. The implement of the palm-climber, however, is simpler and more effective, as it leaves both hands at liberty when desired.

The man cuts a long piece of one of the tough and almost unbreakable creepers which festoon the trees of tropical climes. He passes it round the trunk which he wishes to climb, and fastens the ends firmly together, so as to form a large loose hoop. He then passes the hoop over his head, until it presses against his back, as seen in the illustration, and serves to support him as he leans against it.

Taking the hoop by the two sides, he lifts it up the trunk as far as he can, places the soles of his feet against the tree, and so walks up it, hitching the hoop upwards at every step. When

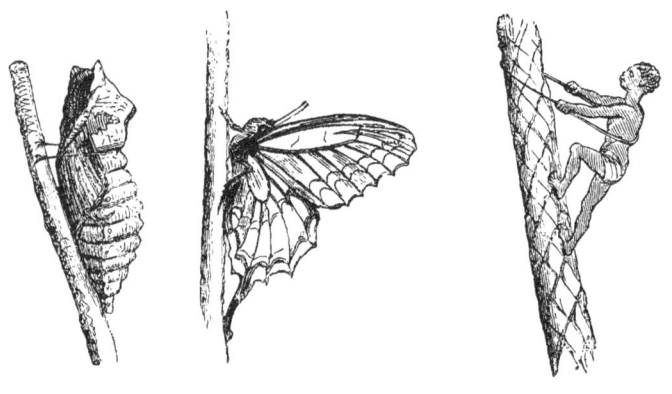

GIRT PUPA AND BUTTERFLY.   CLIMBING PALM-TREE.

he has reached the top of the tree, he supports himself entirely by the hoop, while his hands are at liberty to be used in getting the cocoa-nuts.

IN the insect world there are many examples of support being given by a belt passing round the body.

Among the Butterflies, for example, there are many which, in their pupal stage of existence, are attached to upright stems. They are fixed to the stem by a few threads at the tail, answering to the feet of the tree-climber, while the body is kept in position by a stout silken thread passed loosely round it.

The illustration represents the pupa of the common Swallow-tailed Butterfly, while in the centre is the same insect in the perfect state as it appears when resting. It really seems as if the ancient habit of the pupa had been remembered by the perfect insect, the long ends of the hinder wings taking the place of the pupal tail, and the legs that of the belt.

THE WHEEL.

YET another aid to locomotion is found in the WHEEL, a contrivance for diminishing friction.

When man first learnt that heavier weights could be dragged than carried, he simply placed them on flat boards to which ropes were attached. The next step was necessarily the invention of the sledge, the burden resting on two parallel runners, the ends of which were slightly curved so as to prevent them from hitching against any small obstruction. In some countries—such, for example, as in Esquimaux-land—the sledge is the only vehicle practicable, and even Europeans, when

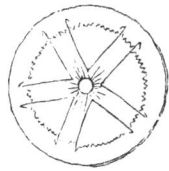

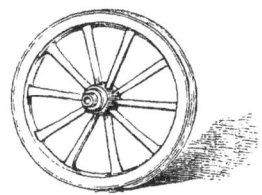

WHEEL-SPICULE OF CHIRODOTA.   CART-WHEEL.

they visit that country, are fain to adopt the sledge if they would live.

But, in more temperate zones, the Wheel is paramount. In its earlier stages the wheel was a very simple business. It was simply a section of a tree-trunk, dubbed roughly round, and with a hole in the centre, through which the axle passed. Such wheels are still in existence in many parts of Europe; and, owing to the want of regularity of outline in the circumference, and the utter absence of grease, the wheels keep up a continuous shriek, almost deafening to those who are unused to it, but perfectly unheeded by those who own or drive the vehicle.

The next improvement was to make the circumference of the wheel as perfectly circular as the art of man could devise, and, instead of having the wheel solid, to fill up its interior with spokes, thus gaining lightness and strength at the same time.

Of all locomotive wheels, I suppose that the modern Bicycle affords the best example. The driving wheel is larger than the hind wheel of an ordinary coach, and yet the spokes are not nearly so thick as the porcupine quill with which this account is written.

If we look at the ancient sculptures and paintings of Egypt and Assyria, as preserved in the British Museum, we shall see that either kind of wheel was used according to the work which it had to do. The solid, uneven, squeaking, wooden wheel was devoted to agriculture, while the light, spoked wheel was sacred either to warfare or hunting.

Let us hope that in the two latter cases some modicum of grease might have been used, as the outcries of tortured and unlubricated machinery are enough to drive away all wild beasts which come within the range of its complaints, while the nervous system of hunter or warrior must have been seriously damaged by it.

EVEN in such a structure as the spoked Wheel, Nature has anticipated Man.

My readers may remember that, when treating of nautical matters, I mentioned the singular anchor-shaped spicules that are found upon one of the sea-slugs, called Synapta.

There is another group of these creatures inhabiting the Mediterranean, in which the skin-spicules take a different form. Like those of the Synapta, they are too small and translucent to be seen without the aid of the microscope and carefully adjusted light. But, just as the spicules of the Synapta resemble the ancient anchor, so do those of the Chirodota resemble the ancient wheel, the similitude being in both cases absolutely startling.

Not only that, but, as all readers must be aware, if they have studied practical mechanics, there are many machines which are toothed on the inner, and not the outer, side of the circumference. Here, in the Chirodota, the inner toothing is manifest.

What purpose it serves we know not. The Chirodota's wheels (of which there are thousands) never revolve, neither do the anchors of the Synapta hold the ground. Yet the very fact that such exceedingly minute objects should be so carefully constructed tells us at once that they must have some important purpose to serve, though at present that purpose is a mystery which no one has attempted to solve.

I have little doubt that when the hour and the man arrive, as arrive they surely will, we shall find in these tiny and almost unrecognised spicules the keys to treasures of wisdom which at present have been opened to no human being.

The whole history of the progress of the human race shows that facts have been allowed to accumulate, fought about, and turned in all directions, before the generaliser comes who pierces to the heart of everything, reduces apparent discrepancies to harmony, and usually is rewarded by finding some one else assume the credit of his discoveries, and receive all the honours and emoluments.

USEFUL ARTS.

CHAPTER XIV.

ART.

WE will now touch lightly on the subject of Art.

In the present day one of the most indispensable accessories to art is Paper.

It is a curious fact that we have no records as to the time when paper was first invented. The Egyptian papyrus we do not consider, as it was not paper in our sense of the word, although we have retained the name.

Paper is a vegetable fibre carefully disintegrated, made into a pulp with water, and then dried in thin sheets. As is the case with many arts, China seems to have taken the lead in paper manufacture, and we are even now indebted to that country for the "India Paper" on which the finest proofs of engravings are taken. This paper is made from the inner bark of the bamboo. "Rice Paper," so called, is not paper at all, but only a kind of pith cut spirally, and flattened by pressure.

There is scarcely any vegetable fibre of which paper cannot be made, and various plants have been suggested for this purpose, such as the stinging-nettle, cabbage-stalks, hop-bines, the waste of sugar-cane, sawdust, &c. Straw has already been successfully used, and so has Esparto grass.

Some years ago, when there was a scarcity of material for paper-making, the well-known Grass-wrack of our shores (*Zostera marina*) was brought into partial use. I believe, however, that the experiment was not a successful one. The Chinese make their paper of bamboo, macerating and pounding it until it is reduced to a pulp, and then shaken into fibres in a mould.

With us, white paper, such as is used by the writer, printer,

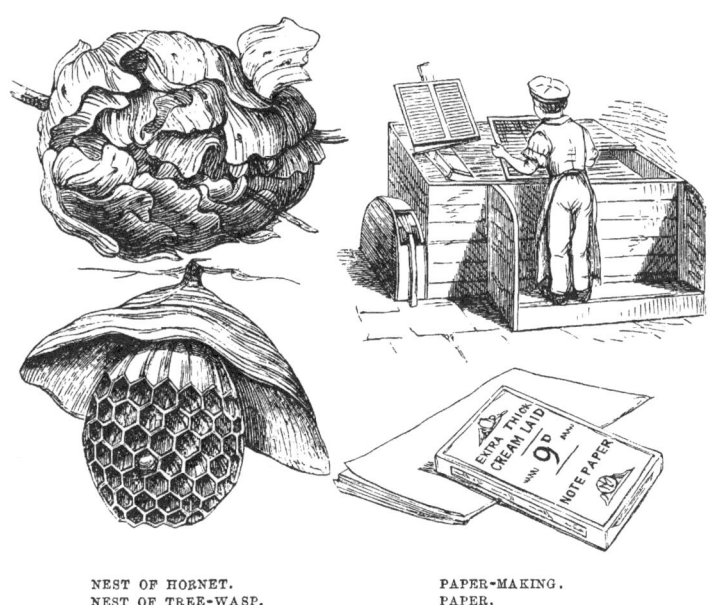

NEST OF HORNET.
NEST OF TREE-WASP.

PAPER-MAKING.
PAPER.

or artist, is made almost exclusively of cotton or linen rags. Upwards of a hundred and twenty thousand tons weight of rags are annually consumed in this country for the manufacture of paper. After being bleached, they are torn and ground into a pulp, which is then handed over to the actual maker.

The illustration represents paper-making by hand, a process which is now rarely used, except for special kinds of paper. Omitting technical details, the mode of paper-making by hand is as follows :—The pulp being prepared, the workman takes a "mould," *i.e.* a frame with a bottom of closely woven wire.

Having put into the mould a sufficient quantity of pulp, he shakes the mould so as to spread the pulp evenly over the surface. The water runs away between the wires, the sheet of pulp is transferred to a piece of felt, and when it is dry it becomes paper. If a sheet of ordinary note-paper be held up to the light, the marks of the wires are plainly perceptible. The so-called "water-mark" is due to wires twisted into the requisite shape.

The Chinese workman makes his paper exactly on the same principle, but the bottom of his mould is made of bulrushes instead of wires.

As for machine-made paper, the process seems absolutely magical. Endless bands of felt and wire are substituted for the hand frames, and, the pulp being poured in at one end, the finished paper is poured out at the other, and self-wound on rollers. Without any exaggeration, paper is now made by the mile, the only limit to its length being the size of the rolls.

WHEN I mention Paper-making in the world of Nature, many of my readers will at once know that I am about to refer to the Wasp tribe.

These insects were paper-makers long before even the Chinese had invented the art, and, so exactly similar is the mode of action, that man might well have copied from the insect.

The Wasp gnaws a bundle of vegetable fibres, mostly of wood, sound or decaying, according to the species. It masticates them until it has reduced them to a pulp, and then, by means of its jaws, spreads the pulp into sheets of various shapes and sizes.

With some of the pulp it forms hexagonal cells like those of the bee, and with some it makes the roof-like covering which defends the cells. Not only that, but it can make a sort of papier-mâché, which it uses for the flooring, if we may so call it, of the different strata of cells, and for the pillars which bind them together.

Like our own paper manufacturers, it is economic of material, will re-masticate any superabundant paper, and is only too glad if it can get hold of any paper made by man. I have seen a wasps' nest which was made entirely from the empty blue and white cartridges that were thrown away by soldiers.

Then there is as much difference in the papers made by wasps as in those made by man. In this country all wasps' nests are made of very fragile material, but in South America there are some wasps which make the external covering of their nests as hard and white as the stiff cardboard employed by artists.

HAVING now got our paper, we will glance at one or two modes of using it for Art. Papier-mâché has already been mentioned, and it is worthy of notice that there are now in existence many decorated ceilings which are made of this material, on account of its great strength and its non-liability to fire.

The first invention which we shall notice is that which is

FERNS IN COAL.   NATURE-PRINTING.

known by the name of Nature-printing, and which has been so successful in transferring to paper an exact representation of vegetable foliage.

One simple tolerably efficacious mode of Nature-printing has long been known. A piece of paper being rubbed with lamp-black and oil, the leaf was laid upon it and gently rubbed, so as to transfer the lamp-black to the nervures. It was then laid on a sheet of white paper, and again rubbed, when an impression of the leaf was left upon the paper.

The present system of Nature-printing is far in advance of this rather rude method, and amounts to an exact reproduction of the plant, not only in form and detail, but in colour.

In order to illustrate this beautiful process, I cannot do better than transfer to these pages the following account of Nature-printing as given in Ure's "Dictionary of Arts," &c. It is an abstract of a lecture delivered by Mr. H. Bradbury at the Royal Institution.

"Nature-printing is the name given to a technical process

for obtaining printed reproductions of plants and other objects upon paper, in a manner so truthful, that only a close inspection reveals the fact of their being copies; and so distinctly sensible even to touch are the impressions, that it is difficult to persuade those unacquainted with the manipulation that they are an emanation of the printing-press.

"The distinguishing feature of the process consists, first, in impressing natural objects—such as plants, mosses, seaweeds, and feathers—into plates of metal, causing, as it were, the objects to engrave themselves by pressure; secondly, in being able to take such casts or copies of the impressed plates as can be printed from at the ordinary copper-plate press.

"This secures, in the case of a plant, on the one hand, a perfect representation of its characteristic outline, of some of the other external marks by which it is known, and even in some measure of its structure, as in the venation of ferns and the ribs of the leaves of flowering plants; and, on the other, affords the means of multiplying copies in a quick and easy manner, at a trifling expense compared with the result, and to an unlimited extent.

"The great defect of all pictorial representations of botanical figures has consisted in the inability of art to represent faithfully those minute peculiarities by which natural objects are often best distinguished. Nature-printing has therefore come to the aid of this branch of science in particular, whilst its future development promises facilities for copying other objects of nature, the reproduction of which is not within the province of the human hand to execute; and even if it were possible, it would involve an amount of labour scarcely commensurate with the results.

"Possessing the advantages of rapid and economic production, the means of unlimited multiplication, and, above all, unsurpassable resemblance to the original, nature-printing is calculated to assist much in facilitating not only the first-sight recognition of many objects in natural history, but in supplying the detailed evidences of identification, which must prove of essential value to botanical science in particular."

Many plans have been tried with only partial success, but that which is now in operation produces the most wonderful results. The plants are laid upon sheets of lead, and then passed through rollers, so as to leave an impression in the soft metal. The electrotype then comes into play, exact copies of the impression being taken by it. As the face of the electrotyped plate is covered with a slight deposit of some hard metal, usually nickel, a great number of copies can be taken without damaging the plate.

A WONDERFULLY exact parallel to Nature-printing is seen in almost every coal bed. In the coal are found impressions of various leaves, mostly ferns, and so exact are they, that the different species have been determined and named with as much accuracy as if, instead of mere impressions, they had been the fern-leaves themselves.

Indeed, if it were needed, it would be perfectly easy to take electrotype plates from these impressions, and to treat them in exactly the same manner as those obtained in the way which has already been described.

STIPPLING.

WE now come to another branch of Art, namely, the production of shadow in an engraving by means of Stippling, *i.e.* the insertion of dots instead of lines. At one time the Stipple was in great favour. Then it was almost wholly abandoned in favour of the line, and now it is much used in conjunction with the line, especially for the delicate shading of flesh tints, such as faces, female arms, &c.

In the illustration a little stippling of a cheek is shown, the dots being purposely exaggerated.

A singularly beautiful modification of the Stipple is now in use. When the engraver wishes for exceptional softness of shading, he does not content himself with mere dots, but, with the aid of his magnifying-glass, converts each dot into a tiny star with three or more rays. Thus the dots seem to melt into each other, and the requisite softness is obtained.

A very good example of this star-stipple is seen in the well-known print called "Coming of Age." If the face and neck of the girl in the foreground be examined with a magnifying-glass, the apparent dots will be seen to be stars, so beautifully arranged that the projecting rays of one fuse themselves, so to speak, with those of the surrounding stars, as is shown in the illustration.

WHETHER the engraver who hit upon this singularly effective plan took it from Nature, I cannot say, but he well might have done so, had he examined the petal of a flower through a good microscope. We all know the peculiar rich softness of a petal, and how our very best floral artists feel the impossibility of transferring it to paper.

The real reason for this special beauty lies in the star-stippling of the petal. The whole surface of the petal is covered

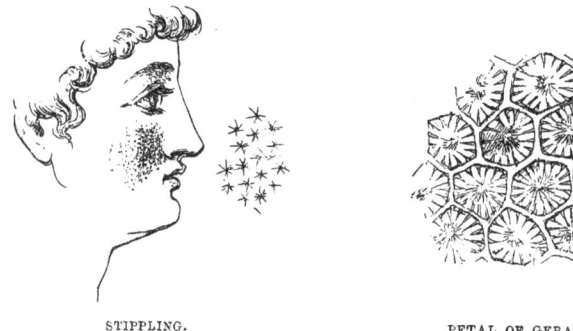

STIPPLING.   PETAL OF GERANIUM.

with multitudinous projections, which are, in fact, undeveloped hairs. These projections are wrinkled down the sides, and so, when viewed from above, they present the curious star-like appearance shown on the right hand of the illustration.

The drawing is taken from a petal of Pelargonium prepared by myself.

There is yet one point in the petal which the star-stipple has not touched, and probably cannot touch. I mean the slight projection of the stipple-hairs, which give an effect of light and shade as well as mere flat softness.

PLASTER CASTS.

WE have already mentioned the electrotype, and may now come to a branch of art which is much associated with it, namely, the Stereotype.

As many of my readers may know, types are very valuable articles, and must not be wasted. If, therefore, a book should be thought likely to have a steady sale, much of its value would be lost if the types were kept standing, inasmuch as they could not be used for any other work.

In such cases the Stereotype is employed. Omitting minute details, the process is as follows :—

The type, ready set up, is carefully oiled. Plaster of Paris mixed with water is then poured into a shallow trough, and the type pressed into it. In a short time the plaster hardens, and the type is withdrawn. The plaster mould is then baked, to

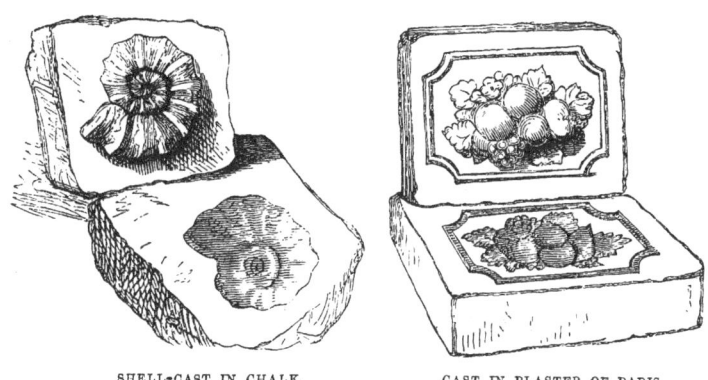

SHELL-CAST IN CHALK.   CAST IN PLASTER OF PARIS.

drive off all moisture, and type metal is poured into it. Thus a solid mass is procured, instead of a number of separate pieces, so that there is no danger of disturbance, and the whole block can be multiplied *ad libitum* if needed. This process sets free the types, which can be broken up and used again.

The ordinary method of taking plaster casts is nearly the same as that which has been described. The object to be cast is oiled, and plaster of Paris carefully applied to it. When it is "set," the plaster "mould" is removed and dried. The process is then reversed, the interior of the mould being oiled, and plaster poured into it, so as to produce an exact reproduction of the original.

In Nature we have almost exactly the same process, although it is necessarily conducted in a much slower manner.

All who have tried their hand at practical geology must be aware of the multitudinous casts of perished beings which are found in various strata. Sometimes the casts are those of vegetables, the original material having been decomposed, and stony matter taken its place. Sometimes there are casts of fishes or echini, while shells, and even insects, are found to have been cast almost as perfectly as could be done with plaster of Paris at the present day.

As might be anticipated, the chalk deposits are peculiarly rich in these casts, the fine particles of the chalk taking the place of the plaster of Paris.

In the illustrations are shown examples of casting in Art and Nature. On the right hand is a cast of fruit and leaves, which may afterwards be reproduced in plaster, wax, papier-mâché, or electrotype. On the left is shown one of the shells so common in the chalk, the upper figure representing the shell itself, and the lower the mould that has been formed around it.

CORRUGATED IRON.

WE have already seen that the Wasps are paper-makers. We may now see how some of the Wasps have anticipated a

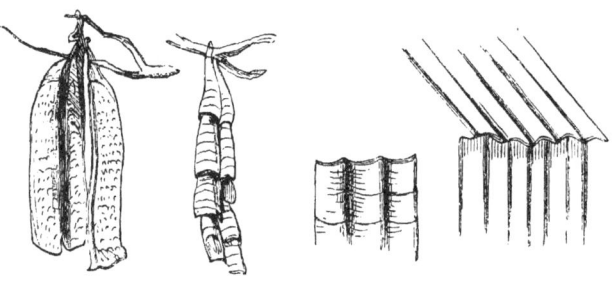

NESTS OF POLISTES.   CORRUGATED IRON.

valuable invention of man, namely, the principle of corrugation, whereby a thin plate gains strength.

Even a sheet of paper gains great strength by corrugation, as is seen in those paper covers which are so much in use for the decoration, or rather the concealment, of flower-pots. But the best example that can be given of this principle is the

Corrugated Iron, which has come so much into use for temporary buildings, such as schools, places of worship, reading-rooms, &c.   It is very light and very strong, and can be used either for roof or walls with equal success.

By means of certain wasps belonging to the genus Polistes, Nature produces corrugated dwellings, which are made of very thin materials, but which are marvellously strong in proportion to their weight.

The insects belonging to this genus are all exotic, but are spread over a very large surface of the earth.

So strong are the nests made by some of these species, that they need no external covering, the corrugated paper supplying at the same time strength and warmth, the latter element being furnished by the air which is entangled between the corrugations.

There are many species of Polistes, mostly belonging to Australasia and tropical America, the latter displaying the greatest variety of form and structure in the nest.

USEFUL ARTS.

CHAPTER XV.

Electricity, Magnetism, and Galvanism mutually convertible.—The Force co-extensive with Nature.—Uses of Thunder-storms.—Languor from Want of Electricity.—Frictional and Voltaic Electricity.—Origin of the Name.—Structure of the Voltaic Pile.—A simple Example of the Pile.—Nerves of a Frog's Leg.—The Electric Shock, and how to produce it.—The Electric Jar and Battery.—Animal Electricity.—The Torpedo and Electric Eel.—Structure of the Electric Apparatus.—The Electric Spark obtained from both Fishes.—Channels of Electricity in the Body.—The Will and the Muscles.—Electricity the conducting Agent.—The Human Body permeated by Nerves.—Telegraph Wires and the Nervous System.—Lightning and the Electric Spark.—The Electric Light and its Power.—The Fire-fly, the Glow-worm, and the luminous Inhabitants of the Sea.—Magnetism and Diamagnetism.—The Electric Telegraph and the Compass.—The Principle identical in both Instruments.

ELECTRICITY AND MAGNETISM.

IT has long been known that Electricity, Galvanism, and Magnetism are but different manifestations of the same force, and that one can be converted into the other at will. It is also known that this wonderful and most important principle lies latent in everything, and only needs the proper machinery to evoke it.

The few following illustrations are intended to show its prevalence in Nature, and that human art does not create, but only makes manifest a power that exists, but lies latent until called forth.

Without going into details, which would occupy the whole of such a volume as this, I may mention that Electricity saturates all the material creation, and that even man himself is not only a reservoir of electricity, but that he feels positively ill if the normal amount be not supplied.

Take, for example, the hours that precede a thunder-storm. We feel languid and depressed. We cannot bring our thoughts together. We are almost incapable even of bodily labour. The reason is, that the portion of the earth on which we live has parted with some of its electricity, and has drawn it out of our bodies.

Then comes the welcome thunder-storm; clouds overcharged with electricity come to restore the balance. The lightning flashes from the clouds to the earth as soon as they are near enough; the rain falls, carrying with it stores of silent electricity; and in an hour or two all seems changed.

The air, which hitherto seemed to afford no nourishment to the lungs, is bracing and invigorating. The nervous system recovers its tension, and the brain can act without a painful effect. All Nature seems to put on a different aspect, and brightness and vigour take the place of dulness and languor.

By a strange coincidence, there is just such a lack of

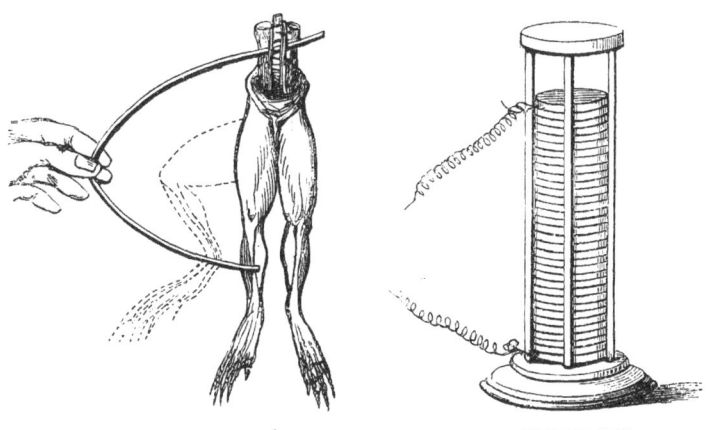

GALVANISING A FROG'S LEG.   VOLTAIC PILE.

electricity as I am writing, and the barometer has rapidly sunk to such a degree that a storm seems inevitable.

One of the chief difficulties in dealing with such a subject as this is to know where to begin. We will, however, do our best to take a general view of it, without going into details.

Many centuries ago it was well known that amber, if rubbed with a dry cloth, would first attract, and then repel, various small and light substances. Indeed, the Greek word for amber, namely, *Elektron,* has given its name to the modern science of

Electricity. Many other substances, such as glass, sealing-wax, &c., possess the same property.

This frictional electricity is but transient, the electric fluid, if we may be allowed to use the term, being driven out by main force from the material in which it was latent, just as fire is procured by the friction of two dry sticks. There is, however, a form of Electricity called Galvanism, from its discoverer, Galvani, who, somewhere about 1790, discovered that the limbs of a dead frog might be excited to action by electricity applied to the nerves.

Afterwards, Volta of Pavia, from whom the Voltaic Pile is named, took up Galvani's discoveries, and produced electricity without friction, by the contact of differently conducting substances.

The right-hand figure represents the Voltaic Pile. It is composed of a series of plates arranged in the following manner— Zinc, Silver, and Cloth, the whole being moistened with diluted acid. Copper will answer the purpose nearly as well as silver, and is not so costly. A very simple mode of demonstrating the presence of electricity is by taking a piece of zinc and a silver coin, and placing one below and the other above the tongue. If the two be then brought together, a very peculiar taste is perceived, and a sudden flash of light seems to pass across the eyes.

The illustration represents on the right hand the Voltaic Pile as at present made, and on the left are the two hind-legs of a frog, with the upper part of the nerves made bare for the purpose of experimenting. The dotted lines show the extent of the movements of the leg when the galvanic current is passed through the nerves.

Now we come to a plan whereby electricity can be accumulated, or locked up, so to speak, and be discharged at once with a definite shock instead of being poured away by degrees. This can be done in many ways, the most common being that which is known by the name of the Electric Jar. It is a glass vessel coated within and without with tin-foil, and having a metal rod passing through the cork in such a way that while the lower end is in contact with the inner coating of tin-foil, the other end is guarded by a ball.

Electricity is now poured into the interior of the jar, and, when contact is made between the inner and outer coatings, a sudden discharge takes place. If a number of persons hold each other's hands, and those who form the two extremities touch the outer coating and the ball which communicates with the inner coating, a sharp discharge is at once made, passing through all the bodies, and inflicting a smart shock, especially at the elbows.

Similar effects can be produced with the Voltaic Battery, but, as that instrument has already been figured, the Electric Jar has been selected. Of course any number of such jars can be connected together, and the shock will be proportionately increased in intensity.

In Nature we have several-parallels. Putting aside the obvious one of a lightning-flash, which has already been

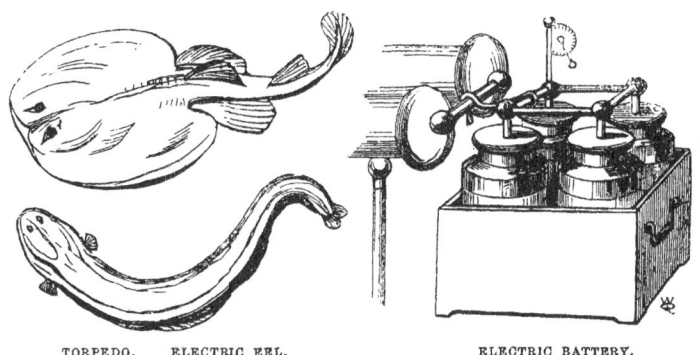

TORPEDO.   ELECTRIC EEL.   ELECTRIC BATTERY.

mentioned, we pass to two remarkable examples of the capability of animal structure to produce electricity, to store it up, so to speak, and discharge it at will. Both these creatures are fishes, one belonging to the Skates or Rays, and the other to the Eels.

The upper figure on the left-hand side of the illustration represents the Torpedo, sometimes called the Cramp-fish, Numb-fish, or Electric Ray. Fortunately for us, it is but seldom found on our coasts, but it is tolerably common in the warmer parts of the world.

The electric organ in this fish is double, and so large that its shape can easily be recognised even through the skin. It is made up of a vast number of discs arranged upon each other in columns like the metallic portions of the Voltaic Pile, and separated from each other by delicate membranes, which take the place of the cloth. When I mention that more than eleven hundred columns have been found in a single Torpedo, and that each column contains several hundred discs, it may be imagined that the shock which such a creature can give must be a very powerful one.

The object of this power seems to be analogous to that of the venomous serpent, i.e. to enable the creature to secure its prey by either killing it or rendering it temporarily insensible by an electric shock. As if to show that the delivery of the shock is achieved by an exertion of will, observers have noticed that just before the shock is delivered, the eyes are depressed in the head like those of a toad when swallowing a large insect.

A STILL more powerfully electric animal is the Electric Eel of Southern America. It sometimes attains a length of six feet, and its electric organs are four times as proportionately large as those of the torpedo.

There is no doubt as to the object of the electric power of this eel, as I have often seen it kill fish, and then eat them.

When about to deliver its shock, it curves its body towards the intended victim, stiffens itself, and with a sort of shudder the electric fluid is emitted. The fish at which it is aimed never seems to escape, but, simultaneously with the shudder on the part of the Electric Eel, turns on its back and lies motionless until it is picked up by its destroyer.

Neither the Torpedo nor the Electric Eel has unlimited stores of electricity. If irritated into delivering repeated shocks, each discharge is less powerful than its predecessor, until at last the creature is almost wholly powerless, and must rest and recruit itself before it can lay up fresh stores of the electric fluid.

I may add that the electric spark has been obtained from both these fishes. It was only a small spark, but in such experiments a small spark is as satisfactory as a large one.

WHAT are the channels by which the electric fluid is transmitted through our bodies?

They are the nerves, which convey from and to the brain a subtle fluid, if it may be so called, just as the arteries and veins convey blood to and from the heart. If any of these nerves be electrified, even after the death of the animal, or after the separation of a limb from the body, muscular movements are induced, and the limb moves as if instinct with life.

Without these nerves we should be unable to feel the severest shock, but they permeate the body so completely, that not a part of the skin can be pricked without a nerve being wounded.

It is by means of these conductors that the will is made to act upon the limbs. The mind, for example, desires the legs to walk, and they do so, the order being transmitted to them through the nerves.

As a rule, we are unconscious of this process. But, when paralysis takes place, and the nerves refuse to perform their functions, the will is absolutely useless, and, however desirous a man may be of walking, he cannot move a step if the nerves of his legs are paralyzed. In cases where the paralysis comes on slowly and in detail, the patient mostly becomes conscious of the part played by the nerves, and feels that his will can to a certain degree rouse the expiring powers of the nerve fluid.

This in its turn is but the conductor for another and infinitely more subtle fluid, of which our space will not allow us to treat, but which forms the connecting link between body and spirit. Perhaps some of my readers may have seen those curious preparations of the human form, when the arteries have been injected with red wax, and the veins with blue wax, and then the fleshy portions dissolved away by chemical means.

The result is a perfect human form, and even to the very tips of the fingers and toes the blood-vessels follow the contour of the body. Did we have means of injecting the nervous system, we should arrive at similar results, except that the nerves would be found infinitely more intricate than the veins and arteries. Thus a human being is a series of human forms, interwoven with each other, and mutually dependent on each other.

It is curious to see how the great discoveries of modern days have but copied Nature.

Take, for example, the network of telegraphic wires which is day by day spreading itself over the surface of the earth, and

the parallel will at once be visible.   Just as the brain transmits
its message to the limbs by means of the nerves, so does the
same brain transmit its message through thousands of miles,
by utilising the wires which are but the rough and coarse imi-
tations of the wonderful nervous system of the human frame.

THE illustration shows the parallelism as well as can be done
by a mere chart.

On the left-hand side is shown the manner in which a nerve-
group is distributed to different parts of the body.   On the
right the railway telegraph wires are seen, and, as the reader
will probably remember, branch wires are carried into the

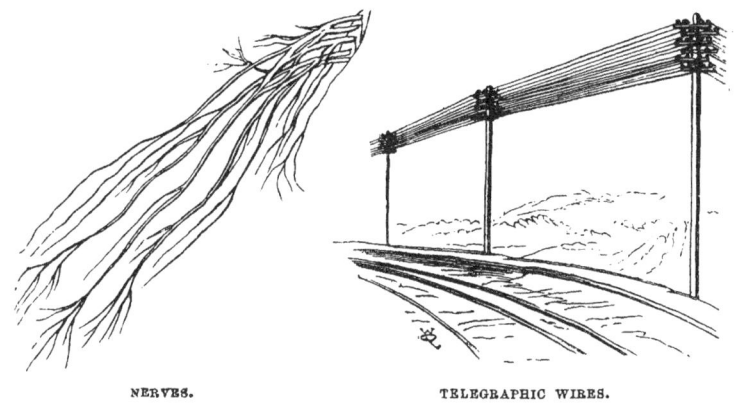

NERVES.   TELEGRAPHIC WIRES.

signal boxes, just as branch nerves are carried to the most
distant parts of the body.

I HAVE already mentioned the Electric Spark, and that it is,
in fact, a miniature lightning-flash, the little crackling report
being a miniature thunder-clap.   It can be produced by frictional electricity, or by the voltaic pile in its many variations,
or by animal substances alone, as in the case of the torpedo and
electric eel.

We now come to a modification of the spark, whereby a continuous current of electricity is sent through two charcoal
points, and inflames them with such intensity that the eye cannot look upon its dazzling whiteness.   There is none of the
yellowness about it which is so great a drawback to our artificial lights, whether they be gas, candle, or lamp, and which

makes ladies' dresses that are really beautiful by day look dull and almost ugly by night.

It is wonderful to see how the Electric Light kills all other lights. The brightest gas becomes dull, and its shadow is thrown on the wall which it formerly illuminated, and the most delicate tints of silks and satins suddenly display themselves in the blinding whiteness of the Electric Light.

At present it is too costly to be brought into common use, but its intensity is so great that serious ideas have been formed of dispensing with street lamps altogether, and illuminating towns with a few electric lamps placed at a considerable height, and having their beams reflected downwards.

London is thought to be a specially fit subject for this mode

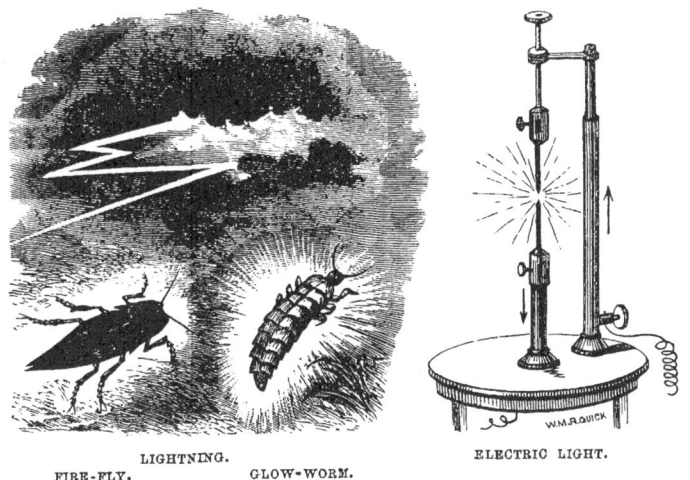

LIGHTNING.

FIRE-FLY. GLOW-WORM.

ELECTRIC LIGHT.

of lighting, as the electric beams can pierce the fogs which the gas-lamp only augments, and give the traveller some hope of finding his way through the most familiar streets.

In the illustration the right-hand figure represents the Electric Light as at present in use. The upper portion of the left-hand side represents the forked lightning, whose dazzling whiteness is so familiar to us, even in the noon of a summer's day.

Below are shown the Fire-fly of warm climates, and the Glow-worm, which, in our comparatively cool country, cheers the summer evenings with its pale lamp. As to the source of this mysterious light, which burns without producing heat sufficient to be recognised by our most delicate instruments, we know but little.

There are instruments so infinitely more sensitive than the best thermometer, that they will record instantaneously an increase of heat if a human being passes in front of them, though at several yards' distance. Yet no effect is produced on them by any of the Fire-flies or the Glow-worm. The spectroscope itself gives little or no information, the spectrum of the light being without bands or bars, and being what is technically called a " continuous " spectrum.

Last year I tried numbers of Glow-worms with the spectroscope, and always with the same result. I never saw the Fire-flies alive, but, no matter what may be the colour of the light, the spectrum, whether of the Glow-worm or any of the Fire-flies, seems to be always continuous, and so to give but little information as to its source.

There appears, however, to be little doubt that animal electricity is the real cause of this curious phenomenon, and that the force which is expended in the torpedo and electric eel, in giving shocks accompanied by slight electric sparks, may develop itself in these insects by producing a continuous light. And just as the electric fishes can emit or withhold the shock as they please, so can the Fire-flies and Glow-worms give out or retain the light by which they are so well known.

Then we come to the multitudinous luminous inhabitants of the sea, which, as many of my readers have probably seen, convert the waves into rolling masses of living fire.

MAGNETISM.

Now we come to another condition of electrical force, called MAGNETISM.

One form of it is strongly developed in the Loadstone, an ore of iron. This ore has the property of turning east and west when suspended freely, it attracts any object made of iron, and can communicate its powers to iron by merely stroking it. There is in the Museum at Oxford a splendid specimen of the Loadstone, which has imparted its virtues to thousands of iron magnets, and has lost none of its virtues by so doing.

All bodies are now known to be magnetic in some way or

other. Several, such as iron, nickel, and one or two other metals, turn north and south when suspended on a pivot, but the great bulk of other bodies turn east and west, and are called Diamagnetics.

As we all know, the property of turning north and south has been utilised in the Compass, without which modern science would be paralyzed, and travel rendered impossible.

It is worthy of notice that although the magnetic needle of the compass turns to the north, it does not do so because it

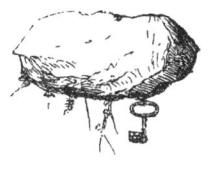

LOADSTONE.

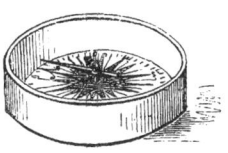

COMPASS.

is attracted by the north pole, but because it is repelled from the east and west.

We have long known that if a current of electricity be sent round a magnetic needle, the latter at once turns at right angles to it. On this principle depends the Electric Telegraph. When communication is made by using the handles, a current of electricity is sent round the needles, and causes them to turn at right angles until stopped by a little ivory pin, which prevents them from overshooting themselves.

There is a perpetual stream of electricity passing over the earth from east to west, and in consequence all magnetic bodies are forced to turn at right angles, just as is the case with the magnetic needle.

USEFUL ARTS.

CHAPTER XVI.

TILLAGE. — DRAINAGE. — SPIRAL PRINCIPLE. — CENTRIFUGAL FORCE.

Systems of cultivating Ground.—The Fallow System.—Manuring the Ground.—Custom of China.—Nature's Abhorrence of Waste.—What becomes of Dead Animals.—Burying-beetles.—The Scarabæus-beetles and their Work.—Drainage *versus* Sewage.—Clay Soils and Drains.—The Mole, the Earthworm, Rats, Mice, and Rabbits.—The Flexible Drain and the Lobster's Tail.—The Turbine Pump and the Ascidian.—The Spiral Principle.—The Smoke-jack, Kite, and Wings of Birds.—Centrifugal Force.—Revolution of Planets.—The "Governor" of the Steam-engine.—The Sling, Amentum, and Mop.—The Gyroscope, the Bicycle, and the Hoop.

SEVERAL times, in the course of this work, we have touched upon man's dealings with the earth, such as mining and tunnelling. We will now take another side of the same question, and, in connection with Tillage, consider Drainage, whereby superabundant moisture is removed from the earth, and Manuring, whereby the exhausted soil is renovated.

We will take this subject first.

It has long been known that it is impossible to get more out of the ground than exists in it, and that when the soil has been so worked as to become unproductive, there are only two remedies. The one is to allow the ground to remain uncultivated for a time. It must be ploughed in deeply, as if it were to be sown with a crop, and must be left to recruit itself from the air. This is the now abandoned "fallow" system, which used to be in full operation when I was a child.

As, however, population increased, and with it the perpetually increasing demand for food, land was found to be too precious to be allowed to lie fallow and idle. Then came the system of rotation of crops, potato following wheat, clover following potato, &c. But, above all, agriculturists learned that in the long-run there is nothing so cheap as manure, *i.e.* the return to the soil by animals of the elements which these animals took out of it.

On the right hand of the illustration (page 495) is shown the simplest mode of enriching the soil, namely, by spreading the manure on the surface of the earth, and then digging it in. Any mode of thus enriching the earth is a proof of civilisation. No savage ever dreamed of such a thing, and I doubt whether barbarians recognised the principle at any time.

Nowadays we have recognised the necessity of returning to the soil in one form the elements which we have taken from it in another. As usual in such arts of civilisation, the Chinese have long preceded us. They waste nothing, carrying, perhaps, its principles to an extent which scarcely suits our European ideas.

They even utilise the little clippings of hair, to which every Chinaman is almost daily subject, if he wishes to keep up his self-respect in public. The barbers carefully preserve these clippings, and sell them to gardeners. They are too precious to be used in general agriculture, but the flower artist, when he plants the seed, puts in the same hole a little pinch of human hair, knowing it to be a strong stimulant to growth.

WITHOUT multiplying examples of artificial manuring, most of which are too familiar to need description, we will proceed to the methods by which Nature has for countless centuries achieved the same work that Man has lately learned to undertake.

Nature abhors waste, and in the long-run will prove it, however wasteful may be the ways of her servants. Take, for example, the case of an ordinary tree, such as an elm, an oak, or a birch. In the autumn the leaves fall. In the next summer scarcely a dead leaf can be found. They have been decomposed by rain, dews, and gases, and have thus returned to the earth more than the nutriment which they took out of it.

Here man is apt to interfere. Knowing the invaluable productive powers of decayed leaves, he removes them as they fall, and stores them in heaps so as to form the costly, but almost indispensable, "leaf mould." In so doing, however, he deprives the trees of their natural nutriment, and by degrees they dwindle and die.

Nature, in this case, shows her superiority over Art.

Then we have the remarkable fact that millions of animated beings die annually, and no vestige of their remains is found. Hyænas and vultures might account for a few bodies, the remnants of which have been found in ancient caverns. But there is no hyæna which could crush the leg bones of an adult elephant; and yet I suppose that neither in Africa nor Asia has any one discovered the body of an elephant or rhinoceros that had died a natural death.

In the first place, there is the curious point, which I have already mentioned, and which is shared by nearly every race of human savages, that when an animal feels that it has received its death-stroke, it accepts the conditions, withdraws itself from those who yet have life in them, and yields up its life as calmly as if it were but sleeping.

But what becomes of the body? As to such enormous beings as elephants, the various species of rhinoceros, and whales, which are as large as several elephants, rhinoceros, and hippopotamus put together, I cannot say from practical knowledge.

Still, as size is only comparative, the rule that holds good with a small animal may hold equally good with a large one. It is my lot to walk very often upon the banks of the Thames. It is a charming walk at high water, but at low water there is too much odoriferous mud, and there are too many dead dogs and cats to make it an agreeable resort, except for enthusiastic entomologists, who seem to swarm in this neighbourhood.

Scarcely has such a carcass been stranded than it is beset by Burying-beetles of various kinds. Hundreds upon hundreds can be shaken out of the corpse of a dog or cat, and, before the next tide has come up, there is scarcely any flesh left on the bones, it having been dug into the earth by the Burying-beetles.

THEN there is that wonderful family of Scarabæus-beetles, which do us invaluable service as scavengers and agriculturists. They follow the path of the caravans, and effectively cleanse the course which has been traversed. Even man is obliged to utilise as fuel the droppings of the horses, cows, and camels ; but the Scarabæus goes further, collecting all that man does not need, and burying it in the earth.

The instinct of the female Scarabæus urges it to gather together the rejecta, to form them into balls, placing an egg in the middle of each ball, and to bury them in the ground. Thus a double object is attained, the offensive substances being removed from the surface of the ground, where they do harm, and being transferred below the surface, where they do good.

Even the curious instinct of the dog, which leads it to bury bones, &c., which it cannot consume, and which it often forgets, if well fed, leaves them to be consumed by the all-absorbing earth.

It is evident that, in the end, the earth *must* receive back again that which has been taken from it. If, for example, we follow the present most wasteful plan of drainage, and fling into

SCARABÆUS-BEETLES.　　　　MEN MANURING GROUND.

rivers everything which ought to be utilised on land, it only gets into the sea in the end, and in the course of years is decomposed, and returns to the earth in the form of gases. Meanwhile, however, we have robbed the locality, deprived it of the nourishment which it required, and forced ourselves to supply it elsewhere at a costly rate.

So runs the cycle of creation. Sooner or later, Nature will have her way, and the more we help her, the better it will be for us.

Of course I do not mean to condemn Drainage, which is an absolute necessity in agriculture, and a matter of life and death in households. But, when rightly conducted, it only signifies that water is removed from a spot which is overstocked with moisture to one where it is needed. Wet clay lands, for example, which were unproductive in point of crops, and injurious in point of human health, have been converted by judicious drainage into fertile and healthy grounds.

This, as it will be seen, is a very different business from removing from the soil the elements which rightly belong to it, and which sooner or later, in some form or another, it will claim and recapture.

Still, it is evident that in the progress of civilisation there must be accumulations of all kinds of refuse, which savages utterly disregard. Then we come to the question of the Drain combined with the Sewer, and are enabled to see how the hand of man, if properly directed, only follows the course of Nature.

So we undermine our towns with a complex system of drains which are understood by only a very few people. For example,

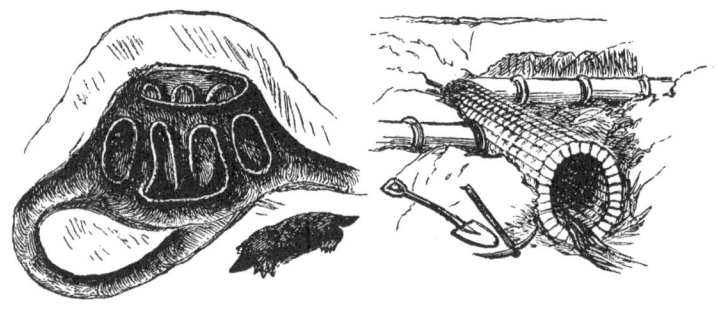

TUNNEL OF MOLE.   SEWER.

just as a tree is only half visible, the roots being about equivalent to the branches, London is only half visible, the subterranean architecture being little, if at all, inferior to that of the surface.

Here, again, we are met by Nature. Very few of us can appreciate the extensive subterranean works which underlie us, even where the hand of man has never been placed. Putting aside a multitude of tiny creatures, there are, in our own country, the earth-worms which pierce the ground in all directions, at the same time draining and manuring it. They penetrate it with their little burrows, thus admitting the air, which the earth needs as much as we do, and allowing moisture to take its right place. Then there are the moles, that are perpetually travelling after the earth-worms, and making drainage galleries of wonderful extent. Then there are the numerous other burrowers, such as rabbits, mice, and rats, which are common everywhere, besides the less plentiful foxes, badgers, and various burrowing birds, all of which assist more or less in the drainage of the earth.

Even bees and wasps of different kinds assist in this work, the hardest soil yielding to their small, though powerful, jaws and feet, and so being made, if only temporarily, able to carry off the superabundant moisture.

ONE of the most ingenious modes of Drainage was that which was invented by Watts, and was avowedly based on Nature. He had engaged himself to carry a drain tube through, or rather over, an extremely irregular bed of a river, where the pipes must accommodate themselves to existing conditions.

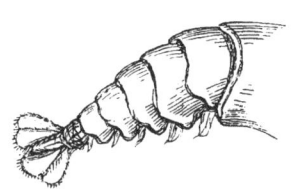 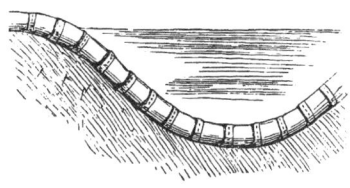

TAIL OF LOBSTER.   FLEXIBLE WATER MAIN.

The modern system of pipes not having been brought into existence, Watts had to adapt himself to circumstances, and did so by making his pipe on the model of a Lobster's tail, as shown in the illustration.

We have already seen how the same object has been utilised in warfare as a pattern for armour, but it does seem rather strange that it should be employed in the tranquil arts of peace.

ANOTHER method of removing superfluous water is by the TURBINE PUMP, by which the water, instead of being cast up in successive jets, was flung out in a continuous torrent. Some of my readers may remember the sensation which was created at the first Exhibition of 1852 by the then extraordinary powers of the Turbine Pump.

Yet this is, after all, nothing but an imperfect copy of the now celebrated being to which human beings have been supposed to owe their origin, namely, the Ascidian, popularly known by the name of the Sea-squirt, and with very good reasons.

As a rule, it keeps up a rotation of tentacles, such as is shown in the illustration, acting exactly on the principle of the Turbine Pump, and drawing in and discharging water with a power that is perfectly astonishing in so small a being. Beside this, it has the power of flinging out at once the whole of its watery contents, and any one who has incautiously handled a mass of Ascidians, and been drenched by them, can answer with more truth than satisfaction as to the water-absorbing power of the Turbine.

Then the Ascidian can do what the Turbine cannot do. In the Turbine the water which is taken in must necessarily be

ASCIDIAN.

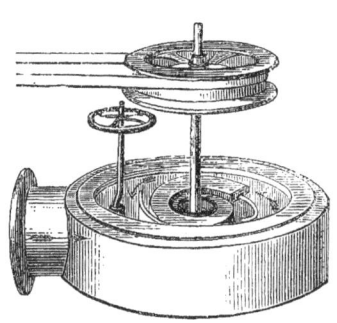

TURBINE PUMP.

ejected in equal proportions. With the Ascidian the same thing takes place, but with the additional power of ejecting all the contained water, and then beginning afresh.

There is now no doubt that the Circular or the Turbine Pump is the most powerful in such cases as emptying mines of the water which, in spite of all precautions, will make its way in, and destroy the labours of the miners. But I merely wish to carry out the object of this work by remarking that the invaluable Turbine Pump is only a very inferior copy of a natural pump, which existed, as far as we know, centuries before Man could find his place upon this earth.

THE SPIRAL.

IN an early portion of this work the Spiral or Screw was touched upon, mostly in connection with the propulsion of vessels. We will now extend it a little further, and see how it is modified so as to perform other offices than those which have been described.

Allusion has already been made to the Spiral or Wedge principle, but some of the illustrations were accidentally omitted. I therefore introduce them here, this being a chapter of miscellanea.

The Windmill has previously been described, as has also the ship's Screw, another form of which is here given.

In the centre is shown the mechanism popularly known as the Smoke-jack, though it really works by means of hot air,

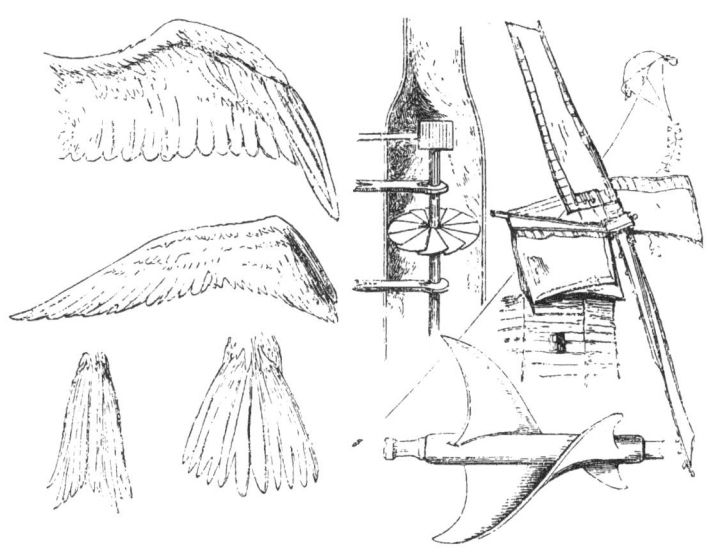

BIRDS' WINGS AND TAILS.   SMOKE-JACK. SHIP'S SCREW. WINDMILL. KITE.

and only becomes gradually choked by the soot which the smoke by degrees deposits upon it. It is, in fact, nothing but a windmill working horizontally instead of vertically, the vanes being moved by the rapidly ascending heated air. So powerful is the spiral pressure of this air, that in my old college days at least a dozen rows of heavily laden spits were perpetually turned by a single Smoke-jack. It is many years since I visited my old college, and I cannot say whether the Smoke-jack still exists, but, as it did its work well so long ago, I presume that it does so now.

Then there is the well-known spiral ventilator set in the

windows of workshops. Perhaps its revolution may not assist the air-current, but it does, at all events, show how much exhausted air has to be expelled from the room, and consequently how much fresh air needs to be brought into it.

PERHAPS the reader may be surprised to see that the Wings and Tail of a bird and a boy's Kite are placed among the examples of the Spiral principle. Yet such is the fact. If the reader will move up and down the wings of any bird which will not bite him, he will find that there is in them a peculiar screwing motion, difficult of description, but very observable.

It is mostly for want of this movement that all our attempts at fitting wings to human beings have been such utter failures. We can make the wings work up and down well enough, but we cannot as yet impart to them the all-important spiral movement.

THAT very well-known toy, the Kite, is another example of the same principle which drives the screw steamer. Its " tail," which need be nothing but a piece of string with a proportionate weight at the end, keeps the Kite in a slanting position, providing that the " belly-band " be properly arranged. The consequence is that the pressure of the wind acts on it as on a wedge, and so drives it upwards until the combined weight of itself and the string counterbalance the upward pressure.

Indeed, the only object of the string is to keep the Kite at a proper inclination; and, if that object could be attained by the force of gravity alone, the Kite would ascend to a height nearly double that to which it can at present attain.

CENTRIFUGAL FORCE.

CLOSELY connected with the spiral principle is Centrifugal Force, that marvellous power which gives to our whole solar system its ceaseless movements, and may extend, as far as we know, to other and vaster systems yet unknown.

Tie a ball to a string, and swing it round, and it will be an exact, though rough, representation of the double power by which the movements of the heavenly bodies are governed, our earth being included among them.

The string represents the force of attraction, which binds all our planets to the sun, and their satellites to the planets, while the force that is employed in swinging the ball represents the mysterious power that issues from the sun, and gives motion to the planets. The metaphor is a very homely one, but it is nevertheless correct.

In the accompanying illustration are several examples of Centrifugal Force as found both in Nature and Art. On the left hand we have diagrams of some of the heavenly bodies, showing the revolution of their offspring, so to call them, while

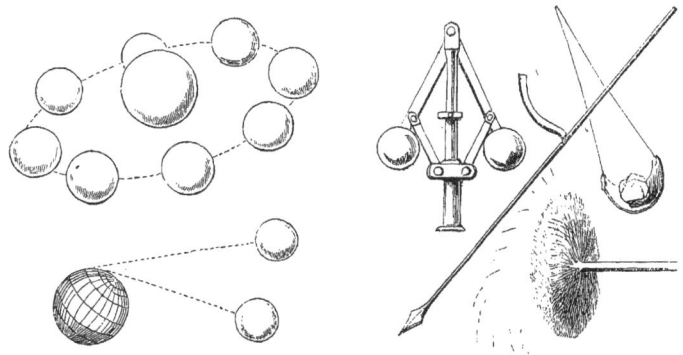

CENTRIFUGAL FORCE OF HEAVENLY BODIES.   CENTRIFUGAL FORCE OF "GOVERNORS" OF ENGINE.
SLING. AMENTUM AND MOP.

on the right are seen examples of Centrifugal Force as applied to human use. For convenience' sake, the illustrations have been separated into two portions.

In the first of these illustrations we have the "Governor" of the steam-engine, that wonderfully ingenious and simple piece of mechanism which controls the force of the steam, and, without the superintendence of man, acts almost as a living being might.

It is composed of two heavy metal balls, hinged, as shown in the illustration, to a movable collar which slides up and down the central rod. When the engine is at work the Governor revolves, and the harder it works, the more rapid is the revolution. Consequently, as it revolves, the balls diverge and draw the sliding collar up the rod.

Here lies the whole beauty of the invention. The sliding

collar is connected with the safety-valve. Thus, if the engine should be working beyond its proper powers, the Governor draws up the collar, and releases sufficient steam to take the undue pressure off the boiler. Thus the engine may be left, so to speak, to manage itself.

NEXT are shown two examples of Centrifugal Force as applied in ancient warfare, namely, the Sling, which is now retained merely as a boy's toy, and the Amentum, which was practically a sling attached to a spear. Both weapons have been superseded by the modern firearms, but the Sling is really a more formidable offensive weapon, in skilful hands, than is generally suspected.

A good slinger is as sure of his aim as a good rifleman, and can send his missile to a wonderful distance. Were I to be armed with the best pistol hitherto invented, I should be sorry to fight an accomplished slinger, unless under cover.

The really tremendous power of the Sling is obtained by Centrifugal Force, the weapon, with its missile, being whirled in the air, and then one string being loosed with a peculiar knack something like the "loose" of a good archer. In consequence, the centrifugal force is converted into direct force, and the missile flies directly forwards.

The Amentum is simply a cord tied to a javelin, so that the thrower has the advantage of a lever, which, after all, is only the conversion of centrifugal force.

The very familiar Mop, flinging off its moisture to a considerable distance, needs no description; but I have introduced it to show the action of centrifugal force in small as well as in great things.

THE next illustration shows how this very same power acts upon the greatest as well as the least of objects, and enables them to maintain positions which otherwise they must of necessity fail to do. Take, for example, our own Earth, and its peculiar position of being tilted on one side, so as to give us the alternative seasons as it flies on its annual course.

This is simply due to its own rapid revolution, which, on the same principle that keeps the arrow and the rifle-ball straight on their course, prevents it from altering its position.

The very same principle acts on the boys' Tops, and is shown in a really remarkable manner by the professional Japanese top-spinners, who will place several tops upon each other, as shown in the illustration, and make them sway backwards and forwards in the most extraordinary manner, sometimes being all upright, and sometimes leaning almost at right angles to each other.

A favourite mode of illustrating this power of Centrifugal Force is by the Gyroscope, a figure of which is given on the right hand of the illustration. The interior wheel is made to revolve rapidly, and the effect of the revolution is to enable the instrument to maintain a horizontal position, even when suspended on one side, as shown in the engraving.

The power of this revolution is quite wonderful, even in a small Gyroscope which can be purchased for a few shillings.

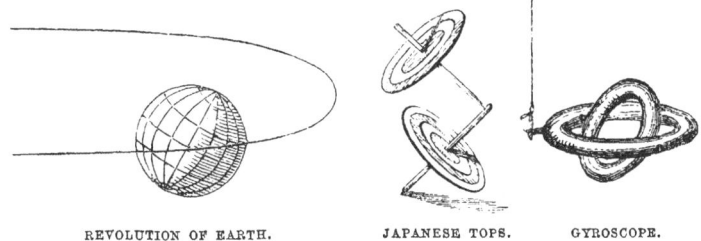

REVOLUTION OF EARTH.   JAPANESE TOPS.   GYROSCOPE.

It almost seems to be alive, and to insist on retaining its position, in spite of all efforts to the contrary.

This principle is used in the swinging cabin of the Bessemer ship, and is also employed by quoit-players in keeping their missile steady as it flies towards the mark. Even the now fashionable Bicycle is managed on the same principle.

As is well known to all bicycle riders, it is comparatively easy to maintain the balance when the pace is rapid and the wheels revolving quickly. The difficulty is, to do so when the pace is slow, and the rider is deprived of the centrifugal force which keeps him on his balance almost in spite of himself. It is just the same with a child's hoop, which runs straight and upright when it is driven rapidly, or when, for example, it runs downhill. But, as soon as the centrifugal force is expended, it begins to waver, loses its direction, and soon falls to the ground.

USEFUL ARTS.

CHAPTER XVII.

OSCILLATION.—UNITED STRENGTH.—THE DOME.

A PORTION of our last chapter dealt of Centrifugal Force. We will now proceed to another well-known power, which seems to be a variation, or perhaps a division, of the same power. I mean the principle of OSCILLATION, which has done so much for the present state of the world. I mention the connection of the two principles because it is evident that, if Oscillation were continued in one direction, it would be converted into centrifugal force. In fact, it can only be considered as centrifugal force interrupted.

The chief point in this subject is the equal time occupied by the oscillating body, no matter what may be the "arc" distance through which it sways, provided that the length of the line remains the same. The discovery of this principle by Galileo in a church at Florence is too well known to need repetition.

This principle may be observed by any one, and at almost any time. The Spider at the end of its line illustrates it, and so does a stone tied to a string, both of which objects are shown in the illustration.

In various departments of Art, Oscillation is absolutely invaluable. We will take, for instance, the best known of these examples, namely, the Pendulum, by which the movements of clocks are regulated. Without some mode of regulation, the works would run down rapidly, and the clock rendered incapable of measuring time. But, in the Pendulum, we possess a means of making a clock go at any desirable rate, and be faster and slower at pleasure; a long Pendulum working slowly, and a short one rapidly.

How the Pendulum affects the working of a clock may be seen by reference to the right-hand figure of the illustration.

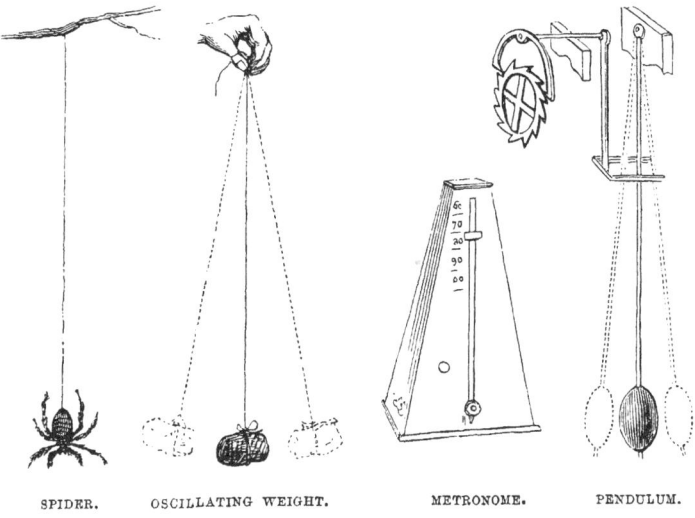

SPIDER.   OSCILLATING WEIGHT.   METRONOME.   PENDULUM.

The movements of the clock are connected with the Pendulum by means of an ingenious piece of mechanism called an " escapement," because it only allows the wheel shown in the illustration to move one cog at each swing of the Pendulum.

Now, as in the latitude of London a pendulum which is a trifle more than thirty-nine inches in length swings once in a second, it is evident that, by lengthening or shortening the Pendulum, we have the rate of the clock entirely under command.

For example, if a Pendulum be required to swing once in

two seconds, it must be four times as long as that which swings once in one second, while to swing once in three seconds it must be nine times as long, the length being measured by the square of the time of vibration.

We are thus able to "regulate" clocks by lengthening the Pendulum if they be too fast, and shortening them if they be too slow. The reader will probably have remarked that the conditions of the atmosphere—such as heat, cold, moisture, or dryness—must have an effect on the length of the Pendulum, and thus alter the rating of the clock. So they do, and in consequence the Compensating Pendulums have been invented, some of them being made of metallic rods of different powers of expansion, mostly brass and steel, while others carry a quantity of mercury in a glass tube near the bottom of the Pendulum.

ANOTHER familiar example of the Pendulum is the Metronome, which is simply a Pendulum with a weight at the top instead of the bottom, the weight being movable up or down so as to decrease or hasten the pace. Generally a bell is added to it, which is struck at the beginning of each bar.

The exactness of its beats is perfect, as is known to all musicians, and is calculated to take the conceit out of players who are apt to disregard their time. I knew one lady, a really good pianiste, before whom I placed my Metronome. Before she had played many bars she broke down, exclaiming that the horrid bell always said "ting" in the wrong place. However, she soon acknowledged the value of the instrument, and was glad to use it.

A very good Metronome may be made by fastening a bullet to the end of a piece of tape, and swinging it backwards and forwards, regulating the tape according to the time required. Such a Metronome is very portable, and extremely useful where the conveyance of the clockwork instrument would be troublesome. Moreover, its beats can be seen by a great number of persons. I have often used it myself.

Such a Metronome is used in the army, in order to regulate the pace of the soldier's step, it being of the last importance that the pace should always be the same. Otherwise it would be impossible to calculate the time which ought to be consumed in marching a certain distance, and the military calculations on which depends the success or failure of a campaign would be wholly upset, half an hour too soon or too late meaning failure.

THE ESCAPEMENT.

As we are on the subject of the pendulum and Escapement, we will say a few words about the latter piece of mechanism. It is here given on a larger scale than in the previous illustration, so that its action may be more easily understood. Whether in watch or clock, the Escapement is exactly the same in principle.

First there is the escapement wheel, the circumference of which is furnished with a number of very deep cogs, varying as to the work which they have to do. Then there comes the escapement itself, which swings on its pivot, and is regulated

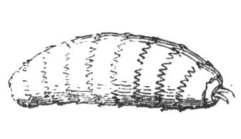
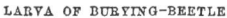

LARVA OF BURYING-BEETLE.   ESCAPEMENT OF WATCH.

in its oscillations by the pendulum. As it swings backwards and forwards, it is evident that only one tooth of the wheel can "escape," and only that in one direction.

We can reverse a steam-engine, but the man has yet to be found who can reverse a clock, *i.e.* enable it to continue going in the opposite direction. The only mode would be to enable one set of cogs to flatten themselves, so as to pass the escapement, and a second set to start up in exactly the opposite direction. Or perhaps there might be two parallel escapement wheels, capable of being connected or disconnected with the clock at pleasure. As, however, a reverse movement is quite needless, no such invention seems to have been made.

ON the left hand is seen an example of the same principle as shown in Nature. It represents a larva or grub of the Burying-beetle. It has no legs available for locomotion, and yet it can get along with tolerable speed.

Many years ago, when living in Wiltshire, I was much struck with this fact. There had been an epidemic among sheep, which killed them off so fast that the farmers would at last not even bury them, but took off the skins, and left the bodies to moulder as they best might.

It was very unpleasant for the farmers, but just the contrary for the Burying-beetles, which simply swarmed in the deserted carcasses. If one of them were tapped with a stick, hundreds of these larvæ came scuttling out, displaying an activity which was really remarkable in creatures practically legless.

In reality this movement is achieved by an apparatus very similar in its action to that of the escapement. The rings, or "segments," of which the body is composed, are furnished with rows of sharp points, arranged very like the cogs of the escapement wheel. By alternately elongating and contracting the body, these points catch against surrounding substances, and force the creature onwards, only allowing of movement in one direction.

Perhaps the reader will remember that in an earlier part of this work it has been mentioned that the various worms propel themselves by the same means. So do the Serpents, the edges of the scales serving the same purpose as the hairs of the worms and the hooks of the grub.

UNION IS STRENGTH.

ON the left hand of the accompanying illustration we have an example of the wonderful power obtained by uniting together a number of comparatively weak objects. It represents a portion of the rope attached to the harpoon with which the natives of some parts of Africa attack and kill the hippopotamus.

Considering that a full-grown hippopotamus weighs several tons, and, in spite of its enormous size, is as active as a tiger, we can infer the strength of the rope which must be needed to hold such an animal when excited with rage and pain.

A few years ago the female hippopotamus at the Zoological Gardens, when deprived of her cub, actually tried to leap over the lofty iron barrier, and so far succeeded as to throw her weight on the uppermost bar. Fortunately it was made of well-wrought iron, and was only bent by her weight. Had it been made of cast-iron, like most railings, she would have snapped it like glass.

Now, the fibres of which the rope is composed are individually feeble, but, when they lend their strength to each other, their strength is amazing. It is well shown by a lasso in my possession, made of the fibres of the aloe-leaf. It is scarcely as thick as a man's little finger, and yet it is strong enough to resist the efforts of the most powerful wild bull. I have some of the separate fibres, and it is interesting to notice how fibres so slight when separate should be so strong when united. Part of the rope has been unlaid, so as to show the manner in which it has been put together.

Towards the harpoon itself, a number of small cords laid loosely side by side are used, so as to prevent the hippopotamus

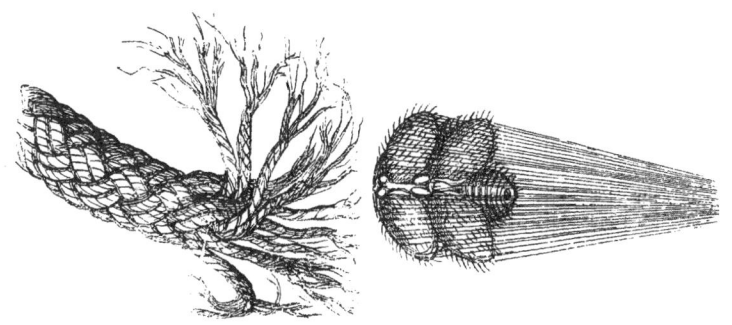

HIPPOPOTAMUS ROPE.   SPINNERET OF SPIDER.

from severing the rope with his chisel-like teeth, which he would assuredly do if it were single. The multitudinous cords become entangled among the teeth, and baffle his efforts; but still their unity is their strength; and, though the animal may sever one or two of them, the others retain their hold until he dies under a shower of spears.

On the right-hand side of the illustration is the Spinneret of the ordinary garden Spider, showing the many orifices from which the silken threads emerge. It is a remarkable point, and one which, I believe, is seldom noticed, that the Spider can at pleasure combine all these fibres into a single cord, or issue and keep them separate, just as is the case with the hippopotamus rope.

The latter operation may be seen whenever a large fly gets into the web. The Spider darts at it, bites it, and then, ejecting a loose mass of fibres, rolls it up in a moment, as in a shroud, carries it off and hangs it in a convenient place, and mends the broken meshes of the web. But both kinds of the cords of the net are made differently from the winding-up fibres, the former being fixed together, and the latter kept separate.

PRINCIPLE OF THE DOME.

WE are all familiar with Domes, especially when the Dome of St. Paul's is the most conspicuous object in our metropolis. Few persons, however, except professional architects and builders, seem to ask themselves the principle on which the Dome is constructed.

The strength of the arch is well known, and the Dome is practically a number of arches, affording material support to each other, and so enormously increasing the strength of the edifice.

A good idea of the Dome principle may be formed by taking two croquet hoops, placing them at right angles to each other, tying them together at the intersection, and pushing the ends in the ground. Even by this very simple arrangement considerable strength can be obtained ; but, if the hoops be sufficiently multiplied to form a close Dome, it will be evident that the strength will be correspondingly increased.

So strong, indeed, is the Dome, that it could be made without mortar or cement, although, of course, its strength is increased by their use. A very good example of a Dome thus constructed is found in the " igloo," or snow-hut of the Esquimaux, which has already been described.

As to the example which I have selected, it would have been easy enough to have chosen one of the great Domes of the world, such as St. Peter's at Rome, St. Maria del Fiore at Florence, St. Paul's of London, or St. Geneviève or the Invalides of Paris.

I have, however, selected the present example on account of the thinness of its walls, the fragility of its material, and the enormous pressure which it has to undergo. This is the " Receiver " of the Air-pump. It is made of glass not thicker than an ordinary tumbler, and yet, even when exhausted of air, it will resist the pressure of the atmosphere for days together.

When it is remembered that the Receiver is deprived of its internal air, and therefore has to resist a pressure equal to fifteen pounds on every square inch of its surface, it may be imagined how strong the Dome is. Were the top or either side to be flat, it would be crushed as soon as a vacuum was formed sufficient to deprive it of the support of the air within.

A GLANCE at the illustration will show how the Receiver is modelled on the same plan as the Human Skull, the outlines being curiously similar. It is this formation which imparts such strength to so thin a set of bones as those which com-

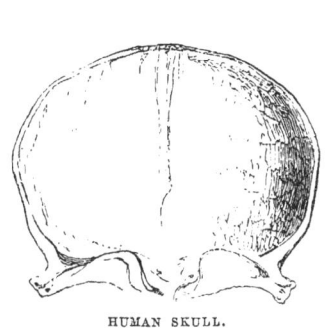 

HUMAN SKULL.   RECEIVER OF AIR-PUMP.

pose the human skull as enables them to protect a sensitive organ like the brain, on which both reason and life itself depend.

Eggs also form good examples of the wonderful strength obtained by this principle, their thin shells protecting the yelk and the white, as well as the chick through its progress to maturity.

THE last subject in this chapter is a curious example of an evidently accidental resemblance in form.

The figure on the right of the accompanying illustration will at once be recognised as one of those Salad-dressing Bottles which try to conceal by their shape the small volume of their contents.

That on the left represents one of the many forms through which the Medusa passes before it attains its perfect form. It was long thought to be a separate creature, and was known under the scientific name of Strobila. Modern researches have, however, made the discovery that it is one of the transitional stages between the creature known as the Trumpet-hydra (*Hydra tuba*) and the Medusa, popularly known as Jelly-fish.

The former almost exactly resembles the Hydra of our fresh waters. It is a tiny transparent gelatinous bag—so transparent as to be scarcely perceptible, and with some thirty or forty long and delicate tentacles hanging from its open end. These tentacles are used in catching the minute creatures on which it feeds. It is fixed, and, to use Mr. Rymer Jones's simile, looks like a beautiful silk-like pencil waving amidst the water. Its length is not quite half an inch.

That it should be identical with the remarkable form shown

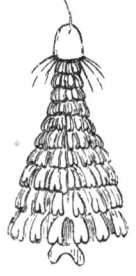

TRUMPET-HYDRA.

SALAD-DRESSING BOTTLE.

in the illustration seems impossible, but such is the case. Its body becomes contracted as if tied with strings, and every segment thus formed develops a set of tentacles, breaks away, and swims off in the form of a Medusa. The upper segment is exhibited as undergoing this process.

The figure is magnified so as to show the structure better, its right length being about one-third of an inch. A full and graphic history of this creature and its manifold changes may be found in Mr. Rymer Jones's "Aquarian Naturalist."

It is not likely that the inventor of the Salad-dressing Bottle ever saw a Hydra, but the resemblance is strangely exact.

ACOUSTICS.

CHAPTER I.

PERCUSSION.—THE STRING AND REED.—THE TRUMPET.—EAR-TRUMPET.—STETHOSCOPE.

The Science of Sound.—Rhythmical Vibrations.—The Drum.—Primitive Drums.
—The Solid and Hollow Log.—The Bass Drum and Kettle-drum.—African
Drums.—Gnostic Gems and the Ashanti Drum.—Tympanum, or Drum of
the Human Ear, and its Mechanism.—An artificial Tympanum.—The
String.—The Bow and the Harp.—The Harpsichord and the Zither.—The
Bow and the Violin.—The Cricket.—The Vibrator, or Reed.—The Jew's
Harp and Harmonium.—The Cicada and its Song.—Harmonics upon Strings.
—The Æolian Harp.—Harmonics upon the Trumpet.—The Trombone.—
Trachea of the Swan.—The Ear-trumpet.—The Sea-shell.—The Stethoscope.
—Savage Food.—The Aye-aye.—The Siren and its Uses.—Echo and Whis-
pering Gallery.

IN a work of this nature it would be absolutely impossible,
not to say out of place, to give an account of so elaborate
a subject as ACOUSTICS, *i.e.* the science of Sound. Suffice it to
say, that all sounds are produced by the vibration of air, and
that the fewer vibrations, the lower is the sound, and *vice
versâ.*

When such vibrations are produced regularly, they form
Musical sounds, but, if irregularly, the sounds can be only
distinguished under the term of Noise. The earliest germ of
music lies in certain savage races, who, as long as they can
maintain a rhythmical beat on any resonant substance, do not
particularly care what it is. A hollow tree is a splendid instrument in their opinion, but, if this cannot be had, a dry log of
wood will answer the same purpose.

Some tribes, more ingenious than others, cut a deep groove
upon the upper surface of a log, hollow it through this groove,
and then hammer away at it to their hearts' content. The

next move was to cut off a section of the trunk of a tree, hollow it, set it on end, and then beat it on the sides.

Lastly, some one hit upon the idea that if the open upper part of the hollowed log were covered with a tightly stretched membrane, and that if the membrane, instead of the log, were beaten, the resonance would be increased. In consequence, the real Drum was invented, and seems to have existed from time immemorial in parts of the world so distant that they could not have had any communication with each other.

Take, for example, the well-known "Bass Drum" of our bands, which is shown on the right hand of the figure. We make it a very ornamental article, with frame of metal, and heraldic decorations of all kinds.

Lying against it is one of a pair of Kettle-drums, such as

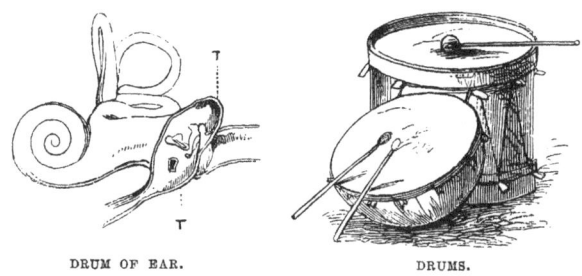

DRUM OF EAR.   DRUMS.

are always seen in mounted bands. They look very easy to play, but, if the reader will try a pair, he will soon find his mistake.

But there are savage tribes of Western Africa who make Drums of such wonderful power that their sullen roar is heard for miles around, as their slow, triple beat summons the tribe to arms like the fiery cross of the Highland clans. As to shape, lightness, and beauty, our Drums are infinitely superior to theirs, but, so far as I can gather from personal and written narratives of African travellers, none of our Drums surpass theirs in richness, depth of tone, and power of carrying sound.

Sometimes these Drums, instead of being mere cylinders, are carved into the most strange and fantastical patterns. I possess one of these curious Drums, brought from Ashanti, and carved out of a solid piece of wood.

The strange point in it is, that it represents a double head carrying, after all negro fashions, a sort of vessel upon it. One part of the head represents a human head (not that of a negro), while the other merges gradually into an eagle's head and beak. It is, in fact, a Gnostic gem, and would pass muster as such if it had been engraved on chalcedony, cornelian, or other semi-precious stones which are employed in the seal-engraver's art.

Upon this composite head is placed the Drum itself, which is also cut out of the solid block, and which, after the fashion of West African Drums, has a hole on one side.

This remarkable instrument was given to me by an old merchant captain, who brought it himself from West Africa, and who, when I made his acquaintance, had actually painted it all kinds of colours, planted it in his garden, and was using the Drum as a flower-pot. Of course, as soon as it came into my possession, I put it in "pickle,"—*i.e.* a strong solution of alkali,—brushed off the paint, and placed it in my museum, where it is now.

On the left hand of the illustration on page 514 is given a sort of map or chart of the human Ear, with its internal Drum, or Tympanum, as it is scientifically termed.

It is by the vibration of this Drum that hearing is made possible, the vibrations of the air being transmitted to the Drum by means of a beautiful bony apparatus, termed the Hammer and Anvil (*Malleus et Incus*). Sometimes the action of the Drum is partially checked, and then the sufferer is said to be "hard of hearing." Sometimes it is broken, or its action totally clogged, and then he is said to be "stone deaf." There have been cases where an artificial tympanum has been inserted, and answered its purpose fairly well.

The String and Reed.

It has previously been mentioned that all sounds are owing to vibrations of the air. But there are many ways of producing these vibrations, and each mode gives a different quality of tone. We have already seen, by means of the drum, how sound is produced by percussion. We shall now see how sounds can be produced by the vibrations of a String.

If the string of a bow be pulled and smartly loosed, the result is a distinctly musical sound, higher or lower according to the length and tension of the string. Perhaps some of my readers may recall the passage in Homer's "Odyssey," where Ulysses strings the fatal bow:—

> " Heedless he heard them ; but disdained reply,
> The bow perusing with exactest eye.
> Then, as some heavenly minstrel, taught to sing
> High notes responsive to the trembling string,
> To some new strain when he adapts the lyre,
> Or the dumb lute refits with vocal wire,
> Relaxes, strains, and draws them to and fro ;
> So the great master drew the mighty bow,
> And drew with ease. One hand aloft displayed
> The bending horns, and one the string essayed.
> From his essaying hand the string let fly,
> Twanged short and sharp, like the shrill swallow's cry."

The Harp is, in fact, nothing but a magnified bow, with a number of strings of graduated length and tension. Some very beautiful experiments have been made on this subject by the Rev. Sir F. A. G. Ouseley, Professor of Music at Oxford, who stretched a string of sixty-four feet in length, and found that although, when vibrating, it must produce a note, there was no human ear that could distinguish it. Yet, if combined with other musical instruments, it would probably do its work well. The theory of the vibrations will be briefly described on another page.

These vibrations may be produced in various manners. The string may be pulled with the fingers, as in the harp, the guitar, the zither, or even the violin, &c., in *pizzicato* passages.

The old harpsichord, now an instrument vanished into the shadows of the past, pulled the strings with little strips of quill, acting like the thumb-ring of the zither-player. The "plectrum" of the ancients acted in the same manner, and the Japanese have at the present day a sort of guitar played with a plectrum. I have heard it, but cannot particularly admire the effect, the notes appearing to be without feeling, and as if they were played on a barrel-organ.

Sometimes, as in our modern pianos, the strings are struck by hammers instead of being pulled by fingers, plectrum, or goose-quill.

The most ingenious mode of causing musical vibration is the Bow, which is too familiar to need a detailed description. Suffice it to say that it really is a modified bow, the place of the string being supplied by a flat band of horsehair, which is drawn over the string, and so causes it to vibrate. In order to enable the bow to grip the string, it is rubbed with resin almost as often as a billiard-player chalks his cue.

Some skill is required even in producing a sound by the bow. It looks as if any one could do it, but a novice, if he extorts any sound at all, never rises above a squeak. When I took my first violin lessons, nearly thirty years ago, I was so horrified at the discordant sounds elicited from the instrument, that I retired to the topmost garret of the house in order not to hurt any one's feelings except my own.

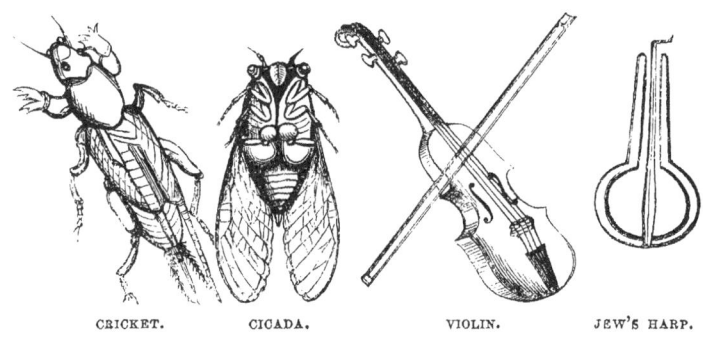

CRICKET.   CICADA.   VIOLIN.   JEW'S HARP.

On the left hand of the illustration is seen a well-known example of the imitation of Nature by Art. This is the common Cricket, whose loud shrill call is more familiar than agreeable.

Some years ago, while engaged on my "Insects at Home," I gave much time to the examination of the structures by which such a sound can be produced. On the under side of the wing-covers, or "elytra," as they are scientifically termed, are notched ridges, which, when examined with a moderate power of the microscope, have something of this appearance ⁓⁓⁓. The friction of these notches produces the musical sound, which, as the reader will see, is exactly analogous to the friction of the bow upon the string.

NEXT we come to the Vibrator, sometimes called the Reed. It is introduced into various musical instruments, such, for example, as the harmonium, the clarionet, the oboe, the bassoon, and various organ pipes.

The simplest form of the Vibrator is shown in the Jew's Harp, as it is popularly called, though it is not a harp, and has nothing to do with Jews.

The word is really a mistaken pronunciation of "jaw's harp," because the instrument is held against the teeth, while its tongue is vibrated by strokes of the finger. These vibrations affect the air within the mouth, and, by expanding or contracting the mouth, the sound is lowered or raised according to the laws of Acoustics. Of course, the range of notes is very small, being limited to those of the common chord, and even they being attainable only by a practised performer. Very good effects, however, have been produced by means of a series

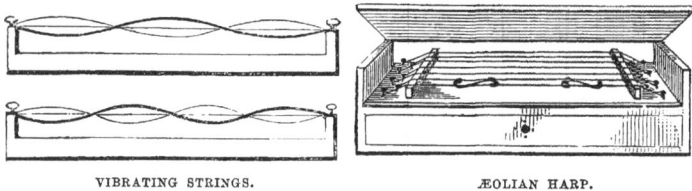

VIBRATING STRINGS.   ÆOLIAN HARP.

of Jew's Harps, set to different tones by loading the end of the tongue with sealing-wax or similar substances.

AN apparatus constructed on the same principle is to be found in the vocal organs of the male Cicada. If one of these insects be examined on the lower surface, two curious and nearly circular flaps will be seen, just at the junction of the thorax with the abdomen. It is by the action of these two little vibrators that the insect is able to produce a sound so loud, that in calm weather it may be heard at the distance of a mile.

THE accompanying illustration is, in fact, a sort of chart as to the vibration of sound.

On the right is shown the ÆOLIAN HARP, with its upper lid raised, so as to show the structure of the strings. These are all tuned to the same note, the present D being generally accepted as being most free from false tuning, and less liable for the errors of " temperament." Several of the strings are an octave lower than the others, but the tonic is always the same.

The instrument is placed in a current of air, generally in a window, with the sash let down upon it, and the air-currents set the strings vibrating in a most wonderful manner.

There is no need for human fingers to touch them, but they automatically divide themselves into the component parts of the common chord, and produce octaves, fifths, and thirds *ad infinitum*.

On the left hand of the same illustration is exhibited a string of the same length and tension, vibrating in two different ways. The upper figure shows it divided into three portions, each of which gives the fifth above the tonic, and all of which, when sounding simultaneously, give a fulness and richness to the tone which could only be attained otherwise by three distinct instruments. All players of stringed instruments know how invaluable are these harmonics, without which many passages of well-known music could not be played, and which are produced by "damping," and not pressing the strings.

So, if the string be lightly touched, or damped at the crossing portion at either end, the result will be that the string divides itself into three portions, and all three resound simultaneously.

The lower string is vibrating in thirds, having divided itself into four portions. If it were damped in the middle, it would divide itself into two portions, and sound octaves.

The subject is a most interesting one, but our space is nearly exhausted, and we must pass to another branch of it.

In all brass instruments furnished with a mouthpiece, and not with a reed, the notes are obtained by vibrations of the enclosed air, caused by the movement of the lips. They are all set to some definite tonic, sometimes C natural, but mostly to a flat tone, such as B flat or E flat.

Taking the ordinary military trumpet or bugle as an example, we have (when we have learned how to play it), first, the tonic. By alteration of the lips we get the octave above the tonic. Then comes the fifth ; then the third, which is, in fact, another octave; and then a few other notes, the truth of which depends on the ear of the player.

Now, all these notes are obtained by means of the lips, which set the column of air vibrating, and divide it into harmonics. The apparently complicated bugle-calls of the army are nearly all formed from four notes only, *i.e.* (taking C as the tonic) C G C E G.

THE Trombone, which is shown on the right hand of the illustration, has the advantage of being lengthened at will, and thus giving the performer a fresh tonic, and consequently

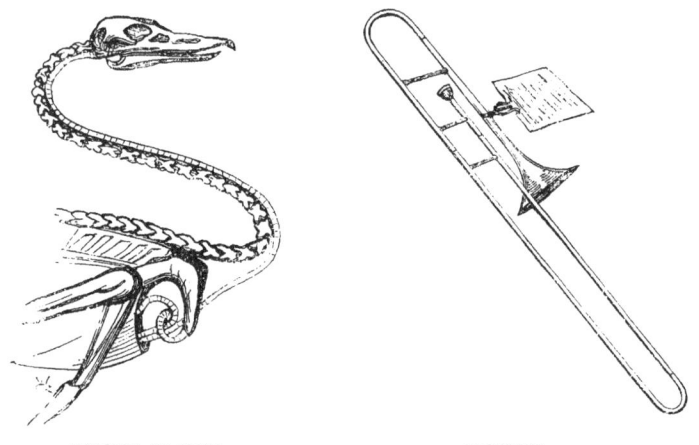

TRACHEA OF SWAN.　　　　　　　　　　TROMBONE.

another series of harmonics. Valved and keyed instruments have a similar advantage, the one acting by lengthening, and the other by shortening, the column of air. The former is infinitely the better plan, as it sets more harmonics vibrating, and consequently gives a greater richness of tone.

A familiar example of this is to be found in the Ophicleide and Euphonium. The former is eight feet in total length, and alters its tonic by eleven keys, which shorten the column of air. The latter is of the same length, but, by the employment of valves, can be made sixteen feet in length. Consequently the euphonium has practically killed the ophicleide, just as the ophicleide killed the serpent. The cornet-à-pistons, the brass contra-basso, the flugel horn, the tenor sax-horn, &c., are all constructed on the same principle.

On the left hand of the illustration is shown the wonderful apparatus by means of which the Swan produces its far-resounding cry. The windpipe, or "trachea," as it is technically named, passes down the neck, protected by the bones, until it reaches the chest. There it leaves them, enters the cavity of the chest, and contorts itself in such a manner as to obtain greater length, just as is the case with the trombone and valved instruments.

ACOUSTICS AS AIDS TO SURGERY.

WE have already seen how the air-vibrations poured in at the small end of the trumpet can make resonant notes. We have now to see how the reverse process can be employed, and sounds poured into the larger end be conveyed to the ear.

The Ear-trumpet is a familiar example of such an instrument, and, as it is shown in the illustration, there is no need

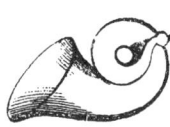

EAR-TRUMPET.

CONCHA OF HUMAN EAR.

of further description. It is rather remarkable, by the way, that the length of tube does not seem to interfere with the conveyance of sound, as may be seen by the speaking-tubes which are now so common in private houses, hotels, and offices.

I know of one church in which there is a special seat for deaf persons. The reading-desk and pulpit are both fitted with the large ends of Ear-trumpets. From them pass tubes under the flooring, and so into the seat, where they can be applied to the ear of the deaf worshippers.

On the right hand is the "Concha," as it is called, of the human ear, which is evidently constructed for the purpose of collecting and concentrating sounds. Instinctively, if we wish to hear any sound more distinctly, we place the open hand behind the ear, so as to enlarge its receptive capacity, and send a greater volume of sound into the ear.

The well-known experiment of holding a shell to the ear so as to hear the murmur of the sea is due to the same cause, the shell collecting, though in a mixed manner, all the surrounding sounds, and making a murmur which really resembles the distant wash of the waves upon the shore.

Then, if we examine the various animals which need acute hearing, either to seize prey or escape from enemies, we shall find that they have large and mobile ears, which can be directed so as to catch the expected sound. The hare, rabbit, and deer are examples of the latter, while the former are well repre-

SAVAGE TAPPING TREE.   SURGEON USING STETHOSCOPE.

sented by the domestic cat, whose ears are always pricked forward when she hears the scratchings of a mouse.

ANOTHER most useful appliance is the STETHOSCOPE, which enables the skilful surgeon to investigate the interior of the body almost as clearly as if it were transparent. It is perfectly simple, being nothing but a trumpet-shaped piece of wood, formed as shown in the illustration. Sometimes it is hollow, and sometimes solid, but the result is the same, sound being transmitted through wood in a most remarkable manner.

For example, if one end of the longest scaffolding pole be slightly scratched with a pin, the sound will be distinctly heard by any one who places his ear against the other end, though the person who uses the pin can scarcely hear the sound himself. The surgeon, therefore, places the broad end of the Stethoscope upon the patient, and the other upon his ear, taps more or less lightly with his fingers, and by the sounds transmitted through the Stethoscope ascertains the condition of the internal organs.

On the left hand is an illustration of the mode in which the Australian savage, without the least idea of the theory of Acoustics, utilises the sound-conducting power of wood. If he wishes to know whether or not a hollow tree is tenanted by an animal of which he is in pursuit, he places his ear against the tree, taps it smartly with his tomahawk, and listens for the movement of the animal inside.

So delicate is this test, that it is employed even when the native is hunting for the large beetle-grubs on which they feed, and which are accounted a luxury even by Europeans, when they have once overcome the prejudice attaching itself to eating, without cookery, fat white grubs as thick and long as a man's finger.

The Aye-aye is said to eat in exactly the same manner, tapping with its long finger the trunks and branches of trees, and, if it hears a maggot inside, gnawing it out.

MEASUREMENT OF SOUND.

Of late years we have had an instrument which enables us to measure the vibrations of sound as accurately as the barometer measures the weight of the atmosphere, the thermometer the temperature, and the photometer the power of light. This is the Siren, which is shown on the right hand of the accompanying illustration.

To explain this instrument fully would require ten times the space which we have at command, and necessitate a great

number of drawings. I will, therefore, endeavour to explain
its principle in as brief terms as possible.

The reader will observe that at the lower part of the instrument there is a disc pierced with a number of holes, and that
above these are two dials. Below the perforated disc, and
therefore unseen, is a circular plate, also pierced with holes.
When a pipe is attached to the lower part of the instrument,
and air propelled through it, the disc begins to revolve, every
revolution being recorded by the dials, after the fashion of the
ordinary gas-meter.

As the pressure is increased, the air, passing through the
holes, assumes a rhythmical beat, which soon becomes metamorphosed into musical notes. It is evident, therefore, that,

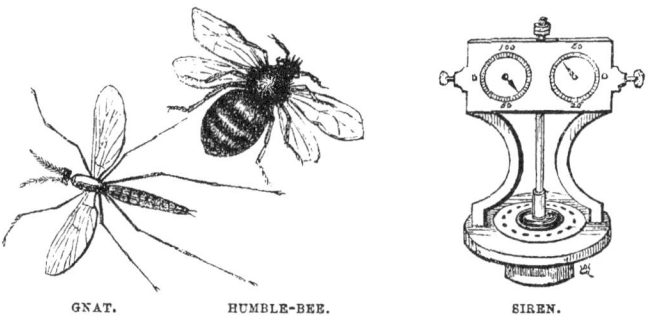

GNAT.   HUMBLE-BEE.   SIREN.

by means of this instrument, the number of vibrations which
produce a definite tone can be measured with absolute accuracy
by any one who has an ear capable of appreciating a musical
note.

It is by means of the Siren that the much-disputed tonic of
C will be settled, the Continental and the English C being
greatly at variance, and even the English C having been
advanced almost a tone since the time of Handel. Much is it
to be wished that Italy, the home of song, and England, the
patron of song, could unite in their tonic, instead of having
systems so widely different that an Italian singer is at a loss
with the English pitch, as is an English singer with the Italian
pitch.

The Siren is even brought into the service of entomologists,
enabling them to measure by the sound the rapidity with which
a flying insect moves its wings. By means of this instrument

we know the origin of the sharp, piercing "ping" of the Gnat, and the heavy, dull boom of the Humble-bee, both of which insects are given in the illustration.

Before taking leave of this subject, I may mention that the instrument is called the Siren because it sings as well under water as in the air, provided that water instead of air be driven through it.

Echo.

OUR last page will be given to the phenomenon called by the name of ECHO, which consists in the power of solid substances, whether natural or artificial, of reflecting the waves of sound thrown against them, just as a mirror reflects the waves of light.

Very often the Echo is naturally formed, as shown in the illustration, by rocks which cast back the sound—waves thrown against them. This is the case in several parts of

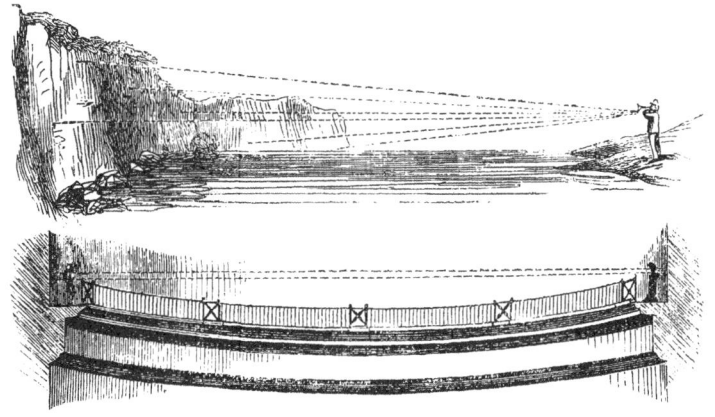

WHISPERING GALLERY.

Dovedale in Derbyshire, where a pistol shot is reverberated backwards and forwards in a most wonderful manner, and a trumpet blast repeats itself over and over again.

At Walton Hall, the residence of the late C. Waterton, Esq., there is a wonderful Echo, nearly half a mile from the house. Mr. Waterton had discovered the Echo, which proceeded from the walls of the house, and, having found its focus, placed on

it a large stone, called the Echo-stone. Any one sitting on
this stone, and singing, speaking, or whistling towards the
house, heard every sound repeated, as if in mockery.

The celebrated Whispering Gallery in St. Paul's Cathedral
is nothing but an ordinary Echo, though so intensified by the
process of radiation, that the sound is transmitted from one
side of the dome to the other, just as light or heat is reflected
from concave mirrors.

INDEX.

Porcupine Ant-eater, 110
Porthesia auriflua, 180
Porthesia˜chrysorrhœa, 180
Portuguese Man-of-war, 46, 372
Pouch-shell, 8
Pressure of Atmosphere, 329
Printing-press, 317
Proboscis of Fly, 379
Processionary Moth, 180
Projectiles, 74
Propolis, 220
Pseudoscope, 287
Ptarmigan, 150
Pucunha, 76
Puff and Dart, 75, 351
Pulley, 452
Pyramids, 216

Q.

Quilt Armour, 126

R.

Radius, 194
Rain-cloud, 429
Ranjows, 109
Rat-tail Maggots, 385
Rattan, 204
Razor, 236
Receiver of Air-pump, 511
Reduvius personatus, 146
Reed, 518
Reverted Spikes, 102
Ribbon Saw, 244
Ring and Staple, 415
Ringed Tissues, 378
Robber-crab, 405
Rocket, 462
Rod and Line, 90
Rolling-mill, 322
Rosemary, 408

S.

Sabella, 218
Saddle-back, 348
Sailing Raft, 5
Salad-dressing Bottle, 511
Sand-paper, 265
Saturnia pavonia minor, 104
Saw, 239
Saw-fly, 241
Sawyer-beetle, 248
Scale Armour, 123
Scales of Butterfly's Wings, 187
Scaling-fork, 133
Scarabæus, 494
Scissors, 228

Screw, 498
Sea-anemone, 8
Sea-basket, 89
Sea-mouse, 353
Sea-urchin, 315
Seed-drills, 336
Sepia officinalis, 167
Serpula, 44, 135, 219, 352
Sewage, 496
Sewing, 406
Shark-tooth Sword, 56
Shears, 228
Sheep-fly, 396
Shell of Tortoise, 188
Ship-worm, 200
Short-tailed Manis, 124, 188
Sialis armata, 275
Siamese Link, 448
Silkworm, 158
Silkworm Cocoon, 179
Siren, 523
Sirex gigas, 252
Skidor, 466
Skip-jack Beetle, 387
Skull, 210, 511
Slates, 188
Sling, 502
Sloth, 398
Slug, 245
Smoke-jack, 499
Snow-house of Esquimaux, 163
Snow-house of Seal, 163
Snow-shoe, 464
Spade, 223
Spear, 58
Spectroscope, 297
Spider, 509
Spider-crab, 147
Spiked Defences, 107
Spiracles of Fly, 357
Spiral, 498
Spiral Spring, 371
Spiral Tissues, 375
Spirit-level, 271
Spokeshave, 236
Spout-hole, 434
Sprat-sucker, 71
Spring, 430
Spring-bow, 142
Spring-gun, 142
Spring-jack, 386
Spring Solitaire, 371
Spring-tails, 388
Spring-trap, 95
Squirrel, 456
Stag-beetle, 248
Star-fish, 332
Steam-blast, 443
Steelyard, 450
Stereoscope, 286
Stereotype, 479

THE END.